Women in Law

Books by Cynthia Fuchs Epstein

*Woman's Place: Options and Limits
in Professional Careers*

The Other Half: Road to Women's Equality
(EDITED WITH WILLIAM J. GOODE)

*Access to Power: Cross National Studies
of Women and Elites*
(EDITED WITH ROSE LAUB COSER)

WOMEN IN LAW

Cynthia Fuchs Epstein

Basic Books, Inc., Publishers

NEW YORK

Library of Congress Cataloging in Publication Data

Epstein, Cynthia Fuchs.
  Women in law.

  Includes bibliographical references and index.
  1. Women lawyers—United States. I. Title.
KF299.W6E65   349.73'023'73   80–68954
ISBN 0–465–09205–5   347.3002373   AACR2

Contents

IV

OUTSIDERS WITHIN

V

MANAGING AND COPING

VI

PRIVATE LIVES

Preface

A GREAT DEAL has happened in the world and in my life since the time when this book was conceived. This is also true for the many people who had some hand in the development of its ideas or the gathering of its evidence. We have grown and changed roles in our personal lives; we have also worked for change in society, engaging in efforts, small and large, quiet and loud. Some of us bear scars, but most of us seem to have profited from living through this past decade and feel that this is probably true for our society as well, although efforts to better ourselves and our society, like woman's work, are never done. Because of these efforts and changes, this book took a long time to produce. However, the process of studying and writing, putting the work aside for a period and taking it up again, had its rewards. Colleagues, mentors, friends, and students (and for many of the people involved, these categories overlap) became part of the intellectual and emotional process.

At various stages and from various sources I also received "guineas and locks," the money and space that Virginia Woolf maintained all writers need. Actually, I had more guineas than locks, many offices but few rooms of my own, involved as I have been through the years with teaching and activity in the women's movement. Life also included involvement in other social causes, work in professional organizations, and time with friends and family.

Through the years I have become indebted to many people. Some of them will appear immediately to readers of this book as their names crop up as contributors to both a theoretical and an ideological framework; others are there more in spirit, their contributions not of the kind evidenced in footnotes.

There is no order: some whom I will mention were active in my work life at times and faded out, only to come back in again; others are spiritually constant. In the revelatory nature of my generation, I will confess that there were those whose influence, at one time inspiring, was at another confounding; such are the workings of our minds and emotions. But if others helped and hindered, I too was responsible for permitting obstacles to get in the way of opportunities. A firm believer in the impact of structure, I found that one must also learn to seize the day. I have always been enormously grateful for support and encouragement, and if I have succeeded in some tasks in my life, I know how much it came from those sources. I know how much structural conditions of opportunity count. There were times when I was given a room of my own, yet preoccupations crowded out the writing that was my mission. Good counsel and age have tempered my preoccupations, and experience has enabled me to make better use of resources.

Certain resources stand out as being of strategic importance. On four occasions, I have been a fellow of the MacDowell Colony in Petersborough, New Hampshire, the last time in June 1980, when a good deal of the writing of this book came together. In this idyllic setting, away from normal responsibilities and contact with the outside world, I was able to write and think, unimpeded by anything but the task at hand. A year as a fellow at the Center for Advanced Study in the Behavioral Sciences at Stanford, 1977/78, also provided an opportunity to work intensively in an environment rich in intellectual stimulation and generous with services. Support from the National Institute of Mental Health, the Spencer Foundation, and the Rockefeller Foundation made the year possible. A Guggenheim fellowship also provided time away from teaching and funds for research in 1976/77, as well as the chance to explore new directions and changes in the law which turned the book around in my mind.

In addition, the Center for the Social Sciences at Columbia University has been my research home for some time, and I have benefited from the assistance I have received there through the supportive colleagueship of its director, Jonathan Cole, and because of a Ford Foundation grant for the Program in Sex Roles and Social Change, for which Miriam Chamberlain was largely responsible. The opportunity this past year to co-direct a training program on the economics and sociology of women and work at the Graduate Center of the City University of New York, funded by the National Institute of Mental Health, also provided a forum to test certain ideas with a group of able colleagues, especially Gaye Tuchman, who shepherded the program into existence and handled many crises single-handedly when the deadlines drew near. The support of the Department of Sociology at Queens College in permitting flexibility of time arrangements is also appreciated. Early research was also supported by the U.S. Department of Labor and the Research Foundation of the City University of New York.

Certain people were important to me in the course of this project because of their encouragement. Emanuel Geltman first backed publication of my early study of women lawyers, which now appears in this vastly changed form. My son, Alex Epstein, not only insisted that I work more efficiently when my efforts diffused, but served as a model by his own behavior.

Certain people were also my critics and sounding boards, providing intellectual and editorial assistance. William J. Goode, teacher and friend, read and discussed many drafts of the work in its early stages and was strong in his urging to move it toward completion. Robert K. Merton's notes on the original manuscript of my first study of lawyers in 1968 were still useful in the writing of the sections of this book which draw on that study. His meticulous editing of a draft on the changing context of law and "outsiders within" was both a challenge to critical thinking and an instructive exercise in the elements of style. Ruth Bader Ginsburg, a source of information and assistance over the past ten years, also read and commented on sections of this work in its various stages, particularly on the law schools and on Wall Street firms, as did Martin Ginsburg. Barbara Underwood and Carole Joffe also read sections of the book and provided helpful comments and information. Donna Fossum provided valuable assistance and offered data for use in the chapter on women law professors. Jeylan T. Mortimer also read critically a version of the chapter on Wall Street lawyers.*

Kai T. Erikson made valuable comments on parts of this work, but my keenest gratitude to him is for the model he provided of exemplary qualitative analysis and dramatic expository writing. Alone with my data and stuck in the language of jargon, I would often turn the pages of his book, *Everything in Its Path,* for inspiration and instruction in the sociologist's craft. Although I could not hope to achieve the level of performance of his exceptional talent, those lessons did move me toward writing some of the sections of this book of which I am especially fond.

But inspiration and enthusiasm often swept me away, and I have been fortunate to have the in-house counsel of Howard Epstein, who contributed once again to my work his considerable talent as an editor and insightful intellect as a critic as he pruned many drafts of this manuscript.

I have also been particularly fortunate in having as research assistants on this book people whose taste and judgment were resonant with my own, and with whom I had many hours of rich discussion. Susan Wolf, who came to the project as a sociologist and is now in the legal profes-

*This appeared as "The New Women and the Old Establishment: Wall Street Lawyers in the 1970's," *Sociology of Work and Occupations,* 7, no. 4 (August, 1980): 291–316.

sion herself, remains my good friend and intellectual compatriot. Mary Murphree and Judith Thomas were research assistants who became true collaborators. Others have also helped in the research over the past years: Allan Grafman, Susan Ogulnick, Carol Finkelstein, Thomas Sykes, Diana Polise, Zeev Gorin, Mark Johnson, Flora Solarz, Claire Morgenstern, Harriet Goodman Grayson, Beth Stevens, Gwyned Simpson, Marlene Warshawski, Mary Berry O'Neil, Simonetti Samuels, and Athena Economides. Those who participated in the last stages of production with spirit, vitality, and good will are Rachel Ovryn, Amrita Basu, Susan Roberts, and Lyn Perlmuth. Helen Danner provided much of the final typescript.

I am also grateful to Julia Strand of Basic Books, who polished the final draft, and to Jane Isay, my editor, for her remarkable efficiency, wisdom, and cheer.

I would also like to express gratitude personally, and as a woman of these times, to Betty Friedan, who provided much of the impetus for change that is considered in this book and whom I have had the good fortune to work with on many projects over the past decade.

Of course, this book could not have been written without the generosity and good will of the scores of lawyers who gave of their time and experience.

CYNTHIA FUCHS EPSTEIN

New York City, 1981

Women in Law

Introduction: Encountering the Legal Establishment: The New Women of Law

ONLY ten years ago, in my introduction to a book on women's equality,* I wrote that "the quiet ones are in revolt." The popular image of women as subdued and passive changed rapidly as they pressed collectively and individually for broader participation in American society during the decade. It was the beginning of wide-ranging changes in women's roles that have affected most institutions of society, including the family, the economy, education, politics, religion, and law, the focus of this book.

What women are, what they want, and what they ought to do and be are now questions in the hearts and minds of most people.

There is no question that women have been poorly represented among those groups that receive rewards and privilege in this society, and whose work is regarded as the most interesting. Few woman are to be found among the managers of industry, among the ranks of professionals in medicine, law, and the sciences—the manipulators of symbols and of power. It once was not believed that women wanted such positions or were competent to hold them. Yet over the past decade more and more women have chosen and fought to enter these spheres, and they are demonstrating competence in many fields that were exclusively male domains before.

*Cynthia Fuchs Epstein and William J. Goode (eds.) *The Other Half: Roads to Women's Equality* (Englewood Cliffs, N.J.: Prentice-Hall, 1971).

3

This change in attitudes and aspirations and acquisition of competence came as more and more women entered the labor force and the growing woman's movement proposed an ideology of equality underscoring women's commitment to work. That ideology had been reinforced by the focus on justice and equality which came to American society's consciousness through the civil rights and student movements of the 1960s. Finally, the impact of legislation was of primary importance, especially the passage of Title VII of the Civil Rights Act, which provided the legal basis for curbing discrimination on the basis of sex in work and education and for outlawing such discrimination in its most blatant forms.

TABLE I.1

Proportion of Women in the Profession of Law, 1910–80

Year	Number	Percentage
1910	558	1.1
1920	1,738	1.4
1930	3,385	2.1
1940	4,447	2.4
1950	6,348 (7,000*)	3.5 (4.1*)
1960	7,543	3.3**
1970	13,000	4.7
1976	38,000	9.2
1980	62,000	12.0

SOURCES: U.S. Bureau of the Census, *Census of Population, 1960*, vol. 1, table 202, pp. 528–33. Statistics for 1900 to 1950 from *Changes in Women's Occupations, 1940–50*, Washington, D.C.: U.S. Department of Labor, 1954, Women's Bureau Bulletin No. 253, p. 57. Table 8-1 "Occupation of Employed Persons 14 Years and Over by Sex: 1960 and 1970," *A Statistical Portrait of Women in the U.S.*, Current Population Reports Special Studies Series P-23, No. 58, U.S. Department of Commerce, Bureau of the Census, April 1976. The census figures may be a slight exaggeration of total number and percent of lawyers since many lawyers *not* in practice may be reporting law as their occupation. There is also some discrepancy of figures and percentages in these charts reported by the government.
*TABLE 8, "Employment of Women in Selected Occupations, 1950, 1970, and 1976," *U.S. Working Women: A Databook*, U.S. Department of Labor, Bureau of Labor Statistics 1977, Bulletin 1977, p. 9.
**Table II, "Employment of Women in Selected Occupations, 1950, 1960, 1970, and 1979," *Perspectives on Working Women: A Databook*, U.S. Department of Labor, Bureau of Labor Statistics, October 1980, p. 10.

TABLE I.2

Year	Number	Percentage
1948	2,997	1.8
1951	5,059	2.5
1954	5,036	2.3
1957	6,350	2.7
1960	6,488	2.6
1963	7,143	2.7

Sources: The small amount of change in the percent of women lawyers in the United States from 1948 to 1963 is even more striking if one uses the figures of Faye A. Hankin and Duane W. Krohnke, *The American Lawyer: 1964 Statistical Report* (Chicago: American Bar Foundation, 1965), based on lawyers listed in the Martindale-Hubbell Law Directory. The Directory's listing, although unofficial, probably also is subject to error but provides an estimate close to the number actually in practice.

Introduction: Encountering the Legal Establishment

The law became a mechanism for change as it was implemented by the concerted action of women's movement activists, by feminist lawyers, and by the acceptance of legal methods as effective tools for winning women's entry into the male-dominated establishment and guaranteeing the possibility of their success.

There were many heroes, both women and men, in the early stages of this saga of change. A good proportion were lawyers. Many had fought for other excluded groups and civil rights causes; some had experienced discrimination in their own careers; and others simply caught the excitement of the successes achieved. Yet the early involvement of lawyers in the campaign to end sex discrimination was not in any sense a measure of women's position in the legal profession. There were few women lawyers in the United States in the 1960s— women constituted 4 percent of the profession, about 7,000 according to the U.S. Census and even less, only 3.5 percent or 6,348, according to statistics published by the American Bar Foundation (see table I.1). Although obviously there had been an increase since the first woman was admitted to practice in 1869, it was very slight over the years.

Ten years later, the picture had totally changed. By 1970 there were 13,000 women lawyers, and they were to increase in numbers even more dramatically. In the 1970s, as the barriers to women's legal education were lifted, as more and more young women looked to law careers as a way of acquiring competence and effectiveness, and as older women encouraged and provided models for younger women to emulate, an extraordinary development occurred: law progressively became a favored field for women. After decades of virtually no movement, the number of women lawyers grew radically in the decade of 1970 to 1980, from 13,000 to 62,000 (from 4 percent to 12.4 percent)[1] and the proportion of women in the law schools rose from 4 percent in the 1960s to 8 percent by 1970, and then to 33 percent by 1980.

I first studied women lawyers in the early 1960s, focusing on them as a "deviant" group—not, of course, a disparaging term, but a sociological and statistical one. These were women who had managed to enter a profession in which most members objected to their presence, at a time when it still was difficult to get admitted to law school (Harvard Law School admitted its first women students only in 1950), and when it was difficult for a woman to find legal work. I was then interested in learning by what paths women managed to become lawyers and how the general aura of exclusion affected their later careers. The interviews with the sixty-five lawyers who comprised the basis for that study were done in 1965 and 1966, the years in which the women's movement was begin-

5

ning to grow, although most people, including the lawyers interviewed, were still unaware of it.

I was struck, in carrying out those first interviews, by how wrong the stereotypes of women lawyers were. Like other women who had dared to enter male preserves such as medicine, large corporations, or competitive sports, they were thought to be tough, aggressive, and generally manlike. But the lawyers I studied were as different from each other in appearance and style as any other group of women, or men. Their similarities were in the kinds of work they found and the legal activities they engaged in. Women were guided toward (and sometimes restricted to) legal work in government agencies or research and routine aspects of case preparation. They rarely went to court to argue cases or were brought into conferences with important clients.

These women as a group had done remarkably well in law school and were hard workers. Many seemed successful, but often that meant only "successful for a woman." A few were doing outstanding work by the standards of the profession, but not the proportion that would have been predicted on the basis of outstanding law school performance which was a clear indicator of future success for their male colleagues.

Although I wrote the story of those women lawyers in 1968 as my Ph.D. dissertation, by the time the analysis was completed the picture had already changed. The book I planned to write on women lawyers was put off again and again; under the impact of the women's movement, women's lives were changing and they were discovering—and creating—new career options. More and more young women were becoming lawyers, and the legal profession itself was changing in response to a variety of forces. The federal government was funding new programs for the poor, many talented young men were choosing to work in these new programs, and there was a general reassessment of legal practice and its rewards by a new generation of law graduates. Large law firms were forced for the first time to compete for gifted graduates with law programs geared to social reform. The change in the legal profession, changes in the attitudes of male lawyers, and changes in the numbers of women who were now getting legal training formed a context in which law became a less deviant and more "normal" career for women.

I interviewed successive waves of women attorneys in the 1970s: the "new women of law," I called them. I spoke to radicals who became

Introduction: Encountering the Legal Establishment

lawyers to fight for civil rights issues; I met with attorneys who were trying to use law to win new rights for women and protect battered wives and rape victims; and with lawyers who were concentrating on constitutional issues to establish new precedents in the law. A Guggenheim fellowship in 1977 permitted me to do interviewing on a broader scale, and I followed up those inquiries by looking at women who were now in those exclusive and powerful domains, the Wall Street law firms.

I also tried to locate the lawyers who had been interviewed twelve years before. The questions they were asked centered on how their work and careers had changed and how the women's movement had affected them. It was clear that some of these older lawyers had also become new women of law. Few were unaffected by the events of the 1970s.

This book is an accounting of the new and the old, of change and stability. The way it was for the 2 or 3 percent who were able to overcome discrimination and become lawyers still constitutes recent history and forms the context in which change has occurred. Many old patterns persist; resistance to women lawyers can still be observed and new patterns have developed to replace the outlawed old ones.

Does it matter that women become lawyers? I think it does. Not only is the entry of women into the legal profession an important indicator of their general equality in American life, but it constitutes a broadening of the profession's responsiveness to the needs of all sectors of society.

There are other reasons to study the role of women in the legal profession. Law is one of the most traditional and exclusive professions in the United States. It also provides access to important positions in business, government, and politics. We can, for example, partly explain women's poor participation in political life and certain kinds of business by the fact that such a small percentage of women have been lawyers.*

The problems that women have faced within the legal profession are in many ways the same they face in other occupations, but they take

---

*About half of the members of Congress and a fifth of state legislators are lawyers, although there has been a slight decrease in the proportion of lawyers among those elected to the House of Representatives (one-third of the representatives newly elected in 1978 were lawyers, 194 out of the total of 433 members). An increasing number of lawyers are represented among top managers in big business. A survey of top corporate officers in 1973, at 300 firms, showed that 4 percent had legal backgrounds. In 1980 that proportion had increased to 16 percent. "Mr. Smith Doesn't Go to Washington," *Student Lawyer* 8, no. 6 (February 1980): 8.

a more intense and dramatic form. Highly demanding of commitment and time, law asks much of its practitioners. Women's ability to fulfill the norms of professional life while performing their roles as women is tested by a legal career, as are their intellectual abilities, especially in the profession's top echelons.

Although the profession has been structured to mesh with the lives of men and the norms of society which encourage men's commitment to work, many women have managed to adapt and perform competently, even brilliantly, as lawyers. Society has changed in ways that make it more possible for women to assume professional roles without assuming they must pay the high costs asked of them in the past. Yet professional work still does exact costs of women and of men. Institutional demands are often overdemanding because individuals have many roles that require commitment and time. For women, there are contradictions as well between the role demands of professional life and those normally associated with their "female" status.

In this book, I shall describe and analyze the ways in which women attorneys are treated by their colleagues and their families, the kinds of pressures and cross-pressures they face, and the new and old ways they have dealt with their problems. I shall look at subtle as well as overt forms of discrimination, but I shall also consider the ways in which women have benefitted by being unique in the realms of men. And in examining the different experiences of women lawyers, I shall refer both to life-cycle changes, which are important to women and men when they go through careers, and to societal changes.

It is, of course, most interesting to locate the patterns that seem to affect most lawyers rather than those cases that are interesting because they are idiosyncratic. Although the women I interviewed were of different personality types and from different backgrounds, all were subject to the same cultural views about women and about women who chose to work in a "male" profession. Some of these views remain. And because I found that the kind of practice a woman worked in often was the most important factor in determining how she saw herself and her future, I shall concentrate on the different kinds of law practice in which women seem to cluster and discuss what this has meant in the past and what it may mean in the future.

I hope that this study will not only answer questions people have had about women professionals, but will raise questions about justice and women's equality and personal freedom. The problems women encounter are ultimately faced also by the men they work with and the men they live with. The sorting out and solving of those problems may be aided by defining what they are, showing how some people have solved them, and suggesting what larger social issues are raised by them.

I

BACKGROUND

Law: The Changing Context

THE MOST admired occupations have long been the professions —law and medicine in particular, but also architecture, the sciences, and academia. High in social rank and perceived as engaged in socially useful work, professionals can hope to find personal fulfillment as well as respect. Professionals' prestige is such that they are often regarded as authorities not only on their fields, but on everyday affairs.

Of all the professions, law is the one most central to the functioning of American society. Members of the legal elite preside over power and property relationships. They play a leading role in the legislative and regulative bodies that write the law; they direct the executive agencies responsible for enforcing the law; they rule the courts that elaborate and apply the law; they guide the corporate and financial institutions that constitute the most important property interests. Lawyers are indispensable in modern industrial America. Capitalist enterprise and the governmental procedures linked to it have grown ever more complex and incomprehensible. The rules and processes of economic life require specialists—lawyers—to guide institutions and the uninitiated. Lawyers do not only advise; many are prominent among the ranks of corporate executives and corporate boards of directors.

The United States, in fact, leads other countries comparable in corporate power in the per capita proportion of lawyers.[1] Great Britain, Germany, and Japan, for example, do not depend as we do on a massive legal structure to conduct business.

In the United States, too, lawyers are very much part of politics and government.[2] Roughly three out of four United States senators are lawyers. Representatives and governors are also likely to have been lawyers, and many presidents have been lawyers. In that respect the United States is like many other countries—everywhere, lawyers are over-represented in law-making bodies.[3] It is not surprising, then, that law should be a prestigious occupation.

However, the American bar is not a monolith. It is quite stratified.[4] Some lawyers work in large firms of hundreds of lawyers, have as their clients national and multinational corporations, and are connected with the power centers of government. Others are solo practitioners who handle the ordinary economic and criminal problems of social life, while lawyers in middle-sized firms and in routine government practice fall somewhere in between. The social distance between the senior partners of a large firm and the small practitioner is great.

The majority of lawyers are white males of middle-class background. Those who practice at the prestigious bar tend to be of higher socio-economic status than those who practice below. Class position has generally correlated with rank in the profession, although law has provided avenues of mobility for American ethnic groups. Although ethnic status once was a barrier to access to prestigious careers in law, it has become less important, even unimportant, in the last thirty years. But blacks and women, who trickled into the legal profession in the mid-nineteenth century, met considerably more resistance from the profession's gatekeepers until the last decade than white males of low ethnic status.

Actually, the legal profession has always been cautious about protecting its reputation and its monopoly on its services by limiting the numbers of practitioners admitted to practice and the character of recruits. It has been like other professions in this respect.[5]

Professional prestige is created both by the merit that comes from performing useful and difficult work (requiring years of training) and by the efforts of professional groups to upgrade and maintain the respect they enjoy.[6] This is achieved by public relations work and control of standards of performance and recruitment of personnel. Monitoring of the professions is accomplished by peer review and control, with emphasis on informal as well as formal rules and constraints. It is widely believed that this is an effective system—for who but an expert could assess a fellow professional's activity—by the professionals whose interests are served, by the social scientists who have studied them,[7] and by the lay persons who are their clients or patients.

Professions constitute occupational "communities" in which members share common norms and values and exercise informal controls known and understood by the members of the in-group.[8] Informal controls include exclusionary practices that limit membership to persons who can be depended on to understand the unstated assumptions and understandings of the professional culture.[9] Gatekeepers exercise discretion in admitting recruits from backgrounds that are different from their own because the nature of informal controls requires that members be from known groups who can be trusted to uphold professional standards and relied on to assure ease of social relations on the job.

The most effective control of recruits has always been at the top of the profession.* In medicine and law, for example, the most prestigious specialties in the major teaching hospitals and the largest firms in the past have tended to be limited to men of the white Protestant "establishment."[10] Paradoxically, at the same time that the professions have been exclusionary, they have created channels of mobility by admitting a limited number of aspiring minorities who were able to acquire sufficient education and who could demonstrate the competence to meet professional standards.† Because exclusion has always been rationalized at the top of the profession as necessary to maintain standards and not as a means to enforce social "purity," the accomplishments of ethnic minority professionals were recognized in many spheres, especially professional education, the gateway to recruitment into the world of elite professional practice.

Women and blacks were not affected by these processes. Prejudice against them was of a different quality than that against men of ethnic and religious minority backgrounds because common cultural attitudes held that they had basic incapacities for professional work, and their race and sex statuses—their physical attributes—could not be concealed. Their numbers were kept limited in the law schools by the use of informal quotas (though a few schools officially continued to deny women admission until the 1950s) and the combined effects of the different criteria used for their admission and the reticence of blacks and women to apply in the face of this array of discrimination.

*Control may also be interpreted as sheer discrimination against outgroupers. William J. Goode points out that both (professional) individuals and groups "have mounted private or public campaigns of propaganda, legislation, formal restrictions, quota systems, false documents, violence and threats, blackmail, and bribery to prevent . . . disadvantaged groups from obtaining as much . . . for their qualities or performances as they otherwise might." *Celebration of Heroes: Prestige as a Social Control System* (Berkeley: University of California Press, 1978), p. 283.

†Harold L. Wilensky and Anne T. Lawrence have shown that structural assimilation has proceeded more rapidly for some ethnic groups than for others, but by the second generation most white ethnics achieved an occupational profile which closely matches that of the native white population. They also maintain that the pattern has not been replicated for black Americans or for women. "Job Assignment in Modern Societies: A Reexamination of the Ascription-Achievement Hypothesis," in *Societal Growth*, ed. Amos H. Hawley (New York: Free Press, 1979), pp. 202–48.

The professional communities, for decades stable in structure and practice, began changing in the late 1960s in response to explosive changes in the society around them. The professions were growing in numbers in spite of a generally cautious attitude on the part of gatekeepers; law and medical school deans, together with bar and medical association officers, had long worked to keep recruitment down in the name of maintaining standards, but also, some admitted, to insure that the market would not be glutted. Nevertheless, there was a quantum jump in the number of professionals. In law, a proliferation of legal work served to allay fears that there would be too many lawyers and the price of legal services would be driven down. There was an increasing demand for legal services by a growing economy in the 1960s and early 1970s, triggered by expansion of the major corporations,* more complex and increased government regulation,† and increased public financing to support the needs of clients whose limited incomes kept them remote from the legal arena. Indeed, a general sense of entitlement of citizens concerned with their rights probably also contributed to an increased litigiousness. From 1972 to 1977 the number of class-action suits in federal courts nearly doubled, and in 1976 a total of 5,320 civil rights suits were brought—a 1,500 percent increase from 1970. Furthermore, new legal business was created by new statutes and regulations dealing with ecology, pollution, and consumer affairs.[11]

The number of lawyers practicing in the United States increased by 33 percent in the decade of 1960 to 1969, after growing just 14 percent in the previous decade. Since 1970, new admissions to the bar have increased by 91 percent. Between 1963 and 1978 law school enrollments more than doubled, from 54,000 to 126,000.[12] In 1970, there were 350,000 lawyers in the United States, about 300,000 actually in practice.[13] Today, unofficial estimates put the number at half a million.[14] Every year more than 30,000 new attorneys join the ranks of the legal profession.[15]

Growth in numbers of lawyers was paralleled by the changing shape of the profession. There was a movement away from solo practice to-

*An example of the order of fees paid to the large firms by corporations is the $2,210,000 paid in 1971 by First National City Corporation to Shearman and Sterling in New York. Pennzoil United paid Baker and Botts of Houston $1,221,300 in that same year. Of companies filing 10K annual reports with the S.E.C., at least twenty-five reported annual fees of more than $500,000 to a law firm. Mark Green, "The High Cost of Lawyers," the *New York Times Magazine*, August 10, 1975, pp 8–9, 53.

†Legislative bodies throughout the United States create new laws at a rate of more than 100,000 in some years, and federal agencies generate about 35,000 or more new regulations every year. "Those #*X&!!! Lawyers," *Time*, April 10, 1978.

ward work in law firms or salaried positions with corporations.* "Lawyers who practice by themselves may go the way of the blacksmith," wrote *New York Times* reporter Tom Goldstein in a series on the rapid growth of the legal profession. There was also a movement toward employment in large firms, and the very largest, the twenty firms comprising the "Wall Street" bar, which ranged from 50 to 125 members in the 1960s when Erwin Smigel profiled them in the *Wall Street Lawyer,* increased their average size many times over.[16] The number of firms that joined their ranks increased too.

Although New York City remains the capital of the American legal profession,[17] there has been a significant shift in the national distribution of lawyers in the past decade. With more than 40,000 of the country's half a million lawyers, New York City accounts for more than $2 billion in legal business.[18] But its monopoly on big business accounts handled by the Wall Street firms has been broken as large firms have grown up in other cities to meet the needs of corporations that no longer feel the necessity of having Wall Street lawyers make their legal decisions.

Houston, an increasingly popular site for corporate headquarters and the center of the oil and gas industry, now has three of the country's eleven largest law firms, as does Chicago (each employs more than 200 lawyers). One of the Chicago firms, Baker and McKenzie—the country's largest—is a loose consortium of 418 lawyers in nineteen countries. Cleveland, Los Angeles, San Francisco, and Philadelphia all have at least one Wall Street-type firm sophisticated enough to handle complex corporate problems. Six Washington, D.C., firms now have over 100 lawyers each, and dozens of others have grown in size, meeting the demand for guidance through the legislative and regulatory mazes created by new government regulations.[19]

Even in New York, firms still referred to as "Wall Street" have acquired addresses elsewhere in the city. Many have moved to the elegant new skyscrapers along Park and Madison Avenues in the midtown area or have opened branches there.[20]

Corporations also have been increasing the size of their in-house counsel staffs, partially as a response to the skyrocketing $100 to $150 per hour fees charged by their Wall Street firms. The legal departments at

---

*In Philadelphia, for example, the proportion of lawyers working in a firm grew from 10 percent of the bar in 1925 to 11 percent in 1952 and 32 percent in 1976. Concurrently, the size of the firms grew. Janet Connally, "Law Firms in Philadelphia," *The Shingle,* March 1977.

Mobil, Exxon, and Western Electric, for example, rank in size with the country's largest law firms.[21]

Yet the proliferation of business still has insured prosperity to the large firms, although some pay more attention than others to acquiring new clients.* Partners in such firms easily make a quarter to a half a million dollars a year.[22]

These changes have meant redistribution of the legal power formerly centralized in New York, although the New York firms certainly remain powerful. The East Coast establishment has had to compete with other areas of the country for (among other things) control of the national bar association and for the bright law school graduates. Ten years ago, when there were 300,000 lawyers in the country, one of every seven new lawyers came to practice in New York City. Now, when law school enrollments have doubled, only one of every ten comes to New York. Lawyers are finding it more attractive to work and live in California, Texas, Florida, and Georgia.[23] Law firms have also faced changes in the kinds of informal relationships characteristic of their ways of doing business.[24] (The consequences for the firms and their memberships are further discussed in chapter 11 on the Wall Street firms.)

The incomes of lawyers were also going up dramatically. Although incomes vary by size of city and of firm, average net incomes of lawyers ranged, in 1975, between $40,000 and $50,000 across the country.[25] A Price, Waterhouse study reported a nationwide average of $63,890 in 1978,[26] and 1975 law school graduates started at an average salary of $15,000.[27] By 1980, the large firms were starting young associates at salaries of up to $39,000.[28]

If a growing profession was good news for most lawyers, it was bad news for ordinary people. Complex laws about complex jobs written in indecipherable language, noted a *Time* cover story in 1978,[29] guarantee employment to lawyers but create resentment on the part of the public. However, legal spokespersons such as Stanford Law Professor John Kaplan maintain that the law's complexity is necessary to guarantee rights, and it is this complexity that creates the need for "secular priests" to guide the lay person through the tangle. Today a businessman, small or large, cannot function without an attorney.

The two trends, egalitarian and technical, are creating jobs in all professions, but also forcing a thoroughgoing reassessment of how well

---

*Firms are catering to corporate clients by opening branches across state and national boundaries to accommodate or recapture them. "We're trying to attract back some of our old clients," related *New York Times* reporter Tom Goldstein on the basis of interviews with several New York lawyers whose firms have opened Florida branches in the last few years. "Law, Fastest Growing Profession," the *New York Times*, May 16, 1977.

professionals do their jobs. The ambiguity directed toward lawyers and other professionals is not new;* it is grounded in the expertise of the professional, which is highly regarded by the public, in the public's dependence on the professional for essential services, and in problems generated by the authority of the professional. Robert K. Merton and Elinor Barber, in a classic exposition of the ambivalence generated by professionals, note: "However great its legitimacy, authority is known to have a high potential for creating ambivalence among those subject to it. Authority creates a measure of respect, love, and admiration, and of fear, hatred and sometimes, contempt."[30] They also point out that the ambivalence extends to such structural conditions as the fact that while professionals are defined as those who should subordinate their own interests to that of the clients, at the same time professionals make a livelihood from their clients' troubles. Furthermore, professionals are agents of frustration because they often ask clients to alter behavior, to refrain from doing things they want to do, or to surrender their values. Also, the standards of accomplishment are different for professional and the lay public.[31] The professional is bound to assess "success" by the "performance in terms of what is accomplished in relation to what, under the circumstances, could be accomplished," and the layperson is bound to judge whether the performance solved the problem to his or her satisfaction.[32] But the "normal" ambivalence stemming from the social roles of professionals and laity has changed in character as a result of the political movements of the 1960s and growing cynicism of the public toward the professions and the legitimacy of their authority.

The movements of the 1960s were important in pressing for egalitarianism in all spheres; they questioned the alleged professional standards that excluded entire categories of persons. Indeed, there were (and remain) challenges to the competence of the gatekeepers to establish criteria for recruiting and hiring which were not exclusionary in effect. The passage of the Civil Rights Act of 1964 prohibiting discrimination in employment and education provided the mechanism for these challenges.† Further, the authority of professions to be entirely self-moni-

---

*Plato, in *The Republic,* showed a contempt for lawyers (and doctors) when he wrote "when intemperance and disease multiply in a state, halls of justice and medicine are always being opened; and the arts of the doctor and the lawyer give themselves airs . . . . Is it not disgraceful, and a great sign of want of good breeding, that a man should have to go abroad for his law and physic . . . and must therefore surrender himself into the hands of other men whom he makes lords and judges over him?" *Plato's The Republic,* trans. by B. Jowett (New York: The Modern Library, n.d.) pp. 110, 111.

†E. Digby Baltzell, in *The Protestant Establishment* (New York: Random House, 1964), argues in his conclusion (pp. 380ff) that the failure of the Protestant upper class to "bring in" a true elite, i.e., to recruit the best, has eroded their own authority and prestige.

toring was being challenged in other realms. As a special report in the August 16, 1976 issue of *Business Week* put it:

Watergate cracked the image of the lawyer as one who reveres the law. Scientists and engineers are under attack for unseemly bickering over the safety of genetic research, supersonic transport and nuclear energy. In a sense, the gap between what the professions can deliver and what the public expects them to deliver results from two trends beyond their control: technological complexity and egalitarianism.[33]

One indicator of the law's "cracked image" was the proliferation of legal malpractice suits, virtually unheard of before this decade. Charges of incompetence against a significant proportion of the legal profession were also made public by insiders as illustrious as the Chief Justice of the United States, Warren Burger, and by the former American Bar Association president, Chesterfield Smith.[34]

The legal profession met a number of crises affecting its ability to be self-regulating in the 1970s, including a ruling by the Supreme Court that the practice of law is a business. The high court thereby cut the ground out from under the bar's claim that it was a self-regulating profession beyond the reach of antitrust law, opening the door to attacks on the basic tradition of how law is practiced. The Justice Department also successfully sued the American Bar Association for unlawful restraint of trade in prohibiting advertising (suits were also brought against engineers and physicians).

The legal profession has been declining in public image in recent years, some observers claim. A 1973 Harris Poll found that only 18 percent of the public had confidence in law firms (a lower approval rating than for garbage collectors, police, or business firms), and another Harris Poll taken in 1978, rating public confidence in eighteen institutions, found law firms at the bottom (along with Congress, organized labor, and advertising agencies).[35] These polls, however, probably do not reflect the positive side of the ambivalence laity feel about the professions.[36]

The mood of the 1960s and 1970s was also reflected within the profession as young lawyers attacked traditional practices and profit-oriented law. A growing body of critics, characterized by American Bar Association president James Fellers as "a reform-oriented group of lawyer activists who have great skill in mobilizing press and Congressional interest in their views" have attacked bar practices.[37] The demands from outside the profession for legal services for the poor were heard by sympathetic professionals within who disdained a legal system that they felt primarily served the needs of those with substantial economic resources. Young lawyers, led by such advocates as Ralph Nader, saw the profession as a leading instrument for consumer advocacy, civil rights, and protection of the poor, and sought to use it to accomplish socially oriented goals. Although their numbers were not large, they represented a substantial increase over the "old left" defenders of the poor and politically problematic clients dating from the 1930s. Furthermore, the issues and cases of the liberal left and radical left lawyers of the 1960s and 1970s moved into the limelight. They, together with labor, civil rights, consumer, and community service groups, were responsible for, among other important work, challenges to bar rules guaranteeing set legal fees. The effect of a Supreme Court decision in the case of *Goldfarb* v. *Virginia State Bar* in 1978 provided legitimation for lawyer reform in permitting lower legal fees for clients than the fee-fixed standard set by the Bar Association.[38]

These developments contributed to a new climate of consciousness and conscience; even young lawyers in the corporate firms felt a responsibility to participate in socially oriented law and pressed their superiors to assign firm time for pursuit of *pro bono* cases.*

The more radical of the issue-oriented lawyers also sought to bring the egalitarian life-style of the youth culture to the practice of law, forming law communes or collectives and abandoning the hierarchical structure of the traditional firm. Many of these firms have since disappeared, due to the problems of economic viability and the changing interests and life cycles of their members, although new collectives have been formed. Public interest law has been institutionalized, to a certain extent, in the form of publicly and privately supported law firms and organizations, although the interest expressed by law students in the 1970s seems to have diminished.

The entry of women into the legal profession is as much a part of the changing context of the legal profession as any of the other factors that have contributed to its changing size, shape, and focus. From a negligible 6 percent of law school enrollment in 1968, women today constitute 31 percent of the total, and at some schools the figure is 50 percent or more. Their dramatic increase in numbers is part of the new fabric of the legal profession. Women lawyers experienced severe discrimination in the past; today their presence is met with responses ranging

---

Pro bono publico work is an old tradition, set down in the Code of Professional Responsibility of the American Bar Association, under which firm time is set aside for no-fee work for worthy causes or indigent clients. In the past, lawyers in big firms often performed such work for their favorite charities or cultural organizations.

from receptivity to rejection, with a high component of ambiguity. But public expressions of welcome are more characteristic in a changing value structure that has generally legitimated women's claim for professional roles. Popular and legal newspapers, magazines, and journals over the past ten years give a generally positive view, as typified by the comment by *Time* magazine that "women are raising the standards of the profession."[39]

Thus growth, the changing shape of the profession, the move toward egalitarianism and the legal machinery to achieve it, the undermining of the gate-keeping function of professional elitists, pressure from groups such as black and women's movement activists, and government regulation have created an opening of structure and opportunities in the late 1960s and 1970s.

The opening up of law schools and private and public law practices to women has caused a surge of women's participation in the profession. In fact, the traditional view of law as a male field has been drastically changed as one third of the nation's law students are now women, and women now work in every kind of practice, up and down the hierarchy. Old prejudices are being challenged as women demonstrate their competence in all aspects of the legal enterprise from corporate law to law in the social interest, from legal research to courtroom advocacy.

But forces remain that mitigate against an entirely open structure, ranging from entrenched views about criteria for a true meritocracy, protection of in-groups, and long-standing prejudices against outsiders. Further chapters of this book explore the progress women have made in the field of law as well as the complex problems of true integration they still face.

Where They Came From
and Why They Chose the Law

WHO ARE America's new women lawyers? What backgrounds do they come from, and what influences have formed them? Occupational heritage, particularly of professionals, has been of interest to sociologists for a long time, but we know more about the backgrounds of men than of women. Do women come from the same backgrounds? Are they subjected to the same influences? In the recent past, becoming a professional was unusual for a woman, and those who chose to do so must have been influenced by special social factors. Yet the question remains whether the background profile of women in the professions differs from that of men.

The study of occupational choice is complicated. Puzzles remain even about those men in occupations that received much research attention. Work on women's occupational roles is rarer, but some historical materials and clues exist.

More attention has been given to the social origins of lawyers than those of other occupational groups. Studies of lawyers have been done by social scientists[1] and writers of fiction.[2]

With roots in tradition and history, linked to big business and to government, law was a worthy profession for the sons of the American aristocracy. It has been a route sought by many minority groups, particularly the Jews, in their quest to "make it" in American society. It has recruited from the economically comfortable, the intellectually talented, and the aspiring.[3] Most lawyers have been, and continue to be, the sons of metropolitan families, according to a 1972 survey of eight selected law schools conducted by Robert Stevens.[4] But have the daughters of these same families been motivated by the same aspirations and driven by the same demons?

Parents' Education, Occupation, and Income

Lawyers tend to have well-educated parents, but the parents of the women tend to be somewhat better educated, on average, than those of the men. A 1961 National Opinion Research Center study of college students who chose law as a profession showed that 46 percent of the students opting for law had fathers with B.A.'s or higher degrees, as opposed to 21 percent of the students choosing careers other than the law. Mothers of pre-law students also had more education than those of their schoolmates: 21 percent of the mothers of the pre-law students had B.A.'s or higher degrees, compared with 16 percent of the mothers of their schoolmates.[5]

When allowances are made for the generational differences separating the women lawyers I studied in 1965 (their median age was 47.5 years) from the college-age subjects of the 1961 National Opinion Research Center study (and for the differences in origins of the two groups), the parents of the 1965 sample exhibited an unusual level of educational attainment. About a quarter of their fathers had B.A.'s, professional, or higher degrees, and almost a quarter of their mothers also had college or professional degrees.[6] A comparison of the mothers of the pre-law group and my 1965 group is especially striking: the percentages of mothers of both groups who had college or higher degrees were almost identical, despite the fact that almost half of the older group's mothers were twenty-five years or more older than the pre-law group, and pursued their education at a time when college-level training for women was far less common. Mothers who were college-educated tended to be American-born.[7] However, lawyers whose mothers were foreign-born and had little formal education often pointed out that their mothers were well-read and concerned with educating themselves. The relatively high degree of education of the mothers I interviewed in 1965 seems even more interesting in the light of a finding from a study completed about the same time of eminent women lawyers listed in Who's Who by Rita Lynne Stafford. She found that only one association between background and motivating influ-

24

ences of these women was significant—that of mother's education as related to mother's encouragement of daughter's education.[8]

The women in the Stevens 1972 study came from families with more education than those of the male students.[9] Almost twice the percentage of fathers of the women students, as compared with the male students, had graduate or professional degrees beyond the B.A. Educated mothers seemed to continue to "count." Of the mothers of female students, 16 percent held advanced degrees as against 8 percent of the male students' mothers.

The mothers of the black women lawyers I have interviewed in the past ten years as part of my early study of women lawyers and as part of a study of black professional women* also tended to be highly educated, especially when compared to the average education of black women. Unlike the mothers of white lawyers, many of the black mothers had higher educational levels than the fathers.

Male lawyers primarily come from professional families, and they are even more likely to inherit their professional direction from fathers than are the sons of other professionals. The 1961 National Opinion Research Center study of college students who chose to study law indicated that a lawyer parent is the strongest single predictor of the choice of law.[10] Of the college seniors who had lawyer fathers, 35 percent chose law, contrasted with only 5 percent of the sons of non-lawyers.[11] Apparently lawyer-fathers are strong role-models for their sons, or they are active and powerful in guiding, persuading, and insisting that they follow in law careers. Studies of Chicago lawyers by Dan Lortie[12] and of New York City lawyers by Jerome Carlin[13] have shown, however, that the fathers of lawyers in these cities were somewhat less likely to have been professional men, probably because many came from Jewish or other ethnic minority groups and were educationally underprivileged.

The women lawyers I studied in New York in 1965 followed the latter pattern, perhaps because half of them, like their male counterparts studied by Lortie and Carlin, were of immigrant origins. Only six of the fifty-four women had lawyer-fathers, and an additional two had surrogate lawyer-fathers—brothers with whom they practiced. The women

---

*This, of course, was a function of the general discrimination pattern up until the 1970s, when black women were often encouraged to get college educations to prepare them to become teachers, whereas it was known that black men could rarely aspire to professional jobs. Cynthia Fuchs Epstein, "Positive Effects of the Multiple Negative: Explaining the Success of Black Professional Women," *American Journal of Sociology* 78 (January 1973): 912–35.

matched the Carlin sample of male lawyers with respect to father's occupation; however, 19 percent of the women had professional fathers, in contrast to 10 percent of Carlin's male group. A majority of the women had fathers who were businessmen (60 percent) with incomes described as comfortable or better.[14]

This pattern held for young women in law school surveyed by Stevens in 1972. They tended to come from families with higher average incomes than those of the men. For example, 33.3 percent of the women, but only 19.5 percent of the men, reported average yearly parental income in excess of $30,000.[15]

There is a good reason why women professionals tended to come from higher income families than men. Only these families were willing to support the higher education of women, who were not generally expected to pursue careers after marriage.[16] In the 1950s, although it was common for young women to work to support their husbands who were attending professional school, one rarely heard of the reverse case. (Women were said to have qualified for the "PHT" when their husbands graduated—"Putting Husband Through.")

The spirit of equality that emerged in the 1970s meant that young couples more often took into account the professional aspirations of wives as well as husbands in calculating their financial needs. Families were also more likely to help support married daughters who were choosing a professional option. Many of the women going to professional schools in the late 1970s were in their late twenties or in their thirties. These women could feel more comfortable than previous generations of married women about husbands supporting them while they were in school because they were convinced they would actually practice law.

Professional Heritage

The size of the sample I interviewed in the 1970s for this study did not permit generalizing statements about women following in the footsteps of their fathers. However, it seemed clear from the interviews that the changing mood of the times had affected the fathers, and the lawyers among them had encouraged daughters to follow in their footsteps. (The non-lawyers had also encouraged their daughters' legal aspirations.) Those interviewed included daughters of prominent civil rights attorneys who were practicing in the public interest sector and daughters of prominent Wall Street lawyers who were in large corporate practices.

By the late 1960s I found that parents were reasoning about daughters going into law the way they had reasoned they should be teachers a generation before. An attorney in a federal prosecutor's office described her discussions with her parents this way:

When I mentioned going to law school my advisor in college said he really wasn't sure it was the thing to do, but my father said it would be really good for me to go to law school because I could have a profession and if I got married and had a family it would be good to have. My mother said the same thing, it would give me options later on; you know, to go back.

It was now also more possible for a lawyer-daughter to follow in the footsteps of her mother. James White's study of lawyers[17] showed that the percentage of women attorneys who had a female relative in the law was greater than the percentage for men by a significant margin, although the absolute number was small—only 47 out of 1,300 women had a female lawyer-relative. The mothers of 4 men in his sample were lawyers, but 20 of the women were following the careers of lawyer-mothers.

Of the lawyers I interviewed in 1965, 20 percent had mothers who were or had been engaged in professional occupations. Of these, not surprisingly, 9 were teachers. One mother was a lawyer and had finished law school only two years before her daughter. About two-thirds of the mothers had worked at some time in their lives, but only a few worked all their lives. Most had become housewives after they had children and nearly one-third worked only before their marriage.[18]

Women law students studied in 1972 by Stevens were daughters of working mothers in greater proportion than their older professional colleagues, and certainly to a greater extent than their male peers. The Stevens study of the class of 1972 showed that women were more likely than their male counterparts to have a mother working outside the home: 44 percent as against 32 percent. And only one-fifth of the working mothers of the female students were employed in secretarial or clerical positions, compared with more than one-half of the working mothers of male students. The study implied that working mothers of women students were more often engaged in professional work than the mothers of the male students.[19]

Working mothers seemed to act as models for their daughters in other ways. In my earlier study, the women whose mothers had worked as professionals or semi-professionals all held full-time jobs. We suspect that these mothers had provided their daughters with an example and had imbued them with a positive attitude toward productive work. As far as these lawyers could remember, most of their mothers who had worked seemed to have enjoyed it. At the very least, most of these women lawyers did not receive negative images of work from their mothers' experiences.

But knowing whether a mother worked or was a homemaker only gives a limited picture of the influence she was likely to exert.

To a striking extent the women lawyers I studied tended to have mothers who were "doers," either professionally or avocationally. Their past accomplishments were evident. Unlike men, women can be rated as "successful" by society even though they have never attempted to work. At least this has been true in the past. Although the lawyers studied earlier sometimes said their mothers had never "worked," on further questioning it often turned out they had pursued unpaid or independent activities that clearly could be defined as work. In one case, the mother had managed the family's real estate holdings, providing the main source of income for the family after the death of the father. Five of the mothers had worked in the family business without pay and did not claim to "have jobs." Some of the mothers had very large families (with five or more children) and were housewives, but were characterized by their daughters as possessing unusual vigor and managerial skill. About one-third said their mothers were active in organizations or clubs.

In my study of black professional women, I found that a high proportion of their mothers had worked. This in itself is not uncharacteristic, considering the economic need of blacks in this country. But the black women's mothers typically had jobs higher in prestige than their fathers. Of the thirty black professional women I interviewed, twenty-six said their mothers had worked: seven had been teachers, one a college professor, two were nurses, and one a physician. We found the same pattern among black women lawyers. The mother of one black lawyer was a physician whose career had been outstanding by any standard. This doctor had been a strong influence in the life of a black woman physician interviewed during a study of black professional women[20] and she had influenced the careers of many young women over the years. Now well past retirement, she was still active as a physician in a poverty program at the time this was written.

Although the mothers of the black women may have been unusual for their race and class, they were similar to many of the mothers of the white women who pursued law, not only in educational attainment but in their general attitudes toward the education of their daughters. Black and white mothers all seemed ambitious for their daughters and pushed, prodded, encouraged, or suggested that their daughters pursue a life goal above the norm.

Although fathers often have dreams for their sons, it is mothers who probably seek more vicarious expression of their own dreams through their children. Mothers traditionally push their sons* to occupational accomplishment and their daughters to "successful" marriages to men of stature or promise.[21] Yet some mothers have seen in their daughters a way to achieve vicarious fulfillment[22] of their own dreams or hopes for women as a class.

Women practicing law a decade or more ago remembered how their mothers had kept alive frustrated ambitions to become lawyers or to have other careers, and had encouraged them toward a legal career.[23] Others simply had high hopes for their daughters and were strong-willed in imposing them. Some of the memories their daughters recounted in interviews in the 1960s pointed to their mothers' influence:

If mother had been opposed [to law school] I wouldn't have done it. She had it in her head that my oldest sister become a school teacher ever since she was a little girl and she became one. She wanted me to be a lawyer. She was a very strong-minded woman.

Mother was a suffragette who organized motorcades when the automobile first came out. She believed that a woman could take any rank, status, or level that a man could.

My mother was one of the first women to be a political leader when women got the vote. . . . I think my mother just pushed me into the law. . . . She was a frustrated career woman. And my brother wouldn't be a lawyer so I had to become one. In those days you did what your parents said you should do.

The lawyers we talked to in the years after the 1965 study often talked of the impact of their mothers too. One or two generations later they told similar stories:

My mother did work in my father's company but she didn't feel she had a real job. I don't think she's really been able to achieve in South Carolina. So I guess she always figured her daughters could if she couldn't. She always told us we

*An early study of achievement motivation by Strodtbeck found that for male high achievers their mothers were important in setting high standards and holding values. Fred L. Strodtbeck, "Family Interaction, Values and Achievement," in David C. McClelland, A.L. Baldwin, Urie Bronfenbrenner, and Fred L. Strodtbeck, *Talent and Society* (Princeton, N.J.: Van Nostrand, 1958) pp. 135–94.

29

would be happier if we worked. She always figured she never made it out but she was going to get us out if it killed her.

A daughter of immigrants who had survived concentration camps and whose mother had wanted to be a doctor but couldn't said:

So I was her big chance. I think she probably saw a lot of her in me. But I certainly never became a lawyer to further her dreams. None of that stuff. I did it for me.

We both laughed at that, knowing it was probably both true and not true.

"Who do you identify with?" was a question asked in the 1960s interviews on the assumption it was a variable important in motivation and the socialization process. Somehow, in the 1970s, the notion that a girl's achievement motivation might flow from a sense of identity with a father seemed less important than it had before. It is hard to say whether mothers were more important than fathers; the process of "identification" is difficult to isolate both for the person experiencing it and for the person studying it. Some of the lawyers observed that their recollections of childhood had shifted and that at various times in their lives they had realized that a father or mother had had more or less impact than they once had supposed. Many of the women lawyers I interviewed couldn't or wouldn't specify which parent was more important to them, although more said they took after their mothers than fathers. In Stafford's study of eminent women lawyers, 27 percent of the sample felt closer to their mothers and 24 percent felt closer to their fathers.[24]

The influence of the mother, not as a model but as a "force," may have been more important than any other single factor in guiding some of these daughters into law, although it is difficult to say exactly how their influence was manifested. Many of the older lawyers we spoke to could not clearly recall early influences—in the original study more than half were over fifty years of age and did not appear to have thought much about the connection between their career choice and the other factors in their lives. But even the younger group, born in an era when psychologizing was in vogue, were less concerned with what shaped them than one would expect. Perhaps this has something to do with learning to think like a lawyer, or perhaps law is selected by persons not given to the torment of self-analysis.

Many of the lawyers interviewed saw their choice of law as an "uninfluenced" one. This had been the prediction of Dorothy Thomas, editor of the first *Directory of Women Lawyers*,[25] who had interviewed women lawyers in the course of compiling her book. In a conversation before I began my study, she commented that lawyers, particularly women lawyers, stress "having done it myself," and that one might expect them to recall that they made up their "own" minds and had not been dependent on help, psychic or social, from others. Almost half the lawyers we spoke to said that *no one* had influenced them.

The lawyers we talked to in the years following the original study were somewhat more affected by the focus on "influences" created by popular sociology and psychology. Like many others, they seem to have searched for the persons who were the "role models" in their lives, even when they were hard put to find any. There were also notions about who a role model ought to be. Many women seemed committed to the notion that it ought to be the father (a current view, popular in feminist academic circles and in publications such as *Ms,* perhaps with Freudian inspiration); few could see parallels in the lives of their mothers, and most were hesitant to say that their mother was a role model although she might have been important.

When Stafford asked her sample of eminent women lawyers about fathers' and mothers' encouragement of their professional ambitions, 75 percent said that their mothers had encouraged them, and 65 percent said their fathers had encouraged them.[26] In the later interviews, we found that mothers were more apt than fathers to have actively encouraged their daughters to go into a profession—the fathers tended to have more traditional views about what their daughters probably would do. The women who ultimately became lawyers usually didn't have to fight opposition at home. Their fathers accepted their aspirations and often became enthusiastic supporters, boasting about their daughters to friends. In fact, in recent years most parents of younger lawyers have shown great enthusiasm and pride in their daughters.

Usually, only positive influences are mentioned by researchers and journalists in the current search for "role models." But motivation can also come from negative role models—those persons whom one does *not* wish to be like.[27] For many women in the 1970s, their mothers were negative role models in some ways.[28] They felt their mothers' lives had been frustrating and unfulfilling, and their talents had been subordinated to their roles as wives and mothers. Further, many mothers encouraged their daughters *not* to follow in their footsteps.

Not all mothers were positive about their daughters' careers, however, and cultural ambiguity still weighed on occupational choice. These tensions were expressed by one lawyer whose relationship with her mother was characteristic of the ambiguous messages girls have been exposed to in American society for many years.*

My mother is a businesswoman. But she resents being one. She says she would much rather have stayed home and had three more children. Her attitude about practicing law was that it was unfeminine and I should get married and have children. I'd never find a husband if I graduated from law school.

She says she hates her job, but she is successful at it and really loves it. She functions best on the job, but she will tell you that all she really wanted to do was be a mother and a housewife.

Yet she was so proud when I was admitted to the bar. She came in full flourish, with her mink coat—the works. . . . She'll talk about "my daughter, the lawyer" —but not in front of me.

Another woman reported that her parents' ambiguous behavior took this form:

My mother said it was the happiest day of her life when I called her up and told her I wanted to be a lawyer. But they worry about me. For a while they were very upset that I seemed to be drifting away from my husband. My father said, "Don't do it. Quit your job before you ruin your marriage."

But some women managed to escape ambiguous or negative messages about the roles they should be assuming. Some of the women I spoke to in the 1960s, before the era of "raised consciousness," reported they were not challenged or harassed for choosing a non-traditional occupation for a woman. In fact, among those who were daughters of immigrants, families often weren't aware of the fact that law was not considered appropriate work for a woman, and many thought it was a good idea for a daughter (like a son) to do well in America by acquiring a profession.[29]

Husbands

One set of influences on occupational selection that sociologists of the legal profession have never studied is that of spouses. One lawyer we interviewed reported that her husband decided to become a lawyer after she did. But the most common pattern was for the wife to follow her husband.

*The earliest sociological statement on this is, of course, Mirra Kimarovsky, "Cultural Contradictions and Sex Roles," *American Journal of Sociology"* 52 (1946): 184–89.

Where They Came From and Why They Chose the Law

The first women lawyers in the United States, whose names are remembered for having battled for the right to legal education or to practice at the bar, were married women who sought to join lawyer-husbands in practice. Many of the wives received instruction in their husbands' law offices, and some lawyer-husbands worked to get legislation passed to open the profession to women. One such crusading husband was P. A. L. Smith, who addressed the argument that admitting women to the profession would interfere with their domestic affairs:

This is all bosh, for in the first place she may not have any domestic affairs to look after; and in the second place, she herself is the best judge as to whether it pays her better or suits her better to look after other business and pay someone to keep an eye on domestic affairs. . . . My wife has studied hard. . . . She had studied at my request and anything I can do to secure her admission to the bar will be done.[30]

Myra Bradwell's name lives in legal history because of the Supreme Court's 1873 Bradwell decision upholding a lower court ruling that denied her a license to practice law on the grounds that she was a married woman. She had been asked by her husband to learn the law and assist him in his practice.[31]

In the days when women lawyers' access to jobs was limited by discrimination, the family law firm often provided the only opportunity for women lawyers to practice. It was probably true that through the years many wives have worked in their lawyer-husbands' offices in the same way that wives assist husbands in other businesses, without formal training or recognition.[32] But for those women who were encouraged by their husbands to study law and become lawyers an unique opportunity was created.

Women came to law by other routes as the years passed, although the impact of lawyer-husbands continued to be important to women's law careers. In fact, apprenticeship in the large firm that employed her husband started one prominent attorney on the route that led to partnership in that firm. A husband's example also inspired the career of another law school professor, one whose name was often mentioned in discussions of who might be the first woman Supreme Court Justice. She told me she chose to go to law school because her husband was in law school, and in the 1950s couples thought in terms of "togetherness." She said that if her husband had been in medical school she probably would have considered becoming a doctor.

And another husband fired his wife's interest in a legal career:

I met the man who became my husband the last year of college. He was working at a Wall Street firm and he was very enthusiastic about his law practice at the time. I had no idea this kind of legal work existed. It sounded very interesting to me, technically interesting. You know, a way to use your mind in a conceptual and technical way. It was a different conception of the law than I had growing up in New Jersey, where you thought of lawyers as politicians in small clubhouse practices and actually basically dishonest. So although I worked in publishing for a few years after graduation I decided to go to law school at night, and did so well I decided to go full time.

Peers

Women who attended college in the 1960s and 1970s were more affected by their peers in deciding on law as a career than preceding generations. During that period, at least in many of the elite women's colleges, the perception (often not correct) was that "everyone" was going to law school. At Barnard College, for example, six months after graduation 2.8 percent of the graduating class was attending law school in 1969. In 1971, the percentage had increased to 5.8 percent. Talk of law was in the air. It was becoming the thing to do when a young woman couldn't make up her mind about what else to do.*

It is perhaps sufficient here to note that background characteristics were probably competing with social events as predictors of motivation.† Of course, background may predispose both women and men to become involved in special events, but the drama of the 1960s clearly seemed to affect the ways in which young people thought about careers.

An article exploring the changing mood toward law school in 1974 observed that one-third of a recent graduating class at Smith applied, and one-half of the Radcliffe class of 1971 actually went on to law school.[33] "Were there many women at Smith who planned to go to law school?" I asked a lawyer just starting practice in 1976. "There were," she remembered. "We had a strong group . . . we were doing the applications and taking the boards together."

*Statistics provided by Jane S. Gould, Director, Office of Placement and Career Planning, Barnard College, in a letter dated August 9, 1972. The largest percentage attending professional school were going to law school. This was followed by medical school, where the increase was from 3.8 percent in 1969 to 4 percent in 1971.

†Other influences affecting young women's career choices in the 1960s and 1970s are discussed on p. 36.

34

Where They Came From and Why They Chose the Law

By the end of the 1970s, law school had become a viable a choice for women graduates intent on pursuing higher education as more traditional fields.

Where They Went to Law School

Proximity to a law school, even its location in the city, were important considerations for the New York area lawyers interviewed in 1965. They were affected by the norms that stipulated that young women ought to stay closer to home than the young men in their families, though even young men were often constrained to attend schools near home for economic reasons. Further, it was common for young people to live at home until they married unless they went off to college. But, as sociological studies show,[34] there has always been a tighter rein on women than on men. Some of the New York City women did not apply to Columbia University Law School because it was distant from Brooklyn and the Bronx, where many of their families lived.

Of the women interviewed in 1965, fewer than 10 percent of those who attended New York area schools chose their law school on the basis of academic criteria. Most of them selected a school because of "practical" concerns of the moment—restrictions on admission at the institution of first choice, pressure to be at or near home, or the desire to combine work with their education. Financial reasons aside, the women's selection of schools was geared to short-term, non-career considerations rather than long-range occupational gain, planning, and goals.

The picture changed considerably by the mid-1970s. Although women were choosing law schools for some of the old reasons, many more were making choices based on practical evaluations of the track that would lead them to their goal.

Married women coming back to school after some years were much more constrained to go to law school in the city where the husband worked. But many adult women have their own ties and roots that hold them in place. Many of the single women interviewed were living in New York when they made the decision to enter law school and had not thought about going out of the city for their legal education.

Many law aspirants are not aware of the professional advantages that graduation from a prestigious law school gives them. Law is an avenue for groups without financial resources, although few scholarships were open or available in the past. Women who came into law were affected by both of these factors, but in somewhat different ways than the men.

First, few women were aiming for the high-level law careers for which the elite law schools would prepare them. The men who went to elite schools typically came from privileged families for whom their choice of school was the conventional and desired thing to do, while the same choice would have been thought unconventional for their daughters. Some might proceed from Harvard College to Harvard Law School, but few families hoped their daughters at Radcliffe would follow the same path, even after women were admitted to Harvard Law in 1950.

In New York, Columbia Law School had a more diverse student body. Although discrimination was encountered by many sons of immigrants who wished to go to Columbia, a large number of very talented students did filter in. Among them were a small percentage of the older women interviewed in 1965, those whose families could offer support for full-time study.

In spite of the fact that 90 percent of the women interviewed in the 1960s described their families' incomes as comfortable or better, only 30 percent did not work at all during law school. Almost 40 percent worked full-time and about an additional 20 percent worked part-time, nearly all for economic reasons. Although many of their families were "comfortable," more than half had three or more children and two-thirds of the women had younger siblings. Some of the families had to save for the education of the younger children and could not invest very heavily in further education for the older children, especially if they were girls. Most "knew" they would have to work and never considered alternatives such as loans from other family members or outside sources.

Today, men and women unhesitatingly turn to outside sources of income in case of need—government and bank loans, scholarships and other forms of support—few of them available in the past and even fewer open to women. In the law schools, important fellowships (which carried prestige and assured entrée into the legal world afterward) were limited to men. Considerable activity by individuals and women's groups succeeded in ending these restrictions and most are now open to women.

36

Motivation for Law School

Before the 1970s it was hard to characterize patterns of motivation of women lawyers. The percentages were not large enough to show the predominance of one set of reasons for choosing law over another. Many of the women lawyers in the group studied in 1965 went into law for the same reasons as male lawyers who were of their class, ethnic background, or personality type. Those who were daughters of immigrants hoped to make money and to rise above the status of their parents. Some were pressed to become lawyers by parents who wished them to follow in their footsteps; or by husband-lawyers who needed them in their law practices. Others found law attractive for a variety of idiosyncratic reasons. Women, like men, read about Clarence Darrow and were inspired to become advocates for justice. It is also hard to classify those women who always wanted to be lawyers and who held on to their fantasies despite often harsh reality.

All were different from their male colleagues in that they were choosing a profession thought inappropriate for their sex. Some women were aware of the prejudice they would encounter, but a surprising number didn't anticipate it. Law was not just another career option for women; they had to be especially motivated to face a system that was exclusionary, or especially insulated from knowledge about its discriminatory nature. On the other hand, opportunities existed: most schools admitted some women and jobs awaited some of those who graduated.

Many law school professors and deans did not believe their schools were discriminatory toward women and explained that there were few applications from women. Indeed, most women seemed to agree that law was a male profession and that they wouldn't like law or be good at it. Knowledge about exclusion certainly undermines motivation. Most people do not want to do things that they know they will be hindered from doing, or are told they are unable to do. Tone-deaf men and women do not aspire to become musicians, and short people do not aspire to become basketball players. Men may follow in the footsteps of their fathers, but only rarely in the occupational footsteps of their mothers, since most men and women work in occupations that are sex-typed as male or female.[35]

The socialization experiences of women and men certainly contribute to their motivations for certain kinds of work. Sociologists usually point to the cultural themes and messages from families (the transmitters of those themes) which create the framework of choices within which a young person begins to think about what he or she will "be." Socialization was emphasized in my analysis of limits on women's options for professional careers in *Woman's Place*. But some women's choices and dreams may not be explained by their early experiences or by identification with another person or group in the manner explained by "reference group" behavior. This does not mean they are not affected by the people or events that form their experience; rather, it is often late rather than early experience that creates or changes their notions of an appropriate life course. The accounts of lawyers and other women interviewed in the past decade indicate that socialization is more complex and more ongoing than previously expected in determining motivation. For some, the creation of opportunity opened horizons. Few women were motivated to choose law before the 1970s, but when they became conscious that it was easier to be admitted to law school, motivation was seemingly created overnight. The psychologically oriented thought it would be necessary to wait for a generation of women to be socialized in a new way before we would see them making such radical decisions for careers in male professions.

It was commonly believed in the past that women did not choose law for the same reasons as men. Before 1970, the studies that documented the reasons for occupational choice of male lawyers excluded the women in the samples because they were too small a group. Common lore suggested that women went to law school "to catch a husband"* or to "do good."

Women often express the wish to engage in work perceived as "good" or aimed at the improvement of society. But to suggest that all or even most women are so oriented is incorrect. The assumption has pejorative consequences to the extent that it undermines women's ability to handle the rough and tumble of business law. Women share these views and often confess to feelings of guilt when they express other goals.

Janette Barnes,[36] in a 1970 study of the entrance of women into the legal profession, suggested that law school gatekeepers held pervasive beliefs that women were motivated by desires to be social workers, to help the poor and the oppressed. As a result, they were thought to be

---

*If any chose law for this reason they did not admit it. Although some did find husbands in law school, they went on to practice law. The nature of the pre-1970 sample excluded many women who dropped out of the law, but some probably did so because they couldn't find work as lawyers.

too idealistic and fragile to survive in the rugged and competitive legal business world. What was the truth?

It is hard to know people's motivations for choosing careers. Theories that measure motivational components abound, but none seems entirely satisfactory in determining the "real" factors. Individuals may answer questionnaires with sincerity, but some of their responses are merely their own "theories"[37] or are self-censored or selected memories about why they did one thing or another. They also forget why they made early choices; they may tend to romanticize early influences or pressures and to forget unpleasant occurrences. For example, an older woman lawyer interviewed in 1965 about her reasons for choosing law said she had *always* planned it. Yet when I interviewed someone who knew her as a young girl, I was told that her real love had been physics but that an accident had confined her to a wheelchair and she was unable to go up or down the steps of the university's science building. Her lawyer-father convinced her to make law her profession because, among other reasons, she physically could get into that building. The "truth" may have been in either account or some combination of the two.

Individuals may also recall their motivations in a way that conforms to a currently popular theoretical model. In the late 1960s, there was widespread acceptance of a theory of "fear of success," identified in a study by Matina Horner.[38] Publication of a version of Horner's study in *Psychology Today,* unfortunately titled by the magazine's editor as "Fail: Bright Women,"[39] gave the theory wide currency, and soon feminists and non-feminists alike were explaining women's career limitations as a result of the psychological fear of success. Social structural impediments to success such as systematic discrimination were suddenly less important.[40]

There has always been a great deal of pressure on women to say they want to work to help their families or do social good. Men are permitted to say they want to work for money or the love of their craft (or they were, at least, prior to the 1960s). Today, however, men are somewhat constrained from offering material gain as a prime motivation for work, and they more often suggest humanitarian reasons for choosing a career.

Although there is no doubt truth in the stereotype that women tend toward doing good works,* a close look at the contexts in which they choose to suggests that this choice often results from limited opportunity channels or perhaps from an underdog's sympathy for humanitar-

*If one needs documentation of this, the millions of women performing volunteer work for charities each year is but one indicator.

ian causes. It would be unfair to suggest that compassion is a uniquely feminine attribute. As social influences have acted on women to make them sympathetic toward the public interest, they have also acted on young men.

The stereotypes about women, whatever their kernel of truth, do not capture many of the other goals and dreams of women of the prior or present generations. Surprisingly, "success" was a motivation expressed by women I interviewed through the years, although communicated in different ways. Women interviewed in the earlier study seemed hesitant to admit they went into law to become "successful" because it was considered unfeminine to do so; those interviewed more recently were more willing to admit it.

Many people are uncomfortable about the "success ethic." A student working on this study, who went to a meeting of women law students in 1970, made these rather disapproving observations in her "general reflections" on the gathering:

[These] women are not at all radical or anything like it. Hard-nosed young ladies who want to succeed. I saw many marriage rings. Not many express a wish to go into legal services. I'm beginning to doubt their social goodness. I think these women have caught the professional bug and want to get to the legal department of GM or even become Chairman of the Board.

Women are believed to be motivated by social goals, and there is a cultural view that they *ought* to be. Law students often expressed distress at pressures on them to feel certain kinds of motivation.[41] That men and women students do not differ much in their responses to questions about their career motivations may come as a surprise to some. Both women and men define their motivations according to the concerns of the time and what they think is right for their sex or age.

For women law graduates in a 1967 study, "help to society" was an important reason for seeking careers in law, but not much more frequently than men. In fact, twice as great a percentage of women as men stated that monetary reward was "very important" to their choice of law.[42]

Women who replied to a questionnaire circulated at the University of Virginia Law School in 1972[43] said they wished to train for a practical and interesting career with good income and opportunities for advancement in society. A few indicated they wished to escape the pigeonhole of schoolteacher or of housewife. Many of the female students indicated they intended to concentrate on fields such as social welfare, legal aid, or urban planning, but so did many of the men.

Where They Came From and Why They Chose the Law

In the 1972 study of law students' motivations for going to law school, differences were found between selected schools, although all students appeared to reflect attitudes fashionable among their age group. For example, "desire for financial rewards," always relatively low in importance among motivations expressed by students at Pennsylvania and Yale, fell slightly from the previous decade at the two schools, although these students expected higher incomes than their predecessors. Although Yale and Pennsylvania students showed a decline in importance of "prestige" as a motivating factor, the reverse was true for students at Iowa and the University of Southern California. (The vast majority of all students surveyed over the ten-year period indicated that prestige was of some importance to them.)[44] "Desire to serve the underprivileged" showed a steady, and in some cases a dramatic, increase in student motivation over time. At Yale, the percentage of those attributing "great" importance to this motive more than quintupled between 1960 and 1970, rising to almost half the class by 1972; almost none responded that service to the underprivileged was of "no" importance.[45] A similar pattern appeared in answers to the question of whether a "desire to restructure society" played a role in the decision to attend law school. Few students conceded that a desire to work in Wall Street firms drew them to law school, although more students admitted such feelings in 1972 than in 1960.

The number of students who entered law school not intending to practice law at all grew between 1960 and 1972. In response to the question, "When you entered law school, did you intend to become a practicing lawyer?" more than one quarter of the Yale class of 1970 answered negatively. The Yale response, however, was double that of other schools. Alternatives to practice included the wish to "restructure society," to become a polititian, to go into legal education, to go into business, to go into government service, and for some men, to postpone military service.[46]

This information is important to consider when one speculates about the stereotypes of women's motivation for law careers. Not only do many men, including those at prestigious law schools, share women's supposed dislike of certain types of practice, but a substantial number are ambivalent about the law altogether and are uncertain whether they will practice. Although none of the women lawyers I interviewed in the years between 1965 and 1977 claimed to have originally planned a Wall Street career when they entered school, many, including some who intended to practice social interest law, ended up in Wall Street firms once training at a prestigious law school opened that option to them. All of the women interviewed intended to become practicing lawyers when they began law school.

Students at a comparable group of law schools* studied by Spangler and Pipkin in 1976 showed some slight differences in motivation for choosing law. "Success" was the strongest and least varying of all motivating factors; the second reason cited was intellectual stimulation, and the third was service. Men and women similarly attributed importance to prestige and income rewards in choosing to study law. Although both men and women were attracted to law by service commitments, women valued them substantially more than men.

Over the twelve-year period studied by Stevens, the predominantly male student population at every school displayed a dramatically increasing interest in legal aid, civil rights, civil liberties, and public defender work.

Yet as Spangler and Pipkin have also observed in their 1976 study, irrespective of initial orientation, both men and women at elite law schools tend ultimately to gravitate to large firm practices. It is clear that women, like men, seek success and find it possible to attain by many routes.

Career Goals

In the late 1960s and early 1970s, women were increasingly concerned with how to achieve careers. The frustrating and often bitter experiences of their mothers or older sisters made it clear that a B.A. degree, even one from a prestigious college, might lead to a teaching job, but it might also lead to the secretary's desk and not to the editorial office. A Barnard graduate recalled:

I was an English major and I started to wonder what I was going to do, and there were really very few options. Either I could teach or I could try to get into writing or publishing, which is extremely difficult. I began to see myself falling into a very traditional woman's role and I didn't like it. . . . So I made a decision to change my major into something more professional, something harder—not harder in the sense of more difficult, but more career oriented and more of a

*Research was conducted in seven schools: three are elite schools and four others are considered good schools. Eve Spangler and Ronald M. Pipkin, "Portia Faces Life: Sex Differences in the Professional Orientations and Career Aspirations of Law Students," Report from the Law Student Activity Patterns Project, Research Program in Legal Education of The American Bar Foundation, 1977.

marketable skill, so I switched to economics and I made the decision to apply to law school.

For some women who were casting about wondering what career to follow, a shift to an environment that offered career-bound tracks was important.* A lawyer who had gone to Vassar as an art history major spoke about how a junior year in an exchange program at Williams College changed her perspective:

> I had no thoughts about what I was going to do after college. But I hadn't been at Williams a week before realizing that everyone had a plan and was thinking about medical school or law school. It would be impossible to go to medical school so I started sending for catalogues to law schools.

A student at the University of Virginia Law School said that she went to law school because at Princeton, where she was an undergraduate, everyone had success goals and she was ashamed that she did not.

The fear of a dead end and the growing sense of self of young women made them seek to order their lives instead of hoping things would somehow turn out well. They realized that professional schools in medicine and law and graduate schools of business administration offered training toward professional careers and status.

In a larger sense, young women for the first time were planning their lives. One University of Virginia law student took me aside after a class meeting in which I had asked the group why they chose law and said she had chosen it to "have control over my life." She said she had been afraid to admit this in class because it seemed selfish compared to the more idealistic motivations of some of the other students.

Women students were worried about how their career plans would fit those of the men they would marry, but they were not waiting to see what the men would do. Prior to the 1970s, women often thought about work as fitting into family obligations or as an emergency measure in the event a husband or father was unable to perform the primary bread-winning role in the family.† But now women were planning careers rationally, and some were putting career choices before romance. One "new" woman commented:

---

*Although I do not subscribe to Jessie Bernard's view that there are separate men's and women's cultures, there are certainly pockets of separatism that socialize and emphasize separate values for women and men. Jessie Bernard, "My Four Revolutions: An Autobiographical History of the American Sociological Association," in *Changing Women in Changing Society,* ed. Joan Huber (Chicago: University of Chicago Press, 1973), p. 88.

†I discussed this perspective in Epstein, *Woman's Place,* pp. 74–76, as one of the problems in women's socialization that precludes them from feeling a sense of responsibility for becoming economically competent.

When I thought about starting law school the guy I was going with was at the University of Virginia. He was someone I really cared about, and the University of Virginia has an excellent law school. But I did come to Columbia because I thought it was going to be a better experience for me. I also thought it would be more useful to get a degree from Columbia. It opens more doors for you.

Some of the "older women," women in their late twenties and thirties, who entered law school in the early 1970s had left college to become active in the civil rights and peace movements. Many of the women who were active in founding feminist law firms or who went to work for the Center for Constitutional Rights, for Legal Aid and Legal Assistance Programs, or for the NAACP Legal Defense Fund came to the law in an effort to become more competent and to gain credentials to better pursue social change.

Other older students went to graduate schools in the liberal arts or humanities and left them for various reasons, not finding their experience enriching or feeling uneasy about the teaching careers that would follow. One said:

I was going to be a French professor. In fact, I spent my junior year in Paris. But when I saw the ridiculous mess I would have to go through to get my Ph.D., I was so turned off that I thought what can I do with my life, and somehow law just popped into my head. I also realized I wasn't an academic. I wanted to be a practical problem solver and deal with people.
I worked for the welfare department and was so discouraged because there was a futility about what I was doing. Then I worked for the New Jersey Americans for Democratic Action but felt that the work I was doing was not transferable after a campaign. There didn't seem to be a decent job of any kind that I could find. So I decided to go to law school.

Some had families but didn't intend to make motherhood a career. One woman had dropped out of a major law school in the 1960s because she was distressed to be one of eight women in the class and had money problems. When she decided to go back to law school five years and two children later she went elsewhere, to Rutgers Law School. "Rutgers was an easy, comfortable place," she reported. "Forty percent of the students were women, and it seemed to me that most of them were like myself, women who had come back in their thirties."

By 1980 it seemed that law school was a normal postgraduate choice for young women straight out of college. Although the women I studied in the 1960s said they would not have dared to consider law unless they had excellent academic records, more women undergraduates whose performance is average profess aspirations to go on for a law career. And a few of their mothers are also thinking about law school; women in their forties who have returned to college for a degree now feel secure enough to go on to professional school, sometimes following in the footsteps of their daughters.

Men are also changing careers and going to law school as older students. A student at Harvard Law School said that a few of her male classmates have Ph.D.'s in other disciplines. Scott Turow, author of *One L*, a novel about Harvard Law School, left a career teaching English to go to law school.[47] At New York University more than 10 percent of the entering law school class in 1979–80 were out of college five years or more.[48] Prejudice against older students has generally diminished, and some schools now even seek them to make up for declining enrollments. Older students are showing that they return the investment in their education by pursuing their careers vigorously.

II

LAW SCHOOL

Getting Into Law School

3

THE LAW SCHOOL lecture hall has replaced the law office as the entrance to a legal career in the United States only in the last eighty years. The first law school in the modern sense dates only from the 1870s, when Christopher Columbus Langdell was made dean of the Harvard Law School and charged with creating an organized and compulsory program of legal education.[1] Until 1900 the most common route to the bar was through an apprenticeship served as a "clerk" to a working lawyer. Today, aspiring lawyers must attend law school. Where they go to school and how well they do there usually is of interest long after the lawyer is an established practitioner. The most efficient way to make a start in law is to attend a good law school, become an editor of the school's law review, and win a reputation for brilliance while still a student. Until recently, this route, as well as the less prestigious alternatives, were limited to only the small number of women who managed to surmount discrimination and exclusion.

Women were notably absent from American legal education for its first 100 years. Some women had gone the route of self-directed reading or apprenticeship,[2] as did Arabella Mansfield who, in 1869, was the first woman admitted to the practice of law, both in the state of Iowa and in the United States. In that same year, women were admitted to the St. Louis Law School, the first law school in the United States to accept students regardless of sex. Of the two women who matriculated that year one, Lemma Barkaloo, had been turned away from the Columbia University Law School.[3] In 1890, when Columbia denied admission to three more women applicants, a member of the Board of Trustees reportedly said:

No woman shall degrade herself by practicing law in New York especially if I can save her. . . . I think that the clack of these possible Portias will never be heard in Dwight's Moot Court.[4]

49

In 1870 Ada A. Kepley became the first American woman to receive an accredited law degree. She graduated from the Union College of Law (now Northwestern).[5] But other women who sought formal education were being denied admission to law schools (and the bar) in Connecticut, California, Colorado, and Indiana. Even after the turn of the century, when women were admitted to the bar of almost every state, the battle for women's admission to law school continued. Although most elite schools had formally opened their doors by then (Michigan, 1870; Yale, 1886; Cornell, 1887; New York University, 1891; and Stanford, 1895), they remained inhospitable to women students. Other schools (both elite and non-elite) maintained policies of total restriction for decades to come.

Professor Ruth Bader Ginsburg, recalling the history of law school discrimination at the twenty-fifth anniversary of the admission of women to Harvard Law School in 1978, quoted the February 18, 1925 issue of the *Nation* (p. 173) on Columbia's adamant refusal to admit women:

The National Woman's Party wants President Butler to admit women to the Columbia Law School. Many times in years before the National Woman's Party was born, women tried to get into the Columbia Law School, and the walls of the masculine sanctuary always stood firm. President Butler long ago turned decision in the matter over to the Law School Faculty, but a large majority of the Professors resisted imprecations, pleas and demands from candidates, organizations, even from benefactors of the school. This defiance of the laws of change and the tendency of the times would be magnificent if it were wholly a matter of principle. The faculty, however, has never maintained that women could not master legal learning or that they should not be made to endure the frank and shocking language of the law. No, its argument has been lower and more practical. If women were admitted to the Columbia Law School, the faculty said, then the choicer, more manly and red-blooded graduates of our great universities would turn away from Columbia and rush off to the Harvard Law School![6]

The 1925 *Nation* article concluded with the suggestion that Harvard and Columbia agree to "dilute the red blood with a little common sense." But by 1928 Columbia, perhaps bending to the pleas of candidates, organizations, and benefactors, opened its doors to women.[7] Harvard, among the last to surrender, admitted women only in 1950, and as the last holdouts, Notre Dame admitted women in 1969 followed by Washington and Lee in 1972.

The major law schools did not help women much by being "open" to them. Policy, though important, was not carried into practice rigorously; law classes with one, two, or three women in them were typical for many decades. In 1915, there were seven women enrolled in the newly opened Cambridge Law School and sixty-one in the Portia School of Law in Boston, both devoted exclusively to the training of women. This number was a large proportion of the women in all law schools at the time, and for some years to come. The situation was not unlike medicine, in which most women physicians were trained at the Philadelphia Medical School for Women. Blacks have suffered similar isolation in their medical educations—most medical degrees for blacks have been awarded by two black schools, Howard and Meharry. The result of ghettoized education was not only that each minority group was isolated, but that these schools did not possess first-rank faculties or facilities.

For women aspiring to law careers in New York City, restrictions or quotas on their entrance meant that channels to the larger firms served by Columbia, Yale, and Harvard were cut off, and channels to the other large firms that comprised the country's principal legal community and to posts in government and leading corporations were also severely limited. The New York women lawyers I studied in 1965, whose median age was 47.5, all had suffered from this discrimination. Columbia Law School's exclusion of women until 1928 meant that some of the older women in my sample could not enter the only Ivy League school in the city. Frances Marlatt, a 1922 Barnard graduate, recalled:

At the time I was ready to enter law school, women were looked upon as people who should not be in law schools. New York University, fortunately, was more liberal than most of the colleges. I wanted very much to go to Columbia, but I couldn't get in. I went over to see Harlan Stone, Dean Stone, who was later Chief Justice, and asked him to open the law school and he said no. And I asked why he couldn't open the law school to women, and he said, "We don't because we don't." That was final and I didn't get in. But N.Y.U. did accept me and I went there.

The 1965 study and a survey I conducted in 1972 reviewed the existing admissions pattern of major law schools in New York and several elsewhere in the country, and found a constant proportion of women over a twenty-year period.[8]

It was difficult to know whether the consistently small numbers of female law students was the result of restrictions or the small pool of applicants. Predictably, admissions officers denied in interviews that qualified women were barred and stressed women's family commitments and early socialization to more "feminine" fields. Women's apparent disinterest in the law had long been used to justify their poor representation among law students by men who refused to face their own complicity in barring women from law careers. None admitted, although it was common knowledge in law circles, that there were informal quotas limiting women's participation in the schools. Ten years later, law school faculty members often referred to the fact that the quotas on women had been dropped in the 1960s.

Helene Schwartz wrote, in her book *Lawyering,* of her experience applying to Columbia in 1962. She was the only woman in her class at Brown University who applied to law school and had one of the highest averages among Brown graduates who applied to Columbia. All the male applicants had been informed about the status of their applications when she asked her dean to inquire about her own.

"It seems that they've already made their decisions on what men to take and they are meeting this afternoon to discuss the applications from women."

"Why are they considered separately?" I asked.

"I wondered about that myself," she said. "But the person I talked to either didn't know or wasn't saying."[9]

There were some breaks in the pattern from time to time. A woman teaching in a small law school recalled:

"I went to Harvard in 1966. I think it was the first year they decided they would not discriminate between women and men applicants. Because when we entered that year there were thirty-seven women in the class and before that the largest number had been twenty. The dean was very upset. He made a speech at the opening . . . that he didn't know what had happened to the admissions committee that year, but they had admitted thirty-seven women to this class."

"You don't think he was joking?"

"No. He made it clear that you were taking up a place that a man could have and you probably wouldn't be serious about your career anyway."

The attitude of the gatekeepers of these institutions was perhaps most succinctly expressed by Harvard President Nathan Pusey, who was reported to have exclaimed when young men were being drafted at the height of the Vietnam War, "We shall be left with the blind, the lame, and the women."[10]

The effects of discrimination were made clear in statistics published by the *Harvard Law Record* (a publication considerably more liberal than the Harvard Law School) in 1965. The *Record* reported that al-

TABLE 3.1

Percentage of Women Law Students, 1963–80

Year	Total Enrollment	Women Enrolled	Percentage Women
1963	49,552	1,883	3.8
1964	54,265	2,183	4.0
1965	59,744	2,537	4.3
1966	62,556	2,678	4.3
1967	64,406	2,906	4.5
1968	62,779	3,704	5.9
1969	68,386	4,715	6.9
1970	82,499	7,031	8.5
1971	94,468	8,914	9.4
1972	101,707	12,173	11.9
1973	106,102	16,760	15.8
1974	110,713	21,788	19.7
1975	116,991	26,737	22.9
1976	117,451	29,982	25.5
1977	118,557	32,538	27.5
1978	121,606	36,808	30.3
1979	122,860	38,627	31.4
1980	125,397	42,045	33.5

NOTE: Enrollment is that in American Bar Association-approved schools as of October 1.
SOURCE: *A Review of Legal Education in the United States—Fall 1980* (Chicago: American Bar Association, Section on Legal Education and Admission to the Bar) Reproduced by permission of the American Bar Association.

though the proportion of women in each law class had stayed at 3 to 4 percent, "applications have skyrocketed."[11] The *Record* said that 36 women had applied to Harvard and 19 were accepted in 1951, the second year in which the school admitted women. In contrast, 139 women applied to Harvard in 1965, but only 22 were reported to have been accepted and registered.

The impact of wartime conditions on women's admissions is illustrated by Columbia Law's pattern during the Korean conflict. The ratio of women in Columbia Law School's graduating class rose from 4 percent in 1953 to 8 percent in 1954 and 10 percent in 1955, coincident with military draft calls. The next year the proportion of women in Columbia's class returned to 4 percent.

But the Vietnam War was another kind of war for many Americans. The discussion over the moral nature of the United States involvement in Vietnam and the widespread opposition to the war by college students raised other issues of social justice. Many of the women who supported the dissent, who marched in protests, and who gave encouragement to young men who resisted the draft also became involved in the women's movement. Law seemed to them a powerful tool for protest and they made use of the places created by drafted or draft-deferring young men. A surge in admissions of women by the New York University Law School in 1967 was the harbinger of a new era, although at the time it was believed that when normality returned women would be excluded as they had been in the past. (The only law school that had a higher proportion of women students was Howard University, but black women have long constituted a substantial percentage of blacks receiving degrees in higher education and the prestige professions, although a greater proportion went to black men and the total number of blacks was tiny.)[12]

New York University was one of the first to admit law classes made up of 25 percent women, in the early 1970s. Rutgers University followed after a spate of activity by women students brought the admission of enough women to comprise 40 percent of the law school. Rutgers' faculty included Ruth Bader Ginsburg, who helped women law students there organize one of the first conferences on women in the law, in May 1970, and was one of the first advocates of women's equality in the law schools and in the courts. Professor Ginsburg, an elegantly mannered, attractive woman who tied for first place in her graduating class at Columbia Law School, commanded a reputation as a scholar and an advocate for the ACLU and was to become a major figure in constitutional law litigation on behalf of equal rights and the first woman law professor at Columbia.

Soon most major law schools were admitting women in far greater proportions than in the past. For the first time a new image was being created: that of the *woman* law student. These law students also formed the first women's law student organizations locally and nationally; it was clear from the direction and ideology of these students that the women's movement was largely responsible for their progress as it was for the progress of women in other male-dominated professions in which similar events were taking place. In 1960, women made up 4 percent of incoming law classes; by 1972, the proportion of women in legal training was 12 percent, an increase from 10 percent the year before and a radical change from the tiny percentages of the previous decades.

Between 1965 and 1975, while the student population of accredited law schools nearly doubled, from about 60,000 to 117,000, the number of women in these schools increased more than ten times, from about 2,500 to 27,000. Both the rates of applications and admissions of women in these schools increased dramatically year by year. By 1977, more than one quarter of all law students were women, and by 1980, 33.5 percent of the 125,397 law students. (See table 1.2.)

It is interesting to note that although the percentage of women was also growing in medicine, it was at a somewhat slower rate. Women in the legal profession had lagged behind women in medicine for decades, either because they were more successfully kept out of the law schools than of the medical schools, or because medicine, in some ways, was more congruent with women's role stereotypes as healers and tenders of the sick. In the medical schools, the proportion of female students almost doubled between 1970/71 and 1974/75, from 9.6 percent to 18 percent, while the actual number of women more than doubled, from 3,878 to 9,661. During the same period, the proportion of women enrolled in law schools more than doubled, from 8.5 percent to 19 percent,[13] and their enrollment tripled, from 7,031 to 21,800.

Activity was generated within the law profession during this period to encourage women to seek legal careers. The legal profession, like other professional communities, was pushed and prodded by the government and women's groups to put its house in order with regard to overt and subtle forms of discrimination against women and minorities. There were gestures such as the Equality of Opportunity provision adopted by the Association of American Law Schools (AALS) and a resolution voted by the American Bar Association (ABA) recognizing "the need to encourage more women to enter and realize their full potential within the legal profession." The resolution urged that law schools recruit women students and professors and refrain from sex discrimination.

With enthusiasm by some and foot-dragging by others, more than 40 percent of the nation's law schools developed programs to recruit female applicants in the 1970s. It should be noted that none of these programs involved preferences given other minority students. The programs were to elicit applications and show that women were welcome. Of the law schools responding to a 1972 AALS questionnaire, 40 percent reported they had programs to recruit women applicants. These included actively seeking out women applicants, having women students accompany recruiting officers, and holding conferences for female undergraduates interested in the practice of law. The law school bulletins sent to prospective students now included pictures of women as well as men bent over law books in the library and in class.

Women's groups at more than 70 percent of the law schools also worked vigorously to attract women to their schools. They distributed posters and pamphlets and sent speakers to colleges and high schools. (A few law schools funded their women's groups for such activities.[14]) The cumulative effect of active recruitment and challenge of old practices and the surprising cultural swing that for the first time was providing encouragement to young women was staggering.

While the total number of law school applicants tripled between 1969 and 1973, the number of female applicants increased fourteen times in the same period. Because of the dramatic rise in the number of applicants, the overall acceptance rate at law schools fell in those four years, from 44 percent to 18 percent, but the female acceptance rate remained a comparatively high 43 percent.[15]

The recent debate over the fairness of affirmative action programs may cause discontinuance of some of these practices. However, the problem raised by preference for women is unlike the problem of other minority group preferences because women applicants have generally been better qualified than men. A 1972 survey of eight elite and "semi-elite" law schools revealed that over 53 percent of the women, compared with only 38 percent of the men, graduated in the top 10 percent of their undergraduate institutions.[16] The average law school admission test (LSAT) score did not vary significantly by sex.

The Law School Admission Test Council reported in a study of LSAT scores for 1973/74 that the mean test score for both men and women was 527. Of registrants for the LSAT in 1973, 75.2 percent were men and 24.8 percent were women. In 1974/75 the LSAT survey revealed that women had a slight edge over the men in law school admission test scores. The mean test score for the male registrants was 522, while the same score for women was 524.[17]

Women have done well on LSATs by the standards set for law students. A 1972 study commissioned by the Law School Admissions Council to determine bias of sex in the tests showed that 1,150 males used as a comparison group with 1,165 females scored approximately 10 points lower, and had a mean writing ability score approximately 7 points lower than the women. Women did better on four of the six sections in the LSAT: reading comprehension, reading recall, error recognition, and sentence correction. Men did better on one section, data interpretation, and the two groups scored about equally on the principles and cases section. A monitoring of the total score statistics for the past several years revealed that the differences have decreased and men and women scored about equally on the LSAT in 1974/75. At the same time, the proportion of female LSAT candidates steadily increased.[18]

But the general attack on affirmative action may produce diminution of activity on the part of the schools. However, women's groups still seem anxious to encourage their younger sisters to pursue law careers. For example, as women's enrollment at Harvard Law School neared 30 percent in 1978, the school maintained that there was a sex-blind admissions policy. The increase was a reflection of a larger applicant pool (10.3 percent of the women who applied were accepted as opposed to 10.8 percent of the men). Reporters for the *Harvard Record*, however, pointed out that the increase in the number was attributable to efforts of the Women's Law Association (WLA), which went all out to recruit female applicants and to personally encourage admitted women to go to Harvard. The WLA used funds budgeted by the Law School to make recruiting trips to colleges and universities. Each woman accepted also received a phone call from a WLA member. In many cases, the WLA member supplied the name of a woman already attending Harvard Law School who shared some special experience similar to the applicant —for example, she might have been married or been of minority status. Thus, special effort was made to give newcomers a sense of identity, a linkage to a network.[19]

Perhaps it is too soon to assume that because law has become legitimated as a career goal for women there is also less need for procedures devised to achieve equitable ends by full-scale recruitment programs. It does seem to be true that where the activity of recruitment has diminished, no matter how institutionalized some of the procedures, there does seem to be a diminishing trend with regard to the rate of increase in women's enrollment in the law schools, although generally the numbers are still on the rise, as table 3.2 shows.

The total number of women in approved law schools increased in 1972 by 36.5 percent, and in 1973 by 37.8 percent. In the fall of 1974, the increase dropped to 30 percent, and in the fall of 1975 to 22.71 percent.

Perhaps it wouldn't have been "normal" for the rate of increase to continue as dramatically as it had in the past. But there had been important shifts in the distribution of women students, and there may have been an overall benefit for women as a result. As table 3.2 shows, the proportion of women in the major law schools (those that "produce" law teachers) was at least one quarter, and in 40 percent one third or more. One major school, New York University, had a record 41 percent of women in 1980.[20]

There were also a few other schools where women had become a majority of the students. Thus, for the first time in history women outnumbered men in some schools open to both men and women. In

TABLE 3.2

Percentage of J.D. Graduates from Twenty "Producer" Schools Who Were Women, 1976–79

Law School	1976–77		1977–78		1978–79		1979–80	
	Men and Women	Percent Women	Men and Women	Percent Women	Men and Women	Percent Women	Men and Women	Percent Women
Harvard University	525	18	544	21	525	20	528	25
Yale University	177	20	174	22	163	28	173	24
Columbia University	297	24	295	24	279	32	278	33
University of Michigan	353	24	373	20	375	25	358	24
University of Chicago	161	18	166	32	168	25	170	28
Subtotal	1513	21 (N*=318)	1552	21 (N=326)	1510	25 (N=378)	1507	27 (N=407)
New York University	361	31	371	27	355	36	355	39
Georgetown University	452	19	477	27	490	25	482	35
University of Texas	570	15	534	22	504	26	511	32
University of Virginia	332	17	355	19	356	20	355	28
Berkeley (UC)	292	34	284	32	279	33	295	33
University of Pennsylvania	200	24	183	28	200	30	211	38
University of Wisconsin	291	24	284	32	281	25	305	36
Northwestern University	185	28	170	35	178	31	183	31
Stanford University	148	18	143	24	161	27	198	21
University of Iowa	208	20	200	17	276	26	202	25
University of Illinois	187	9	203	16	207	20	212	25
University of Minnesota	233	21	210	33	220	33	230	28
Cornell University	257	15	172	25	170	24	155	25
Duke University	145	14	155	20	143	22	196	24
George Washington University	349	30	274	32	311	36	312	37
TOTAL	5623	21 (N=1180)	5567	25 (N=1392)	5541	28 (N=1552)	5709	30 (N=1713)

SOURCE: Elizabeth A. Ashburn and Elena N. Cohen, *Women's Integration into Law Teaching* (American Bar Association, Section on Individual Rights and Responsibilities, November 1980), p. 60.
*N denotes women.

the fall of 1974, Northeastern University reported that women outnumbered men in the first-year class 66 to 61. Again in the fall of 1975, the women in the first-year class at Northeastern outnumbered men, this time 68 to 65. The University of California at Davis had 84 women and 67 men in the fall of 1975 and at the Antioch School of Law there were 61 women and 59 men.[21] It was probably not irrelevant that these three schools were all new schools, without traditions of prejudice, without powerful alumni to protest the admittance of women, but dedicated to teaching law geared to the public interest and the needs of general practitioners. None were geared to corporate practice. But even where women were 20 percent of law classes this was a significant enough proportion to make their impact felt. Their ever increasing numbers in the elite law schools meant they had attained serious visibility in the profession and had access to important jobs in business and government.

Going to Law School

4

LIKE many other institutions, the nation's law schools have endured dramatic and significant changes in their life-styles in recent years. There has been a significant change in the general ambiance that pervades the law schools. Once the bastion of traditionalism and conservatism in the universities, the law schools devoted themselves to what they believed was the honing of the legal mind and the inculcation of the profession's values and principles. A style of education geared the students to think fast, on their feet, and to be prepared. Compulsory class attendance, mandatory jackets and ties for men and dresses, stockings, and high heels for women were the norm. The professor conducted classes from a raised platform, calling on students by name as listed on his map of assigned seats. The method of teaching, which still endures, is Socratic—the professors challenging students to analyze cases assigned to be prepared in advance. It was not uncommon for professors to be harsh in their questioning and sarcastic and biting in their comments. The student was expected to learn to stand up under harsh pressure in preparation for the courtroom. The intent was for the weak and unfit to fall by the wayside and for the survivors to develop a "legal" mind.[1] Professor Soia Mentchikoff, some years ago, warned members of the student wives association at the University of Chicago that they might expect their husbands to change in the course of their law school careers: to become more critical, cynical, and harsh.[2]

If there were no student husbands organizations to be also forewarned, it was perhaps because there were so few women students; and perhaps because those few were not experiencing the same education, even at the same schools. Most male students banded together in study groups to share notes and discuss cases that they might be asked to present in class the next day or on exams. The women, however, tended to study alone because they couldn't easily join in groups that developed in the quarters where the male students lived. Women students usually had to find their own quarters outside of the school. In a few places, like Brooklyn Law School in New York, women were physically segregated in the classroom.[3]

Women students were not participants in the male culture of the law schools, yet they were immersed in a male world in which there was no separate women's culture. Law students tend to live together and eat together. Professors stress the importance of talking over the day's work with fellow students. "A lone wolf in law school," Karl Llewellyn told his Columbia students, "is either a genius or an idiot."[4]

Some women undoubtedly were geniuses. Most were simply superior students. Yet some dropped out of law school, and one can only surmise it was because law school was a formidable milieu. Some law school deans supposed it was because the dropouts married; they could not see how punishing an environment their schools were for the few women students.

The Strength of Numbers, the Weakness of Few

If the growth in the number of lawyers has had an impact on the changing structure and culture of the legal profession over the past twenty years, numbers have been important as well for the changing experience of women in the law schools.

It is in the nature of social experience that the existence of few among many has consequences for the relationships between them. It is not inevitable that the few come off badly, but it is often so. Certainly when the few lack legitimacy, they are in a poor position. For women going into the legal profession, law school was the first place where they would learn what it meant to be a minority in an inhospitable work world. Women's place in the profession was such that no one needed to demonstrate unpleasantness in order for them to be aware that by simply being there they were engaging in a deviant activity.

Deviance may have nothing to do with acting differently from others or even possessing different norms or values. Deviance comes from *being* different, and thus, even when it is contradictory to the experience of majority members of the group, the assumption is made that the minority acts differently. Sometimes the assumption of deviant behavior becomes a self-fulfilling prophecy. Women were not forced to act differently. They went to classes and did their work. Perhaps a few more women than men dropped out; perhaps, as some have observed, more women avoided participation in class (but that was true for many men as well); perhaps some worked especially hard and some had the disadvantage of not being included in study groups. But none of these traits stood out as particularly different from those of some men, or as strange or unusual.

As Erving Goffman has pointed out,[5] some members of groups stand apart on the basis of one or more characteristics that the group chooses to focus on. Sometimes the trait may be a physical impairment that makes it difficult for those afflicted to act like the other. Sometimes the trait is irrelevant to functioning—it is skin color or national origin—and is the basis of stigmatization. And though the stigmatized devise ways of coping, they are not supposed to be able to and are placed in a no-win situation. The black who "passes," if unmasked, is considered to be guilty of falsehood and of overstepping bounds of propriety. So the competent woman law student was guilty of trying to be something different than what she was. Like others who are stigmatized, women have been self-conscious and concerned with their special situation, even if determined not to let it bother them.

A woman professor at one of the elite schools talked candidly in an interview with me about her experience today as a teacher, comparing it to the past:

The numbers are very different. The great change in the proportion in the class came about in 1972. This made a huge difference, all having to do with how conspicuous women were as students. When women are very rare in a class there is something special about calling on them. It can involve bias or it can even involve non-bias, you are just very aware. In a class today I am very aware of calling on a black. I wish it weren't so, but you can't help but know that there is something special going on. If there were lots of black people, as there are now lots of women, you just wouldn't be so aware of the special significance.

She was especially sensitive to her own reaction because she remembered the problems she faced as being one of few women among many men:

I had the sense of being exceedingly conspicuous. The things people experienced as being offensive didn't happen to me, but when I was in law school . . . the chances were that you would be called on more often, which is either good or bad, but different anyway. Chances are you'd be called on if you weren't in class; you'd be noticed; somebody would ask why you weren't there; in fact, if you were looking glum, or if you were not understanding, or if you were scratching your head, it would be noticed.

62

The effect of numbers was also commented on by a young male professor at Columbia:

It used to be you knew every woman in the class. In a class of 300 there would be 10. Two hundred ninety were undifferentiated and 10 women, they stood out. I met a woman student who graduated in 1972 and I realized how surprised I was that I didn't know her.

An older male professor at Columbia perceived his women students this way:

When I started teaching thirty years ago there were women in the law school and we did expect the women to be very good, because they were chosen .... I don't believe there was a quota system, but in effect there was a relatively small amount and so the competition being what it was, only the very top young women were able to get in, with the result they were very good in class, and the impression I had was they used to show up the men. When I called on two or three men who reported unprepared or just didn't answer at all, then I would call on one of the women and I knew she would be prepared, because they were in a defensive position. They had to prove they were good.

But a younger male professor, told of his older colleague's comments, responded:

That surprises me because frequently you hear the other story. If someone called on a woman she would be terrified, she might not be able to speak. She might have known the answer but you would never have known it.

Women law students faced similar treatment by their male classmates, whose reactions ranged from indifference to hostility, although they were sometimes supportive. Some male students saw women as uncommitted. A 1969 Harvard Law School alumna remembered:

Practically every male student I met wanted to know "what is a nice girl like you doing in a place like this?" They would interrupt my studying in the library to ask inane questions, confident that I couldn't really be seriously concentrating.[6]

Hostile male peers sometimes accused women of being in law school to find a husband, or of taking the place of a male who, as a future breadwinner, needed the education more than she. One woman recalled being chastised by a male classmate during the Vietnam War for taking the place of a man whom she was probably sending off to his death. It depended on the group one chanced upon. Judge Florence Kelly remembered her male classmates at Yale Law School in the 1940s as being helpful and supportive, a "tight" group. But Virginia Hall, a 1970 Harvard Law School graduate, reported in the *Harvard Law School Bulletin:*

There were many men at HLS who saw me as a threat, as unnecessary competition, and many others who applauded what I was doing intellectually, but would say, "good for you, but I want my woman to stay at home."[7]

Women I interviewed were subjected to comments implying that if they did well it was because of favoritism. As this lawyer recalled:

I remember when the first semester grades were posted. There were few A's given out, but if two A's or three A's went to women, the men would say, oh so-and-so likes girls. I don't think it was the view of the majority of the men, but there were enough who were very competitive, and when they did not do as well as the women it was obviously because of the teacher who had lower standards for women.[8]

The sense of competition has always run high in law schools because students are highly aware that their class ranking is dependent on doing better than a fellow student. It has been widely suggested that women have been less socialized for competition. Trying to suggest a rational explanation, some blame it on women's failure to play team sports. However, because it has only recently become fashionable for women to admit to a competitive spirit, and since a lack of competitiveness is sometimes given as a reason for women's disinterest and incompetence in many spheres of law, it is appropriate to look at the role of competitiveness in women's lives.

Although individuals certainly differ in their will to win, culture surely plays a part in setting the context for their competitiveness. Women's entrance into those spheres in which competition is for money and even honor has been severely restricted, but in fact they have been competing in those and other spheres throughout history. It is certainly a commonplace that American women have competed in rearing the most accomplished children, raising the most money for the church, having the most beautifully decorated home, and leading movements for social reform. In elementary school, girls have competed for high grades, and usually do so successfully until the age of puberty, when they seem to respond to pressures to hold back because of yet another competition—winning the male. Many assume that learning how to be competitive as a member of a team prepares men in special ways for the marketplace. But we have no real data that inform us that foot races or baseball games are better training grounds for the combat of the adult world than the competitions that women have engaged in as children.

We do know, however, that many a man feels diminished when a woman does better than he; and in law school, which is so evaluative and where grades count for so much, it is not surprising to find women the butt of antagonism. Furthermore, the women may feel especially vulnerable and uneasy on finding just how competitive they really are and how resentful of this others seem to be.

The women and men who go to law school "want to be somebody or else you wouldn't be here," observed a psychiatrist asked to speak at a student seminar at Stanford. "Out of 4,000 applications, 172 are chosen," he said. "Everyone's sense of self-image and confidence is dependent on whether their grades are good; not only good, but outstanding," as one student pointed out. The high anxiety characteristic of students in law school creates an environment in which one person's success is another's failure.

Although this partially explains peer pressures on women students, it is harder to understand the antagonisms of professors. Perhaps women cannot return the investment male professors expect from the men. Perhaps their visions of molding a new generation of strong, caustic, and brilliant replicas of themselves cannot be fulfilled by the women students. Perhaps not only do students need role models but the keepers of the roles need a constituency for whom they can play model. The law school classroom has for many decades been a theater with professor and students performing at a high level of intellectual exchange and wit. The experience of most men, in their private lives, is that women are listeners, not contenders. Therefore, it was not to be expected that women would contribute to the banter in which the professor demonstrates his verbal virtuosity.

If the behavior of some male students toward the women was disquieting, it was hardly a match for the behavior of the professors. A Harvard Law graduate recalled that in 1965:

One professor announced to the class that he had called on everyone in class and inquired as to whether there was anyone on whom he had not called. I raised my hand and stated that I had not been called on nor had any other woman in class been called on.[9]

Another reported that as late as 1969:

Even the most liberal professors rarely called on women, and when they did, hurried to get on to a man whom they could harass without fear of provoking overt (i.e. feminine) emotional collapse.[10]

Some law professors called on women in their classes with an eye to singularly embarrass them. One woman remembered a professor who called on her whenever a case involving the sordid side of sex came up. She was bombarded with questions regarding evidence of rape, such as the degree of penile penetration required. And all women students knew they would be asked to recite on "Ladies' Day," an institution common in many schools. Harriet Rabb recalled that at Columbia,

I heard of Ladies' Day here when the professor would say, "Will all the little virgins please come to the front of the room." Of course the women didn't know whether to go or not. And when I was in school here between 1963 and 1966 there was one teacher who was known for his "Valentine's Day massacre." The women were obliged at the beginning of the hour to stand up at their seats and remain standing through the hour and get called on . . . and that would be the day when he did all the embarrassing and difficult-to-discuss problems.

At Harvard, Professor W. Barton Leach was known for his special Ladies' Day performances. One of his students, a 1965 graduate, reported that he

sat in the audience and asked questions in a "humorous" tone of the women who were exhibited on the podium rather like performing bears.[11]

A Harvard 1969 graduate recalled Professor A. James Casner's Ladies' Day:

On Ladies' Day in Professor Casner's first year Property class we were quizzed on the intricacies of dower, that is, on the rights women have in their deceased husbands' property. For the professors and the male students, Ladies' Day was an entertainment, a show put on at our expense.[12]

By the mid-1970s, it was becoming difficult for law professors to exhibit such behavior. "One of my colleagues," reported Professor Ruth Bader Ginsburg, at Columbia University, "with the best intention, had a Ladies' Day to celebrate Billie Jean King's victory over Bobby Riggs.* The hissing could be heard down the corridor."

Perhaps the students who sat through Ladies' Day without protest would have reacted if asked to participate in a Jews' Day or a Cripples' Day.

*Riggs is the male tennis player who said that even at his age (fifty) he could beat any woman player. He lost to Billie Jean King, September 20, 1973.

66

All law students were subject to ridicule, but the women were subjected to a special sort. "Better go back to the kitchen," a professor harangued women who stumbled in recitation.[13]

From one year to the next, Ladies' Day became a memory, and no one dared mourn it publicly. Harvard women of the law class of 1968 had helped its demise:

On Ladies' Day, we dressed in black, all wore glasses and carried black briefcases. We totally devastated Leach—knew *all* the answers, and, at the end, when he asked, "What was the chose in question?" (his big punch line—the answer was "underwear" and was supposed to embarrass [*sic*] us), we replied, "we've replevied a few samples," opened our brief cases, and threw fancy lingerie at the "boys." Leach almost had a stroke on the spot, and never had a ladies' day again![14]

In 1971 Dean Robert B. McKay of New York University Law School, writing on "Women and the Liberation of Legal Education," conceded that "What we did not know was that we were sometimes prone to what I call unconscious sexism through jokes or illustrations in class offensive to women; embarrassment of female students by differentiated treatment; and even expressions that might suggest male superiority."[15] A questionnaire sent to all law professors associated with the American Association of Law Schools in 1973[16] indicated how far the pendulum had swung with regard to their public attitudes. When asked whether "there should be equal opportunity for both men and women in law and law school," the mean score for those answering was close to "strong agreement." The professor also agreed, though to a somewhat lesser extent, that men and women are or can be equal as professionals, lawyers, and law students. Agreement, however, about the role of law schools in taking affirmative action to correct inequities was more modest, though still positive.[17] Students generally confirmed in interviews that these public views were consistent with the behavior of law professors, particularly in the elite schools (where gentlemanly behavior was the norm), except for some residual hostility by a few older men.[18]

But the new attitude was not due only to normative change brought about by a conversion to feminism of law professors. It was becoming harder and harder to treat women students in the ways they had been treated before. Students generally were unhappy with autocratic behavior of professors and were pressing for a more egalitarian style. The student revolts of 1968 and 1969 made this message clear. Furthermore, younger faculty members, themselves close to the student environment, rejected the autocratic style. There was also the structural change resulting from the increase in numbers and proportion of women students attending law school. Scapegoating became harder; the students were no longer as passive as they once were in accepting humiliation.

The newer generation of women law students organized to deal with discrimination in the schools. The obvious modes of exclusion were the easiest to deal with. Delegations of women at NYU, at Harvard, and elsewhere, went to see male professors to discuss their sexist behavior (in their offices they were more contrite than in the classroom); students publicized improper behavior in law school newspapers, at public meetings, and "hearings"; formal complaints were lodged with deans, and committees were set up to advance the integration of women students.

Many professors were genuinely unaware of the unfairness and hurt they inflicted. In an interview with me a male professor at an elite law school commented about a colleague:

Years ago we gave out a questionnaire at the law school and students would rate their professors. One student criticized one of the professors because he was very prejudiced against women, that he used to make remarks in class that seemed to be snide remarks. I don't know if this was just a supersensitive group of girls who felt this way about it. He happened to be one of the most pleasant colleagues. He was very active in the work of the Chaplaincy here, and it was hard for me to believe that anybody with this approach, a religious approach to dealing with human beings, would have been discriminatory in his dealings with women. But some of them did. . . . I think it must have hurt him badly to have them criticize him. He was a very sensitive man.

Activity by formal groups was not the only means of professorial consciousness-raising by students. Hissing could be heard in law classes throughout the land at teachers' missteps. Let a professor use "he" repeatedly in referring to lawyers, or address the class as "gentlemen," or use a woman as an example of a person lacking in competence, and the sound of hissing was sure to rise.

Women students, joined by sympathetic men, engaged in effective techniques of social control and socialization. Old dogs learned new tricks, and they learned them fast and often with the dedication of recent converts. Furthermore, many seemed to forget it ever had been otherwise. Not only was misogyny less fashionable, the conservatism and proper upbringing of many male professors served the rapid changeover. As long as it was proper to exclude women there was no reason to change. But when etiquette demanded equality of treatment, and mannerly behavior consisted of treating women and men equally, there was no recourse but to do the "right" thing.

An eminent male professor at a West Coast law school commented candidly to me on the impact the new attitudes had made on his relations with women students and on the sexual element sometimes present:

"Do you think your own experience in dealing with women in law schools, both as colleagues and students, is any different than it is with men?"

"No. [But] I'm less relaxed with women students than I am with men students."

"Why?"

"Because I'm always worried that I'm going to get into some kind of a sexual accusation."

"Has it always been true for you or is it because of the recent publicity about sexual harassment on campuses?"

"Oh no. It's always been true. I'm very leery and careful dealing with women, especially with women in a situation where the power relationships are such that exploitation is really not, unfortunately, uncommon."

"Well, do you think this has some effect on the extent to which women get the same training from you that men do?"

"I don't think it has anything to do with it. As a matter of fact, I go out of my way to take them on and give them a break."

"You do?"

"Oh yes. But I do treat them differently than I treat men."

"How do you treat them differently?"

"Well, I think about whether the door is closed when I'm with them."

"How about intellectually?"

"Well, I'm a little more careful about bruising them because I think they bruise more easily then men or I think they do—and I may be wrong about it—but my fallback position would be that even if they don't bruise more easily, they can shout more and the last thing I want is a hysterical woman on my hands."

In recent years, some women have developed special relationships with women law professors in clinical programs that deal with women's issues such as Title VII employment practices, rape, and so forth. In fact, the clinical program at Columbia under the direction of Dean Harriet Rabb worked on the landmark sex discrimination case against American Telephone and Telegraph Company as well as important cases against major law firms. A number of student participants interviewed had the experience of doing law work of major social significance and also had a chance to work with a lawyer not much older than they, who had a husband and small children and who was also attractive and stylish. Although a few thought her awesome, and some couldn't adjust to her strength of purpose, Rabb inspired respect and love in most students who worked with her, and was especially important to the women.

There are also young male professors active in clinical programs who are working with students, men and women, in a more egalitarian way than the austere professors of the past. However, clinical programs remain somewhat peripheral to the teaching program in most law schools. The national law schools that have prized legal theory over law practice have viewed these programs as experimental and are ambivalent about their utility. As a result, the status of the lawyers brought in to teach in these programs has always been uncertain, and the programs themselves may seem insecure.

Women faculty were important in changing the views of their colleagues. Vivian Berger, while a professor at of Columbia Law School, reported that a male colleague,

whom I regard most highly and respect and who is very open-minded, confessed that until he talked to me (I had just written an article on rape) he didn't think a woman could be "raped" unless a weapon was involved.

And Ruth Bader Ginsburg, then at Columbia, noted that male colleagues who were revising legal case books often asked her to read through them to "see if anything in them is offensive to women."

The new attitudes expressed by many younger deans was also effective in changing the tenor of the law schools. Especially in the more elite law schools, young male deans and a growing number of younger women deans were given responsibility for student affairs, which included reviewing the representation of women students and charges of discrimination against them by the faculty and in placement after graduation. Some of the women took a special interest in bringing about change swiftly, and arranged confrontations between the students who had charged sexist or racist treatment and the person accused. The new standards of "proper," egalitarian behavior became stronger when discrimination was no longer tolerated.

Women's Groups in the Law Schools

The impact of women students' groups on the position and numbers of women in the law schools has been mentioned, but their activity and history bear special notice. The women who organized them became leaders later in pressing for changes in the profession which made life easier for the next generation of law students.

Women banded together for the first time at any law school at New York University in 1968 and formed the Women's Rights Committee.[19] The work of this group became a model for groups elsewhere and their activities spread into other legal spheres. The increased numbers of women at NYU (where there were more women students than anywhere else in the country at the time) brought them visibility and a collective spirit. The committee's first efforts were directed at making available to women the prestigious Root-Tilden Scholarship Program, then restricted to men. They were successful immediately, and women students were able to benefit from the program in 1969/70.

The Women's Rights Committee was instrumental in forming the National Conference of Law Women, which held its first meeting at NYU in April 1970 and has met every year since then, attracting thousands of women. As a result of the Committee's efforts, the American Association of Law Schools established its present Committee on Equality of Opportunity for Women in Legal Education in 1969. The first student member of this committee spurred its successful effort to amend the AALS articles to prohibit sex discrimination by member schools in their admissions, placement, and institutional hiring offices.

In 1969, women at NYU law school also urged the creation of a Women and the Law course, the first in the United States. Diane Schulder was the first instructor for the seminar, and was succeeded by Eleanor Holmes Norton, who was then New York City Commissioner on Human Rights. Inspired in part by the NYU example, similar seminars were offered in the years following at other law schools, including Yale, George Washington, Rutgers, Georgetown, and Michigan.[20]

The Women's Rights Committee worked to establish, in 1971, a clinical program at NYU on women's rights. Columbia also developed a program, and Harvard women students established a research group on sex discrimination issues in 1975. The group went on to develop materials for recruiting women students and persuaded the Harvard Admissions Committee to permit women students to accompany faculty members on recruiting trips. Efforts of women students at NYU, Columbia, Harvard, and elsewhere were an important factor in increasing the numbers of women in the law schools.

The NYU group, together with women at Columbia University, filed suit with the New York City Human Rights Commission against a dozen Wall Street firms, charging them with discriminatory hiring practices. The recruitment patterns at those firms were seriously altered by this effort, as we shall see in the discussion of Wall Street firms.

Still another breakthrough was effected by women students when they acted as plaintiffs in a suit filed in 1971 and won by the American Civil Liberties Union against the State Board of Law Examiners of New York for discriminating against women in Bar Examinations procedures. This overturned the longtime practice of segregating women during the exam.

The dramatic victories won by the NYU group and others mobilized attention and participation of women students. Inevitably, attention diminished when there were no longer the same mobilizing battles to be won. Nevertheless, women's groups have continued to support the Women and Law Conferences, focusing attention on legal issues of specific interest to women, such as sex discrimination, abortion, battered wives and children, and on the problems of women lawyers in practice.* They have also remained active in monitoring the admissions policies of the law schools and have provided help in recruitment and in some cases specific support activity for women wavering in their decisions to become lawyers.

---

*In 1979, more than 2,300 participants, 60 percent of whom were students and 40 percent lawyers and other legal workers, met for the tenth annual conference. In the ten years since its inception, the conference had become a place for women to discuss the broad issues of the status of women in U.S. society and the relationship of law to this status.

The 1980 conference (the Eleventh Annual) met in San Francisco with a program that included 160 workshops. The workshops covered a wide range of topics, among them employment discrimination, the Equal Rights Amendment, abortion, violence against women, and professional issues such as the selection process for judgeships, legal specialties such as divorce and daycare, fund raising and grant proposals, health care delivery systems, housing and tenant issues, female occupations and unionization, and social security law and its effect on older women.

The National Conference continues to serve an important role in its original goal of developing networks of women across the nation. In addition, it supplies important information and even training for women in legal careers. The 1980 conference, for example, had a trial skills training session in which videotaping and critical feedback from a panel of women trial lawyers provided important help for women involved in litigating cases.

Value Conflict

In my travels to law schools through the 1970s, I found women's groups exploring conflicts in values they believed women faced when they entered the first stage of what they regarded to be the male establishment. The conflict in what some students felt were female values as opposed to male values was a theme that came up again and again in law schools and in law practice, reflecting a general concern expressed by women in other occupations.

I became directly involved in one such exploration during a year spent on the Stanford University campus. I was approached by a committee to organize a workshop on values which the women's student organization was preparing. Once again I heard concerns expressed that had been raised at every other campus where I had spoken to women students. Women seemed uncomfortable about what they perceived as pressures, subtle and direct, that were leading them to become less "caring," more instrumental, propelled along a track where only work mattered. They were concerned they would be forced to set aside what they considered to be human values.

They asserted that men shared this conflict and the conference was planned for them too. But women were organizing it and they seemed to feel that men thought less than they did about balance in their lives and the means and ends of professional life. They thought that men would not have to make the kinds of choices women must between the private and professional spheres.

Many women change the direction of their aspirations in law school. Some who went to law school to change the world have found the business side of law fascinating and have made successful careers at it. One of the organizers of the Stanford workshop told me after graduation that the nagging concerns which had led her to become involved in the workshop had changed now that she was practicing. She no longer felt that her sense of identity was under attack or that she was "selling out."

Where one went to law school had a lot to do with value conflict issues. Women at the top national law schools, such as Harvard, Yale, Columbia, like the men, are tracked toward careers demanding total commitment and oriented toward the corporate world rather than social interest. For the best students at such schools there are guaranteed career paths—clerkships, teaching, and the big firms. Women, like men, are learning to want those positions because they are the symbols of success and accomplishment. At other kinds of law schools the track is not as clear, and women go on to serve in spheres that seem more comfortable ideologically. New schools without ties to the establishment, such as Northeastern University or Antioch Law School, also have a larger proportion of women students, 50 percent as opposed to between 20 and 30 percent in the large national schools. Few expect to go to large corporate law firms from schools such as Northeastern, and there is less competition and less agony of decision about the "right" thing to do. Women who attend such schools go into small practices and do traditional women's work in law (matrimonial and real estate work) or move into the public sector in district attorney's offices or as public defenders. A selected few go into foundation supported feminist law practices emphasizing public policy-making legal cases.

Women's Groups Today

The activity of women's groups and the participation of women students in meetings geared to their interests is mixed these days. The women's groups now have their section on the bulletin boards along with those of other groups, for job notices and special meetings. They tend to advertise occasional meetings on subjects such as "Mixing a Career and a Family," "Sexism and the Law," or "Careers in Law." Female faculty members are favorite speakers, and alumnae who are practicing in various kinds of legal work. Harvard Law students celebrated the twenty-fifth anniversary of women graduates at the law school in 1978 with meetings that drew hundreds of judges, law professors, and law partners. It was the usual array that one might expect at a Harvard Law reunion, except that these were alumnae and not alumni.

Berkeley (Boalt Hall) women have also been active, and women students at the University of Chicago have sued the law school placement office to insure equal treatment for women, to give only a few examples of women's groups' activities on law school campuses. Women students have also created ongoing publications that range from news about women alumnae and students to scholarly journals. New York University students published the first issue of *Women News* in March of 1980

74

and the first issue of the *Harvard Women's Law Journal* appeared in the spring of 1978.*

They have started programs to help attorneys in rights-related cases. Harvard women have founded an "issues research organization" to provide assistance to attorneys and rights advocates doing sex discrimination cases.[21] The program, which also included interested male students, planned original investigatory work and analysis of legislative and regulatory decisions.

Besides offering support and counseling on the practical "how to" issues of getting jobs, managing sexist rudeness, and doing clinical work on issues, the women's groups offer a sense of camaraderie and mutual support. There is an informal air at their meetings; speakers know one another; they tend to have attended many conferences together, to have been through the sex discrimination "wars." The network of women throughout the profession is evident in the women's groups at many schools and at the national conferences they have planned and carried out. In a very real sense, women are engaged in the "bonding" that Lionel Tiger wrote about in *Men in Groups*, where he attributes men's success in business and war (and women's lack of it) to a primal, genetically selected set of attributes which brings them together in task-oriented groups.[22] Had Tiger waited a few years he would have been pressed to explain women's links through the very same kinds of groups. Women's "bonding" certainly was instrumental, but its character was somewhat looser than that of their brothers in the profession, observers note.† For example, at the twenty-fifth anniversary meeting of women at Harvard Law School, a reporter noted that the festivities were filled with "merriment, warmth . . . and lots of hugging and kissing. . . . perhaps in contrast with the formal male-dominated gatherings at the school."[23]

---

*A sampling of articles included such topics as "Restrictions on the Abortion Rights of Minors: *Bellotti* v. *Baird*," "Sex Discrimination in Coaching," "Sex-Biased Pension Plans in Perspective: *City of Los Angeles, Department of Water and Power* v. *Manhart*," and "The Law School Admission Test and the Continuing Minority Status of Women in Law Schools."

†Not different from all men, but those of the upper strata. Men in less conservative occupations or from certain groups in lower strata tend also to be less restrained in style.

III

THE MANY PRACTICES OF LAW

5

Breaking In: Dark Days of Discrimination and the Beginning of a New Era

FEW WOMEN LAWYERS found a warm greeting in their chosen profession when they sought their first jobs in the years before the 1970s. Strangely enough, this often came as a surprise. Even afterward, the experience of many women was far from what they had hoped, although the new generation of women lawyers was more alert to expressions of prejudice against them.

The older generation had found that, unlike male law school "stars," women graduates with outstanding records were not courted by firms that professed to constantly seek talent, or even by smaller firms that cannot always bargain for the best.[1] Many women hadn't thought much about discrimination when they chose to go to law school. The older women from immigrant families simply were unaware of the way the opportunity structure worked. Some simply assumed they could get what they aimed for in the individualistic American tradition. Some were thick-skinned enough not to notice prejudice. There was little discussion of the reasons for women's tiny representation in the profession, and talking to them one could easily remain unaware of how pervasive was the prejudice against them. Although accounts of unpleasant experiences were not in short supply, many had sought jobs in sectors where they knew it would be less difficult for a woman. They avoided those areas of practice in which it was common knowledge that women were not wanted.

The situation of women in law was something of a paradox. On the one hand "everyone knew" that law was a male profession and that women were not considered desirable candidates for jobs. On the other hand, the women who became lawyers either remained unaware of this view or disregarded it. James J. White, in his 1967 study, reported that nearly half of the women lawyers in his sample felt they had been discriminated against in their work; but this, of course, meant that over half did not.[2] Discrimination is perceived differently and discrimination is dispensed differently. Some women lawyers were confronted with discrimination head on, some indirectly, and some never met it at all.

Women, of course, were not alone in facing discrimination. "The best talent of the Bar will always muster to keep Ins in and to man the barricade against the Outs," wrote Karl Llewellyn, long before it was fashionable for lawyers to admit such practices existed.[3]

The wise have long sought to explain the instances of "senseless" differentiation between insider and outsider, the lengths to which insiders will go to keep outsiders away, and the animosity and rationalizations that develop in the process. Insiders contradict their own professed ideals in the process, unaware, or uncaring, of the inconsistency or hypocrisy in their actions. Law, based on an ideology of even-handed fairness, should by the canons of its own logic welcome to its practice all those who qualify by reason of intelligence and facility. Yet it is an uncontested truth that lawyers discriminate in ways that violate this standard. Why that is so has been explained by various theories of human nature and group behavior.

One of the theories focuses on the competition for scarce resources and the acquisition of wealth. Limiting the profession can indeed insure a monopoly for the few. Selective recruitment keeps the privileges and other benefits that have accumulated for one's own group.

Another perspective, not contradictory by any means, centers on the nature of groups themselves and the ways in which communities form and maintain identity. Professions are communities,[4] dependent like others on boundary-maintaining mechanisms that define groups as distinct and foster integration of their parts. Kai Erikson has explained how communities tend to make statements about the nature and placement of their boundaries by "locating and publicizing the group's outer edges" through confrontation between outsiders and deviant members on the one hand and official agents of the community on the other.[5] He has shown how events such as wars and rituals can dramatize the difference between "we" and "they."

The legal community has long been successful in maintaining cohesion by creating its own confrontations and adversaries; by welcoming "ins" and repelling "outs." Sometimes that task has been easy—the outsiders literally come from outside; another country, another place. But where the facts are not obvious, justification may come merely from the act of defining certain people as outsiders. Women, forever integrated with the lives and communities of men, have been defined as "outsiders" serving the purpose of reinforcing the bonds of male association. Rationalizations may follow the exclusion rather than account for it; rationalizations change, become dated and absurd, yet new ones arise because what is necessary is that the division remains—that outsiders are labeled by those whose interests are served by creating a sense of difference.

The history of discriminatory practices by the legal profession is well documented, partly through studies made of the inner workings of the professions. The social sciences turned their attention to the professions in the 1950s and 1960s.[6] A burst of exposés followed, documenting exclusion in a far more passionate and evaluative way than had the studies that had preceded them. Those dealing with the legal profession typically were the work of advocates for the public interest.[7] Before the 1970s, attention to the ways in which the legal profession made women outsiders was slim—relegated to a few paragraphs or footnotes.[8]

The history of discrimination against women in law has now been explored by others. A description by Sachs and Wilson traces it from the first decision about whether the "persons" admitted to the bar could be women in 1894 (the Supreme Court ruled that the word "person" meant "male") to the 1970s in both Great Britain and the United States.[9] The dimensions of discrimination, its tonality, the consequences it has had for the patterning of women's experience in law, and the ways it has changed and is changing are complex. Some practices have endured long after the formal rules that supported them.[10] Today, no law or Supreme Court ruling specifies that women are not full citizens or "persons." No rules prohibit married women from practice. Yet sexist prejudice remains, as does the necessity for affirmative guarantees of women's rights to opportunities open to men.

A 1958 U.S. government publication advised women lawyers to concentrate on "real estate and domestic relations work, women's and juvenile legal problems, probate work and patent law," reflecting the wisdom of insiders about areas in which women were apt to find work.[11] But no sphere of legal work could be depended on as a hospitable environment for the able woman lawyer coming out of school. Echoing the culture's prejudice about the impropriety of women

TABLE 5.1

How Employers Rate Characteristics of Law Students

		Female	Black	Protestant	Catholic	Jewish	On Law Review	Upper 1/5 of Class	Upper Half of Class	Lower Half of Class	Harvard Law School	Dad—In Profession	Dad—Blue Collar
Overall Averages	(100)	-4.9	-3.5	.6	.2	-.8	6.8	5.2	.9	-4.5	2.2	.9	-.2
Firms 50 and over	(13)	-3.3	-1.7	0	0	-.1	6.9	5.0	.9	-5.3	1.6	.8	-.1
Firms 25 to 50	(24)	-5.0	-3.5	.6	-.1	-.6	6.6	4.8	1.2	-4.3	2.1	.4	0
Firms 15 to 25	(19)	-5.0	-3.0	0	.1	-.3	6.8	5.5	-1.3	-4.3	1.9	1.1	-.3
Firms 5 to 15	(32)	-4.4	-3.7	1.1	.6	-1.3	7.0	5.4	2.1	-4.6	2.3	1.3	-.2
Firms 1 to 5	(12)	-7.8	-5.5	.8	.4	-1.4	6.3	5.6	2.4	-4.4	3.0	.9	-.6
New York City	(15)	-4.3	-2.3	0	0	-.3	7.3	4.8	.5	-5.1	1.7	.3	-.1
New York City—50 and over	(8)	-3.5	-1.9	0	0	0	7.5	4.9	.1	-4.0	1.9	.4	-.1
Chicago	(8)	-5.4	-2.5	0	0	-.1	6.7	4.6	0	-5.1	.7	.3	0
Boston	(7)	-2.1	-.2	1.8	.2	-.6	7.0	4.4	-.4	-5.4	4.3	.8	0
Washington, D.C.	(5)	-5.0	-4.5	0	0	-.8	7.0	3.0	-2.6	-7.4	1.8	.5	0
Philadelphia	(4)	-6.8	-2.3	0	0	-2.5	5.8	4.0	1.0	-4.5	1.0	.3	0
Los Angeles	(12)	-3.0	-5.2	2.5	1.4	-.2	6.6	5.9	-.5	-7.0	2.3	2.2	0
San Francisco	(7)	-6.1	-8.0	2.2	.7	-1.0	6.8	4.7	-.2	-5.3	4.3	1.6	-1.9
New England	(8)	-2.1	-1.4	1.0	0	-.7	7.7	6.9	4.9	-2.6	2.1	.9	0
Middle Atlantic	(10)	-4.4	-2.9	0	.2	0	6.2	5.7	3.1	-1.3	2.1	.1	-.1
Middle West	(17)	-7.3	-6.9	.6	.2	-1.8	7.1	5.8	2.4	-1.5	2.4	.8	.1
West	(7)	-5.7	-1.4	0	0	-.1	6.6	5.3	.6	-5.3	1.3	.9	0
Large East (N.Y. 50+)		-3.5	-1.9	0	0	0	7.5	4.9	.1	-4.0	1.9	.4	-.1
Same—Student Prediction		-2.6	-6.2	2.1	.1	-2.6	8.2	5.1	1.5	-3.4	2.8	2.1	-.6
Medium-Size Midwest		-5.8	-5.1	0	.2	-1.8	7.2	5.8	1.0	-3.2	2.0	.8	.2
Same—Student Prediction		-3.8	-7.1	2.4	.1	-3.7	8.4	5.6	2.8	-.6	3.6	2.8	-.4

SOURCE: This table is adapted from the chart, "Quality by Quality, How They Rate," in an article, "The Firms—What do They Want?" by Bruce Abel in the *Harvard Law Record*, December 12, 1963, vol. 37, no. 9, pp. 9-11. The chart reports results of a questionnaire which asked "What are you looking for in a job candidate?" The questions called for a rating of factors on a scale of plus ten (definitely does get the job) to minus ten (definitely does not get the job), with zero indicating that the quality is not a factor in evaluating the candidate.

seeking men's work, most male attorneys were no more educated or understanding than anyone on the street about women's capacity to be good lawyers. Furthermore, lawyers were not embarrassed about expressing their views. The *Harvard Law Record* editors informed readers about how their various characteristics and attributes might affect their chances for a job in 1963, after survey results were tallied of law firms' views about what they wanted in a candidate.[12] A special article on women noted special discrimination against them. "Hiring partners admit they do, and the girls agree," wrote the reporter,[13] and the evidence presented in the same issue left no doubt. (See table 5.1 and the appendix.)

The *Record* asked 430 private law firms across the country which characteristics and attributes they evaluated most highly in judging new recruits, as well as those characteristics they judged undesirable. The responses confirmed the profession's norms of achievement in that qualities of initiative and scholarship were rated as most important.[14] However, when asked what kinds of characteristics would weigh against job candidates, other than a poor or mediocre academic record, the responding firms listed qualities quite unrelated to the professional task. Women, blacks, and Jews, and those candidates with rural backgrounds or fathers in blue-collar occupations, all were considered undesirable by law firms of varied sizes in all parts of the country. The survey confirmed the common knowledge that most firms were prejudiced against candidates whose backgrounds were considered to be of low status in American society. Differences in attitudes toward minority groups were evident in the responses, however; some low statuses were lower than others. Catholics were the most acceptable minority group. Only in firms of twenty-five to fifty members—at the time, the larger firms—were they negatively valued.[15]

Jews, blacks, and women were all consistently rated negatively. *But of all the "deviant" statuses, females drew the most negative rating, — 4.9 on a scale of from —10 to +10 for those least likely to be hired.* Only poor scholarship drew greater opprobrium than being female. The intensity of feeling against women varied with the size of the firm, however. Small firms, with from one to five members, found women most undesirable (with a rank of —7.8). The largest firms found them most acceptable and gave them their best rating on a national basis, although still clearly negative: —3.3. But, as we shall see, the large firms had the option to track women differently than the men within the firm, and did.

Law apparently is worse than other professions in its antipathy to

83

women. In a study of perceptions of sex discrimination, Soule and Standley reported that women lawyers perceive more discrimination against them than women physicians or architects, and male lawyers express even more rejection of women than do men in the other professions.[16] The "even hand" of discrimination has been raised against women in all sectors of the legal profession.

As the years have passed, women lawyers whose abilities have been proven in spite of discrimination, or because of opportunities that emerged as discrimination was reduced, have recounted the obstacles they faced early in their careers.* Judge Cecelia H. Goetz of the Bankruptcy Court of the Eastern District of New York remembered that upon graduation from law school no firm was willing to interview her, let alone offer the $5.00-a-week clerkship then common. "If not for the refuge my father's firm offered, I might still be unemployed," she told a group of lawyers not long ago.[17] Goetz was to experience almost the entire array of discrimination to which women were subjected. Government work, considered a refuge for women (and indeed it was for some) was by no means entirely open. Judge Goetz recalled the days of the New Deal in Washington, where

discrimination was rampant against women, against blacks, against New Yorkers. . . . The Antitrust Division of the Department of Justice, then headed up by a leading liberal, Thurman Arnold, made no secret of the fact that it would not hire women. Even where women were not automatically excluded they were unwelcome. At the SEC, I was told that they had hired a woman the year before, and she had had the ill grace to meet and marry a Wall Street attorney, convincing them that it was unwise to hire women. I was upbraided for trying to take the bread out of the mouths of married men by competing with them for what jobs existed.

After World War II, when I wanted to join the attorneys engaged in prosecuting war crimes in Nuremberg, Germany, it was necessary for Telford Taylor personally to authorize my employment because the Department of the Army, which had no difficulty in finding accommodation for women secretaries, claimed to be unable to accommodate women lawyers in occupied Germany
. . . .

In 1953, with the end of the Korean War, I again found myself job hunting. One firm to which I came highly recommended suggested I learn typing, since secretarial positions were available. This, after I had been occupying supervisory positions for several years!

That year, several of my colleagues at OPS [Office of Price Stabilization] holding jobs similar to mine had been hired by one of America's major corpora-

*As this book went to press, President Reagan nominated Sandra Day O'Connor to the U.S. Supreme Court. A law review editor at Stanford Law School, O'Connor's only offer from the major west coast law firms to which she applied in 1952 was the position of legal secretary. *Time*, July 20, 1981, p. 12.

tions interested in beefing up its in-house staff. When my resume was put before the firm's general counsel, he turned to my sponsor and said: "Bring him in. What are you waiting for? This is exactly the type of person for whom we are looking." To which my sponsor replied: "It's not a 'he'; its a 'she.' " The general counsel, I am told, replied just as quickly: "What are you wasting my time for? You might just as well recommend a Negro."

Wall Street was hostile territory for most women lawyers. Judge Nanette Dembitz recalled, in a *New York Times* interview,[18] that she couldn't get a job in a Wall Street firm though she had been an editor of the *Columbia Law Review* and had first-rate family connections (she was a cousin of Louis D[embitz] Brandeis, then an Associate Justice of the Supreme Court). Another lawyer I spoke to was told at her Wall Street interview that the firm would "love to hire me, but they couldn't because what bathroom would I use? Really, they said, you couldn't use the secretaries' bathroom, and you couldn't use the lawyers' bathroom, and they couldn't build another one, so what bathroom would you use?"[19] Another woman applied for a labor law job and was told that a woman couldn't be hired because "union members wouldn't put up with it."

Discrimination showed varying faces to women job-seekers, and each form had its consequences. If some women didn't recognize discrimination, perhaps that was because it was masked. For example, a woman I spoke to who had been third in her Ivy League law school graduating class in the early 1970s reported discovering only by chance that she had been a victim of discrimination:

I remember having lunch with one of the guys I'd gone to law school with . . . . I had just come from a firm which had said "We have nothing against women lawyers. If a woman had higher marks then a man and they both applied to our firm we would take the woman. But we cannot now." My friend and I had lunch, and he told me that he had just been hired by that firm. My marks were about ten steps above his, so I knew that they disliked me. They never did take women in that firm.

Sometimes discrimination was compounded by the other statuses a woman held. For example, a woman of Italian extraction might have two counts against her, or three counts if she were from a poor family as well. The women of my early study had demonstrated superior performance in law school as a group, yet many would have faced discrimination even if they hadn't been women because they came from immigrant families and low status ethnic backgrounds.

It was certainly necessary for women to have the most preferable array of statuses possible to impress the hiring committees of prestigious firms. For example, a former Wall Street lawyer reminisced about being hired by her firm:

The reason I think I gained entrée is because I went to Mount Holyoke College. Mr. T. and the firm . . . would not be very much inclined to take a woman who had not gone to one of the major women's colleges. And its also a firm, I'm quite sure, that would not consider a woman unless she had done very well in law school and had gone to one of the major law schools. I know that a number of girls that I thought would be quite well qualified but that went to obscure schools were not offered jobs. I suspect that they feel this way about men too; they're not very likely to take a man that's not from one of the major law schools, unless perhaps he was first in his class or something like that.

Most of the older women lawyers interviewed who worked in high prestige firms at any point in their careers fulfilled most of the other status-set expectations of the professional elite. Like the men Smigel surveyed in his study of Wall Street lawyers, they were graduates of Ivy League colleges and law schools, and were from Protestant professional families. Smigel pointed out that the prestige firms demanded candidates with "lineage, ability and personality . . . from the right school and the 'right' social background."[20] (These attitudes have persisted to a certain degree, tempered by the changes in the firms and in cultural views about the acceptability of minorities.) Nevertheless, it was not easy even for the Protestant woman with the "right" lineage and the "right" social background to get a job on Wall Street, or for a Jewish woman to get a job in a Jewish firm. Those who did usually knew they were not on the road to partnership, and that their choice of specialities would be restricted to the classic "women's" fields of trusts and estates, or "blue sky" work (the routine analysis of securities laws).

A few professional gatekeepers were frank about the attributes they considered disreputable. Mary Christiano (not her real name), a New York attorney in partnership with her brother, had vivid memories of just how unacceptable she was to Wall Street:

I first tried Wall Street. If you want to know about discrimination let me tell you what I was told six and a half years ago [1958] . . . that number one, I was a girl. Number two, I was too young (I was just twenty-one when I was admitted to the bar), and I was an attractive nuisance because I was supposed to be so pretty I would disrupt the office. Number three, I would get married within a year and they would waste their time teaching me (I'm still here . . . unmarried). Number four, I was a graduate of Brooklyn College, and number five, I was Italian and a Roman Catholic . . . that's what I got told point-blank down on Wall Street. One firm offered to give me a position as a file clerk.

Discrimination Gives Way to a New Era

The biography of the late Judge Francis Dooling,[21] who recently came to public attention for a supportive ruling on a New York abortion appeal case, is interesting to juxtapose against Christiano. Dooling, a Roman Catholic from a poor family, worked as a clerk for the Wall Street firm Sullivan and Cromwell. Identified for his outstanding talent, he was told he could have a future in the firm if he went to law school. Dooling planned to go to St. John's University at night. That wouldn't do, he was told by the partner he worked for, it had to be Harvard. And so it was. Dooling went to Harvard (which was not open to women at the time), did well, and returned to work as a lawyer for Sullivan and Cromwell and became a partner in due course.

In a sense, Dooling's career was the reverse of Christiano's, the opportunity presented to her was not to go up the ladder, but down—to become a clerk after law school. Her experience was not unusual. Of the sixty-five lawyers I studied in the 1960s, ten had first jobs after law school as legal secretaries, and many others in the group were offered such jobs. And it was not only the local law school graduates. An alumna from the first class of women graduated from the Harvard Law School recalled in the Twenty-fifth Anniversary Directory,[22] "My faculty advisor suggested I learn typing and shorthand, for I could, no doubt, be hired as a secretary for *the* senior partner while I might never be hired as a lawyer."

Having one status that is considered inappropriate or unattractive to members of the group one aspires to join is a problem. Having more than one compounds the negative image, as a general rule, as in the case of Ms. Christiano. Yet the rule has interesting exceptions. There are cases in which the negative statuses in some women's "status sets" tend to cancel each other. In the study of older women lawyers, interviews with some black women suggested that by hiring a black woman a law firm could fulfill two debts to society's "current call for equality, without any great threat to the firm's continuing internal structure."[23]

That suspicion was borne out by further interviews with black women in a study conducted in 1972[24] and in interviews with black lawyers in 1977 and 1978. It became clear that many of the successful black women, survivors in a discriminatory society (the odds they would be able to graduate from law school were small indeed) actually got their chance because of their "double impairment." What were the advantages? The strongest—in the 1970s—should come as no surprise. It was summed up in one ironic comment: "I'm a show woman and a show nigger, all for one salary."

When black women interpreted their situation as favorable, they

87

were not using white men as their reference group. They felt an advantage in comparison to white women and to black men. Two negatives didn't make one more attractive than a properly credentialed white male. It might, however, make one more attractive than another disadvantaged person. In the 1972 study of black women, about a third of the women felt they might have an advantage over black men in entering the professional world (usually a white world). Not that they had actually competed with black men for their jobs, but they had asked themselves why they had "made it" and why no black men had. The reason they gave, and it was typical, was that they believed black men were a "threat" to white men, while a black woman was not. Why? Because a black man would have more legitimacy in invading the white establishment; in demanding partnership—as a "man." A woman could be kept in her place.

Some of the black women also felt they might have had an advantage over white women because they were taken more seriously by white men than women of their own group. Their reasoning was that white men suspected white women were looking for husbands and would drop out of the profession as soon as they married. They sensed that white men would not expect a black woman to so frivolously throw over hard-won training, and indeed could not, for there were no eligible black men in the settings in which they were finding opportunities. Furthermore, black women were not seen as potential defectors—it was well known that they "needed" to work. And they were not in competition with the men for partnership. I wrote, a decade ago, "Jewish women, or, today, Negro women, may have a greater opportunity to be hired by an elite Gentile firm because . . . *as women* they are not expected to be, and do not expect to be, partnership material." I felt then that the "sex-status" effect might cancel out objections to "undesirable religious, ethnic or racial statuses."[25] There were a number of reasons why these may be of less consequence when coupled with female sex-status. The primary reason was that women could be kept separate from the "inside" network of a firm with less embarrassment than could a man. It is more difficult, for example, to "hide" the man with the objectionable characteristic because he expects the same treatment as other lawyers of his age. The woman, however, understands that she may not be treated like the man and therefore objects less or not at all. The early study noted that:

The woman . . . understands that she may not be treated like the men and therefore objects less or not at all if she is assigned a less visible role (legal research, for example). Once she is hidden because she is a woman, it may not be so important that she has a lower-class background or other religious or class-associated characteristics. Therefore, a firm can hire a competent woman of poor social background without the risk of trouble that might result from hiring a man of the same background.

The data available when that statement was written only suggested the phenomenon, and by the 1970s it was too late to test the notion because women were becoming intent on working at "visible" jobs and aiming for partnership.

The advantage of the multiple negative was rare enough not to be a "route" of any substance for women, although one hears many comments—bitter, joking, or serious—that it *is* a route. "The best recruit," goes the current wisdom, "is a black, handicapped woman with a Spanish surname." But the statistics don't seem to bear that out.

"We don't because we don't" was the reason offered to Frances Marlatt in 1922 in denying her a place at the Columbia Law School. "We don't hire women" was the only answer many law offices offered to women, and for those male lawyers who felt pressed to provide other excuses, "My wife wouldn't like it," "It would be distracting," and "We don't have the bathroom facilities." These were just a few samples from the litany of reasons offered by those men with the quickest minds, the most facile tongues, and in some cases, the strongest defenders of the "rights" of people and groups around the profession.

In view of the generally negative attitude toward women lawyers it was no surprise that they were not to be found randomly scattered across the spectrum of jobs offered by the legal profession. Only some minds and a few doors were open. Many women retreated from the profession, discouraged by resistance, and took the "easy" and socially legitimate way out by returning to the home. Some took other kinds of jobs, in teaching or business, for example. Nearly all accepted their situation one way or another. There is not much one can do if told how socially unacceptable one is, not without companions to make an issue of it; not without law to support efforts for change.

In order to see how far women have come in the legal profession, it is instructive to see what kinds of work patterns they had in the 1960s and before. Only 20 of the 104 firms that answered the *Harvard Law Record*'s questions in 1965 about female employment had women on their staffs. The women in the 1960s study found jobs that were of three types. There were the "assured" jobs—those waiting in family law firms, for example. There were the "protected" jobs—those known to be relatively open and less discriminatory, government legal jobs, for example. And there were the jobs of the "seekers"—those who went out in the market as men did, looking for the best possible thing. Some who sought, found. Others who sought, settled, a more common experience.

Some women, surprising today but not for their generation, returned to the jobs they held as legal secretaries before law school. They would, in time, become lawyers, but first returned to work as secretaries, without bitterness but with resignation. Others became research assistants for their professors, collaborated with authors on textbooks, or took other ancillary jobs. None of these women identified their situation as the result of discrimination. It was the way things were. Society used women's talents behind the scenes and they rarely expected independent recognition. (Thousands of book acknowledgements attest to the professional contributions of wives to their scholar-husbands' work, or to the contribution of women research assistants content to stay in that position.[26] Judge Shirley Fingerhood some years ago, while still in independent practice, recalled a story that resonates with the sense of the future many women had when they were girls:

I remember that someone gave a speech my first year in law school in which he said "and your highest ambition will be to be an individual practitioner or a senior partner in a firm . . ." and I thought, hah . . . that "highest ambition" . . . was the last thing I could think of. . . . As a matter of fact, I think the image I had of myself at the time was typical for a woman of my generation. That is, I saw myself as the brains behind some public figure. It never occurred to me that I would be a public figure, or that I would stand up in court. I thought of myself as being the person who wrote the speeches, or the person who did the research, or the person who had the idea which somebody else translated into action.

There is much disagreement today about how to take the pulse of prejudice. In what sense is an institution to blame if no woman or blacks apply? To what extent is the "establishment" profile an outgrowth of natural relationships among like-minded and like-mannered people, and to what extent is self-selection to blame for the fact that unlike-mannered people shy away?

When people accept an unfavorable definition of their competence and potential, they may be realistic but they may also be suffering what Richard Sennett and Jonathan Cobbs[27] have so poignantly called "the hidden injuries of class." Hidden injuries may make one numb, too weak to plunge ahead, or may kill one's spirit. Discrimination causes such injuries. And because such injuries don't show, even the persons suffering may deny they exist. Some of the older women, reinterviewed later, described "waking up" to the bias against them or said they "only began to think that they might have been discriminated against" when class action suits began to be mobilized.

Discrimination is a chronic condition that can induce trauma in a way similar to the consequences of chronic disasters, described by Kai Erikson:

> A chronic disaster is one that gathers force slowly and insidiously, creeping around one's defenses rather than smashing through them. The person is unable to mobilize his normal defenses against the threat, sometimes because he has elected consciously or unconsciously to ignore it, and sometimes because he cannot do anything to avoid it in any case.[28]

Is the analogy of disaster too strong to apply to the case of women lawyers, or women generally? Does it overdramatize their condition? The most pervasive kind of prejudice women have faced has centered on the feeling that women are protected and at least some of them are pampered. As a result, their claims to be full persons have been trivialized. One could empathize with the poor or homeless or unemployed, but women, especially middle-class women, seemed to have more than their share of society's comforts and their problems seemed minor. The hidden injuries were rarely recognized, and still are not. Perhaps the disaster analogy is more apt than at first it seems.

The older group of women lawyers, satisfied with the job opportunities then open, felt few pressures to move ahead and certainly were not encouraged to do so. Incidental information that a firm or government office was not hospitable to women was enough to persuade a potential recruit not to apply. Sociologists and anthropologists know and teach the impact of informal social controls—the grimace, the withheld hand or the too-eager hand, termination of conversation without appropriate cues. But, caught up in an orgy of quantitative assessment (for example, comparing the number of applications submitted to law school to the number accepted), they often forget that "open" movement into and out of groups (or in this case, elite professions) is only grossly determined by formal procedures.

Individuals also learn to work within and to manipulate the system that surrounds them. Women lawyers quickly learned to identify the good firms; some were less resistant to hiring women and became known for it. The extent to which women could approach them for employment depended on their access to the relevant information networks and their ability to mobilize resources to get the sought-after job. Two early interviews illustrate the point:

All the members of my class, men and women, had regular nets out. We all knew where everybody was. Another girl in my class had a father who had been tied up somehow with Johnson's firm. . . . Someone from the Bar Association had gone to Mr. Johnson and said "We protest that you don't have any women lawyers." So he thought to himself, "Whom do I know that I trust in a woman lawyer?" and he phoned this girl's father, and she went over. But she wanted to get into a straight downtown firm, and she told the firm that she had a friend, and called me up. So before she hung up, you know, I was winging out the door. We were all helping each other then. I mean, everybody was trying to get everybody else a job, and if they wouldn't take you, you know, you'd say "Just a minute; I know somebody that is just what you want. . . ." We were a very close group, and we were pretty much 100 percent employed, which was fabulous in those days [the early 1940s].

I tried to steer clear of firms where it was pretty obvious that they were not about to hire a woman. We knew this because the girls who try the year before report back. I went to the four firms and there was only one that I really wanted to work for, and that was Firm X, . . . and they took me, so I didn't have a very difficult time of it. I didn't know anyone there, but my husband's a lawyer, and he had graduated from Harvard Law School two years earlier than I graduated from Columbia and I knew a lot of his friends. By that point, you see, we knew people in just about all the firms and also had another way of knowing what was a likely firm to accept a woman and what was the kind of firm that I would prefer to be with.

Women lawyers who are not well placed in the information networks, or whose social background statuses are not considered advantageous —each, of course, is a correlate of the other—did settle for poor or mediocre jobs. These two cases were typical:

I'd like to know what experience other women lawyers have. There's only one that I know of who's really happy with her job. The others have taken mediocre jobs. These are very capable women. Just for a woman to get through all the years of school and work, she has to have a certain kind of mind and they're not giving her an opportunity that she needs. . . .

I was quite lucky in getting that job because I had been to so many interviews, and they all said, "Oh, we'd love to hire you except for one little handicap— you're a female." Three or four different firms each told me something on the order of . . . "Normally if we can't use an attorney we recommend him to another firm . . . but we don't recommend girls." In desperation I went to an agency and I said, "Get me anything!"

Much as rivers change their course because of an overwhelming natural disaster such as a flood or a storm, so did women's channel of opportunity change during World War II. The non-traditional labor force utilization of women was one of many unanticipated consequences of the war. Although it had occurred in World War I, twenty years later there were more women available who had expertise in professional and technical fields. The war not only made a place for women, it also created a small demand for them. A lawyer who was a beneficiary recalled:

And then the war came along. And the downtown firms had to make up their minds whether their senior partners would go and look up law themselves because they were losing all their young men. . . . None of the firms I'd gone to before would take women. Now they called me up; they said, "Come on down, we've changed our minds and decided we'll have to take women." I said, "Well, I don't want to come; I'm very happy where I am." One said, "Oh, come on down anyway," and they offered me so much more money than I was making, I thought I'd better go. That was a big corporate firm on Wall Street.

The opportunities provided by World War II, however, did not bring radical or permanent changes in the law career structure for women. The women who went to work for firms that formerly excluded them, or who were offered jobs by such firms, knew they were filling a gap that would close again when the men came back. Sex-status did not recede in importance; it was as salient as ever. As one lawyer put it:

When the war came and this downtown firm called me, it was because I was a woman. In other words, their young men were all being drafted, and they wanted a woman. They weren't just calling for a lawyer; they were calling for a woman. Someone they wouldn't have to keep.

During and after the war, some law firms hired women to do special jobs. This was in some sense an opportunity; but it was limited. One of the women in the early study warned:

Beware of an old traditional firm who wants a *woman* lawyer. They want her for something they can't get anybody else to do!

Another recalled:

Frankly, I got this job because one of the partners knew my background and again it was one of these accidental things where I was taken in on sort of a temporary basis. I have remained there, but the whole experience of somebody with my background having to find a job through coincidence has made me extremely cynical and disillusioned about the law for women.

Although not meant to be career tracks, these backdoor routes did provide opportunities for women who were present when their firms felt the first pressure to promote women, an unexpected consequence of "exploitation." Like the women in the law schools who became professors after years of service in ancillary positions, on research and in special institutes, some women hired by large firms for special assignments became permanent staff members, even eventually reaching partnership. These clearly were unanticipated consequences of a selection system expected to keep women in a particular place.

The legal profession's treatment of women in practice, in teaching, and in the job market changed dramatically during the late 1960s and 1970s, at least partially in response to legislation curbing unequal treatment of women and minority groups. There is no question that these changes in the law were crucial in bringing about a change in discrimination patterns. Without them, it would have been difficult to fight entrenched modes of discrimination. The laws themselves required time for interpretation and enforcement; it took a decade.

The Civil Rights Act of 1964 prohibited, under its Title VII, employment discrimination on the basis of race, color, national origin, or sex, and also created the Equal Employment Opportunity Commission (EEOC) to aid in enforcement by acting in response to and on behalf of groups whose rights had been violated by discrimination in the workplace. In 1967, Executive Order No. 11375 added sex to the other categories of discrimination prohibited by an earlier order applying to organizations with federal contracts of more than $10,000. Although law firms didn't ordinarily have contracts with the government, the large corporations that employed them did. In 1972, a series of amendments to Title VII were passed that extended its provisions to all employers with fifteen or more workers and gave the EEOC the power to initiate suits. This meant not only that smaller law firms were affected but also that individuals did not have to jeopardize their careers by filing personal complaints against their employers. The Justice Department could also bring "pattern or practice" suits. The coverage was extended to state and local governments, agencies, and educational institutions.

Of particular impact on the employment of lawyers, legislation defined law school placement offices as employment offices and thus made them targets for suits. Their gatekeeping functions were radically altered as a result of suits brought against them by students and feminist attorneys.[29]

The year 1972 was a milestone for women's equality. That year Title IX of the Educational Amendments Act made it illegal to discriminate on grounds of sex in all public undergraduate institutions and in most private and public graduate and vocational schools receiving federal monies. This applied to all major university law schools and affected not

94

only the enrollment of women students but the distribution of scholarship aid to them. Further, it put the universities under the gun to hire and promote more women faculty members (see chapter 12).

Law schools, for example, used Title VII and worked through the human rights commissions of the various states and cities to deny placement service to any law firm or employer of lawyers known to be practicing discrimination. It was applied early in the 1970s by the University of Chicago, the University of Michigan, and Harvard University,[30] and other law schools followed suit. It became illegal for law firms to say they would not interview or hire women. They could not ask whether women intended to marry and have children.[31] But the rules meant nothing without implementation, and that was forthcoming. It was the beginning of a new era of access to jobs within the profession, still tentative in some places, but nevertheless far wider than could ever have been imagined a generation before.

Several key decisions handed down by the Supreme Court also paved the way for clearing out entrenched discriminatory practices. In 1971, in the unanimous *Reed* v. *Reed* decision,[32] the Court held a statute giving men preference over women for appointment as estate administrators was inconsistent with women's equal protection. By 1975, in place of the proposition that women make the best secretaries, the Court, in *Stanton* v. *Stanton*,[33] judicially noted "women's presence in business, the professions, in government, indeed in all walks of life." In 1977 the Court's newest member, Associate Justice John Paul Stevens, (finding a law discriminating against men that made widowers demonstrate dependency on spouses to receive benefits automatically obtained by widows) "wrote that habit rather than analysis or actual reflection made it seem acceptable for the legislator to pigeonhole people by sex."[34] A majority of the Court in 1976 and 1977 openly acknowledged that a dynamic equal protection principle mandates an elevated level of review for explicitly gender-based classification.[35] But, as Ruth Bader Ginsburg, who successfully argued before the Court to achieve sex equality in a number of important cases,* has noted, "The development remains uneven and unfinished, impeded by the unsettled fate of the federal ERA. But it has reached at least a mid-passage state, it has progressed beyond the point of return to old ways."[36]

---

*Ruth Bader Ginsburg argued the following cases before the Supreme Court: *Frontiero* v. *Richardson*; *Kahn* v. *Shevin*; *Healey* v. *Edwards*; *Weinberg* v. *Weisenfeld*; *Califano* v. *Goldfarb*; and *Duren* v. *Missouri*. After arguing the above cases, she also wrote briefs in several subsequent cases but did not argue them.

Patterns of Practice

Kinds of Practice

The range of possible career paths in law is large. Today, with the "law explosion," a new member of the bar presumably can elect to choose among judicial clerkships, criminal law, tax practice, corporate law, work in legal aid or public defenders' offices, a prosecutorial job at the state or federal level, government practice in a regulatory agency, or any number of others. Some of these fields and specialties are new, but the choice was wide in the past too. Yet law is stratified; different kinds of firms and legal work attract and select lawyers from different groups and backgrounds. Subcultures, some clearly defined, some not, can be identified. I shall consider only the element of women's sex-status in locating them among the various kinds of practice.

In the past, women's law career choices were structured by the interplay between discrimination, adaptation to that discrimination, and opportunity. In the absence of comprehensive data, the pattern of women's legal employment, how it differed from that of men, and how it is changing can only be sketched. Nevertheless, distinct patterns can be identified—the specialties that women have traditionally gone into, those specialties that have become indicators of the changing role of women in the law, and the general changes in the legal profession that affect all who practice it.

The broad parameters of the distribution of women lawyers among types of practice can be located in census figures, studies done on lawyers in various regions, and the jobs that recent law school graduates have received. Census figures are outdated because analysis of the 1980 count is still in process. However, past censuses reveal the crude dimensions of lawyers' employment (see table 6.1). In 1960, 65 percent of male lawyers, but only 40 percent of female lawyers, were self-employed

TABLE 6.1

Employed Lawyers and Judges by Class of Worker and by Sex, 1950, 1960, 1970

Class of Worker	Total		Men		Women	
	Number	%	Number	%	Number	%
1950						
Total	180,461	100.0	174,205	100.0	6,256	100.0
Private wage and salary workers (firms)	45,236	25.1	43,296	24.9	1,940	31.0
Government workers	26,428	14.6	24,695	14.2	1,733	27.7
Self-employed workers	108,758	60.3	106,188	61.0	2,570	41.1
Unpaid family workers	38	*	26	*	13	*
1960						
Total	208,696	100.0	201,556	100.0	7,140	100.0
Private wage and salary workers (firms)	43,726	21.2	41,597	20.6	2,229	31.2
Government workers	30,855	14.7	28,848	14.3	2,007	28.1
Self-employed workers	133,874	64.1	131,009	65.0	2,865	40.1
Unpaid family workers	141	*	102	*	39	*
1970						
Total	276,688	100.0	263,506	100.0	13,182	100.0
Private wage and salary workers (firms)	76,440	27.6	72,480	27.5	3,960	30.0
Government workers	53,128	19.2	48,280	18.3	4,848	36.7
Self-employed workers	146,983	53.1	142,651	54.1	4,332	32.8
Unpaid family workers	137	*	95	*	42	*

SOURCES: 1950 data: U.S. Department of Commerce, Bureau of the Census, Census of Population 1950. 1960 data: Table 21, U.S. Bureau of Census *Subject Reports: Occupational Characteristics,* Final Report PC(2)-7A, 1963, p. 277. 1970 data: Class of Worker or Employed Persons by Detailed Occupation and Sex: 1970, U.S. Bureau of the Census, 1970 *Subject Report, Occupational Characteristics,* Final Report PC(2)-7A, Table 43, pp. 693, 704.
*less than 1 percent.

according to the census;[1] and in 1970, 54 percent of the men and 33 percent of the women were self-employed,[2] reflecting the diminishing corps of single practitioners generally and women's greater tendency to go into salaried employment.

A large proportion of women lawyers work for government, and a higher proportion of women work for government than do men. This has held through the years. The 1960 census showed that 28 percent of women lawyers and judges* worked for government, as opposed to 14 percent of male lawyers and judges.[3] In 1970, 37 percent of women lawyers and judges worked for the government, as opposed to 19 percent of male lawyers and judges.[4] A 1976 study of the Chicago bar

*Of course, all judges work for the government, but the census groups "lawyers and judges" as one category.

showed that women were "overrepresented in government employment," but gave no figures.[5] Two surveys of the National Association for Law Placement (NALP) showed that in the class of 1975 (nationwide), 18 percent of the total class went to work for government, but 24 percent of the women did.[6] There was a small overall drop in 1977, as 16.7 percent of the class went into government work, but 22 percent of the women sought employment in government.[7]

Census data indicating where lawyers work show only a slight male-female difference in employment by private law firms. In 1970, census figures showed that proportionately fewer men (27.5 percent) worked in firms as private wage and salary workers than did women (30 percent).[8] The NALP statistics cover a different population of lawyers—namely, recent graduates—and are gathered on a different basis from the census data. The NALP figures for the class of 1975 showed that 51 percent of all graduates were employed by private law firms, but only 38 percent of women graduates were so employed.[9] Two years later, in 1977, the percentage of all graduates entering private law firms had risen to 53 percent, and the proportion of women who found employment in those firms had risen 5.3 percentage points, to 43.3 percent.[10]

Law firms come in many sizes and models. Large and small firms are not only separated by numbers; they are worlds apart in structure and function. It seems that more women are to be found both in larger and smaller firms than in those of medium size. In 1977, 35 percent of women graduates found jobs in firms with two to ten lawyers and 24.3 percent found jobs in very large firms (more than fifty lawyers). In fact, a greater percentage of women graduates found work in large firms than male graduates (15.5 percent).[11]

The entry of larger numbers of women into large firms probably is the result of compensatory hiring to make up for past inequities, although it may also reflect greater merit on the part of female candidates. Those counted as entering small firms probably are engaged in independent and family practices, a pattern found in the past, although it is said to be on the decline.

There are no reliable figures on how many lawyers served in public interest jobs in the 1960s, most of them in legal aid and public defender offices. Furthermore, since much of this work was voluntary, it was not counted by the census figures and since women were more likely to do voluntary work, they might not have been counted at all. By the 1970s, much work in the voluntary sector was being handled by salaried lawyers in public agencies and public interest firms. We know that some women continued in this tradition, in a greater proportion than men, but they were not a major segment of women lawyers. While only 5.5 percent of the 1975 class of law graduates went to legal services jobs, 12 percent of the women graduates took these positions.[12] In 1977, 4.3 percent of all graduates and 7.8 percent of the women graduates took such jobs.[13]

The percentage of women and men going into the academic sector (a catch-all category ranging from post-graduate work to teaching law) was the same (3.3 percent), and almost the same percentages of all lawyers (10.1 percent) and of women (9.5 percent) went into business, a growing sector of employment.[14]

The NALP survey of the class of 1975 proved that women lawyers had won their spurs on the legal job market. The survey reported that 87.5 percent of women law graduates that year were able to find law-related employment within six to eight months of graduation. Yet women and minorities were still being employed at a slower rate than their male counterparts, since 91 percent of all 1975 graduates were able to find employment within the same time.[15] In 1977, women lagged by less than one percentage point in employment—92.9 percent, compared with 93.6 for all law graduates.[16]

A survey conducted by the California Young Lawyers Association on levels of unemployment and underemployment among the 22,500 lawyers admitted to the California bar between 1972 and 1977, called "the first of its kind in California and the first of its scope in California," also showed some differences.[17] More women reported being unemployed (6.5 percent) than men (3.7 percent). It also took women longer to find a job. Asked if the following were true, "Even though I am now employed, I was unemployed for a period of at least three months after my admission to the bar," 21.2 percent of the women answered "yes" as opposed to 18.4 percent of the men. A slightly greater percentage of men, however, reported that although they were practicing attorneys, they did not have sufficient work (15.7 percent of the men; 13 percent of the women). But greater percentages of women than men felt that the employment survey was worthy of substantial attention, would pay a $20.00 fee for a placement service, and would use a placement service if it were adopted by the state bar.*

---

*The percentages were as follows: Felt that issue of lawyer employment was important —male, 78.3; female, 82.7.
 Willing to pay a $20 placement fee—male, 56.7; female, 67.0.
 Would use a placement service—male, 35.5; female, 48.5.

From Tables 35, 36, and 37, California Young Lawyers Association, "Report on Unemployment," 1977.

In the 1970s graduates at elite schools did not have the placement problems experienced by graduates from schools with lower standing. At Harvard, Yale, Columbia, and Stanford, I was told that there were no difficulties in placing either women or men students. It was a distinct change from the past, when some elite schools had told of the troubles women students had in finding jobs. The patterns of employment also had changed. At Yale University in 1969 only two women out of a total of eight women graduates went into large law firms (New York City corporate firms); in 1978 half of Yale's women law graduates went into these firms. A large number of Yale's women graduates were also getting judicial clerkships—nine out of forty women in 1976. Very few went into legal services. The numbers are significant because the Yale Law School's graduating classes are small —between 150 and 200.[18] At Columbia University, between 1972 and 1977, the percentage of women graduates entering law firms rose from 38 percent (ten individuals) in 1972 to 58 percent (forty-four individuals) in 1977. Clerkships also went steadily up in this period, mostly to judges on state courts, and public interest work attracted about 8 percent of the women. Most Columbia women, like their male classmates, got jobs in New York, although by 1974 women increasingly were going to Washington for government jobs or to Washington firms or clerkships. As at Yale, relatively few Columbia graduates took work with the federal government.[19] There was also an increase in the percentage of Harvard women law graduates going to large firms: in 1972, 55 percent did so; in 1977, 63 percent, compared with 68 percent of all Harvard Law graduates.[20]

Lawyers are not equally distributed over the country. Most employed graduates are located in New York City, followed by Washington, D.C. and Chicago. If one ranks the states for numbers of attorneys, New York again leads, followed by California (combining the figures for Los Angeles and San Francisco), Washington, D.C., Ohio, Pennsylvania, Texas, and Illinois. Although local law schools tend to place their students in the region of the law school, the major law schools tend to place them in the big cities of the Northeast, primarily in New York. The factors that affect men's job selection also affect women: when women began attending the elite law schools in greater numbers they began working in the big firms in New York. But following the "woman's pattern" of selecting government work in the past, they located in the big cities and to some extent in Washington, D.C. The areas of greatest economic growth, the Sunbelt states, remain more resistant to women lawyers than the Northeast or Far West.[21]

Specialties

Women have not only been selectively recruited into certain kinds of legal employment but into certain specialties within them. There are *subject* specialties such as probate (trusts and estates), corporate law, litigation, real estate, domestic relations;* and *function* specialties such as research, negotiation, or courtroom advocacy.[22] These overlap as specialties do in medicine: there are specialists for gastrointestinal problems, but surgeons who operate on the stomach may perform other operations as well.

Within practices, there is a sorting out of lawyers with respect to who will do what, and people make their reputations in different spheres. Of the founders and name partners of one powerful Washington, D.C., firm, a firm graduate summed up their respective talents as, "Burling was the genius and Covington the wheeler-dealer . . . an unbeatable combination."[23] The brilliant brief writer is not always the dazzling courtroom performer; the passionate advocate probably won't serve as a moderate arbitrator.

How one finds one's niche and one's legal persona may have something to do with the self formed in one's early years, but it is also affected by where and how one finds a job—in a firm, in the government, or in a corporation. Many an aspiring legal theoretician does years of service as a legal "squirrel": "digging out obscure cases from the firm library for his or her superiors."[24] Many an aspiring Perry Mason bent on saving the disadvantaged ends up devising tax shelters for wealthy businessmen.

Some individuals have more control than others; they select specialties rather than permit themselves to be coded into a slot. Yet men and women in the law profession often find themselves in specialties because of drift, chance, and recommendations from clients.[25] But if that is true, the sociological eye can usually detect the pattern of "chance" and its roots in social sources.

To begin with, cultural definitions determine the "right" specialty for the "right" person. Gatekeepers act as central casting agents in fitting people to their professional parts, based on talent, assuredly, but also on "image" and prognosis of capacity. As performers at auditions, some individuals are better prepared than others, knowing how

*Key specialties in large firms are corporate, litigation, trusts and estates, sometimes real estate and tax. Some firms also have particular specialties such as entertainment, patent law, and copyright.

to speak and dress, but when employers hire, they usually have a "type" in mind—one with specifications of color, age, and probably general appearance.

The professions, like other occupations, use the most elemental casting procedure in society—dividing candidates on sex lines. In the past, those women who managed to get into law found there were jobs they were expected to do although, given the small numbers of women in the profession, no specialty can be said to be "women's work." Rather, women were guided and directed until "by chance" they found themselves in specialties considered to be appropriate for them.

The specialties appropriate for women covered a broad spectrum and were not connected professionally with each other. "Blue sky" work, which has been assigned to women, bears no resemblance to matrimonial law, which women tend to do when they are in private practice or in mid-size firms. The work in law libraries, as librarians or doing research, bears no resemblance to appearing in court on behalf of a juvenile offender in criminal court. Handling wills and probating estates bears no resemblance to working on copyright problems of authors.

Yet this is the work women have done. In 1958, a U.S. Government publication describing careers for women in the legal profession reminded them that:

Women's opportunities seem best in those specialties where their contributions to the field have already been recognized. Some of these are real estate and domestic relations work, women's and juvenile legal problems, probate work (about a third of all women judges are probate judges), and patent law for those who have the required training in science.[26]

"Where their contributions . . . have already been recognized" was a delicate way of putting it. Women in the early study clustered in specialties where their "opportunities seemed best": matrimonial law, real estate, general practice, and trusts and estates. A national sample of women lawyers showed this pattern in 1967.[27] Asked what kind of work they performed, the women responded: trusts and estates (60 percent), domestic relations (50 percent), and tax law (31 percent),* far higher percentages than reported by men for these specialties. The study also found that 45.6 percent of the women handled litigation and 27.7 percent were involved in criminal work. But no information

*Totals are higher than 100 percent because respondents, like most lawyers, performed more than one kind of work.

was sought on the women's level of responsibility in these categories or the specific kinds of work they performed. The study pointed out that "litigation can include an appearance in court for the purpose of procuring a signature on a probate order or obtaining an occasional uncontested divorce."[28] Women lawyers rarely argued a case in court, and much of their criminal work consisted of cases handled under Legal Aid assignments. In 1956, Harvard alumnae were advised that "no women can as yet hope for a niche in the litigation section of a large firm."[29]

It seemed unusual, therefore, to encounter a woman in the 1970s (recently retired from a large firm) who had gone into litigation two decades ago. "Wasn't that unusual?" I asked. "The reason," she answered, was that "I was hired to do the file search on a litigated case. I thought the firm was so very benevolent to take me. I thought it was a great opportunity."

A male associate in the same firm with this woman described her situation as a member of a litigation department who was not permitted to engage in the active role of litigation:

She was of the typical mold of a few years ago, in that she is very bright and they put her in the position where she could work on theoretical and antitrust stuff without ever being an actual litigator and counselor—never going to court; she was in the back room in the way a lot of places used to keep their bright women in that kind of theoretician role and let the men go in and make objections and argue in court. People consulted her on any interesting, or even not so interesting, antitrust problems that came up, but she could not do more, she would always be on the shelf, a resource.

Women lawyers did not go to court. Like law students on summer internships, women did research and brief writing. Another lawyer in a large firm who ultimately rose to partnership via an alternative route explained:

When I came into the firm I said I wanted to be a litigator. The firm had never had a woman in litigation, but they thought maybe it was time, so they said, "fine." It really didn't work. I could see very quickly that I was going to be just a brief writer. I don't know whether it was because I was a woman, but I didn't see myself making a future, because the way to become a top litigator was to go through the courtroom.

Although trial work for indigents seemed all right for women, trial work for paying clients was not. *Time* magazine made it all clear in a March 6, 1964, article that described the legal profession's view of women lawyers as "unfitted for trial work, suited only for matrimonial cases or such backroom fields as estates and trusts."[30]

Have women gone to court in the past few years? A male associate in a large firm felt there was still resistance:

They really don't go to court, they are sent to the library and they write briefs, which is what everybody does for a couple of years, but then people get kind of pulled out of that and told to go at least argue motions and stuff—but the women don't get as precisely a fair shot.

But this was not true for women in the district attorneys' and public defenders' offices throughout the country, as we shall see.

It was characteristic before the 1970s to place women in a form of professional purdah in their relegation to routine behind-the-scenes work, concealed from colleague and client. No studies or records exist to document the percentages of women in the "hidden" specialties, but that women were clustered in them seemed to be common knowledge among the men and women I interviewed, and it is supported to some extent by data.

White suggests that the women lawyers in his 1967 University of Michigan study were poorly represented in firms of from five to thirty lawyers. They were better represented in large firms because the female law graduate's function in large firms "is to do research, mind the library, and perform other specialized tasks which fall somewhat short of practice on the same scale as her male colleagues,"[31] and smaller firms did not have elaborate libraries or research departments.

In a typical description of women lawyers in large firms, one woman respondent recalled a role model:

I even had an aunt—an "Auntie Mame" who studied law and who became a lawyer—and went into the firm in the usual women's capacity. They had her in the library, you know, doing research, keeping records.

The smaller firms that could not afford to hide the woman lawyer could not afford to hire her.

In the past, women lawyers tended to go into specialties or areas of the law in which they assumed handmaiden roles, ancillary to the dominant and highly visible male legal roles. These were of two types: jobs *as assistants* and jobs *of assistance.*

A few of the women in the older group began their careers as assistants to professors and judges. Although these positions are often useful for later career development, and fledgling male lawyers compete to become judges' clerks, women often take assistants' jobs that are not defined as the first stage in a career status-sequence.* They tended not to be clerks.

One woman in the group was awarded a doctorate in law and stayed on at her law school to become a research assistant to her major professor. She held the job for seven years, until he died, and then took on another "handmaiden" job as secretary to the law faculty. Although this may have seemed a good choice because of the opportunities it offered at the time, objectively she was over-qualified for the position. A man probably would have pressed for a more commanding job.

A combination of factors have usually been at work in women's acceptance of such employment. Two women lawyers who took such positions saw them as opportunities at the time they were offered. One woman was offered twenty-five dollars a week as an assistant during the depression years (at the time, men going into practice were being offered five dollars a week). In the second case, the attorney could not get other work. As she described it:

Well, that was a very pathetic time for women and I didn't know it. Nobody had told me that no men's firms would hire you. When I went to the employment service at New York University, they wouldn't even take my application, because they said they had no jobs whatsoever for women.

When one of her law professors asked if she would write a book for him, she replied, "Yes, I'll write a book for you, what's it about?"

The assistant's job also characterized the experiences of an outstanding law graduate who applied to a noted Federal Court judge in the early 1970s. The judge asked her whether she was interested in a family as well as a career in law. At her reply of "yes," he was delighted and offered her a job. He had long hired women with legal training who he believed would have limited ambitions because of family responsibilities and who would be more than content to write his speeches, précis material, and do research on the "cutting edge" in various fields by going straight to the experts and interviewing them. They were usually content to stay with him five years or more (normally a clerkship is one

---

*Women are eminently exploitable as assistants for the reasons noted by a professor of biophysics in advising his colleagues on maximizing their output: "You must have a laboratory assistant, preferably female . . . a female is better because she will not operate quite so readily on her own. . . . " E. C. Pollard, "How to Remain in the Laboratory Though Head of a Department," *Science*, September 4, 1964, p. 1020.

year). This attorney pondered the offer and agreed to work for two years providing she would have a chance to do legal work. On the job, she rapidly found that the demands of her work as "assistant" crowded out any opportunity for legal writing. It was only after taking advantage of all available private time that she was able to write an opinion on a case and submit it to him. He signed and used it immediately. The event established her competence, but she had to carefully structure her other work so that the opportunity to learn by doing law would be sustained. The changing times gave the lawyer confidence to press for the equivalent of a legal apprenticeship, and made the judge feel obliged to provide it (at least partially). He later served as her "mentor" and recommended her for jobs that started her on a track to high-level legal work.

Many of the older women interviewed who were in practice with other lawyers or a husband or father worked in jobs of assistance—performing the practice's non-legal administrative tasks; hiring secretaries, renting office space, and keeping the work calendar. These "housekeeping" responsibilities guarantee that those who accept them remain in obscurity. This was the job description of one lawyer:

I do research, briefs, preparation of legal documents, and office administration. I make sure that the girls are all doing work they're supposed to be doing . . . in general, the management of the firm. I see that the bills are paid and proper allocation of vacations made—this is in addition to my primary work which is the preparation of legal documents and briefs.

The court appearances of most of these women also were limited to administrative tasks: submitting papers, making motions, and asking for adjournments—clearly, work that does not qualify a lawyer for advancement or enhancement of reputation. Husbands or other male members of the firm handled trial work. All of the women in joint practice with husbands professed not to like litigation. Perhaps their objections were a response to pressures on them to avoid direct competition and let their husbands and other male colleagues shine. One lawyer, asked if her husband ever turned court cases over to her, replied, "Yes, especially if it's in criminal court." (The profession ranks criminal law as the lowest form of courtroom work.)

Lawyer wives who were in more prosperous partnerships were less likely to be assigned housekeeping jobs. However, even where there were younger lawyers or assistants to take over these responsibilities, the lawyer-wives generally supervised them.

For some women, the specialization problem is now regarded as ancient history. Aware of the sex-typing of the past, younger women going into law in the 1970s were alert to the danger of being pigeonholed in "appropriate specialties." In large firms they demanded and got assignments in corporate work and litigation. But some women have followed the pattern of the past, and if there is less "typing," there is still patterning. Why do women cluster in certain specialties? Some of the reasons are the same as in the past, and some are different.

The legal system appears to have its reasons for clustering women in specialties that are remote from the high-echelon, high-powered, and money-generating areas of law. But to expect a neat typology of reasons is to assume a decision reflects an internal logic and a sense of reasonableness. Those who made the decisions thought they were "reasonable," and indeed they were. But reasonable for whom? According to the law's own standard of the "rational man," not for women, not for justice.

In the past, many employers did not want women in their offices—period. Women made them uncomfortable. Some employers felt that women lawyers would cost them clients and money. Many women agreed with them. One commented, "I think one of the reasons there weren't more women corporate lawyers was because of client difficulties. I think a lot of corporate clients didn't want women."

Law firms also thought women would not be competent to handle the rough and tumble of negotiation or to participate in the "old boy" camaraderie between lawyer and client. At times the client's attitudes were such that few firms would risk assigning a woman to work with them. One woman recalled:

One firm had a lot of international work. They have a lot of Japanese clients. They said they really wanted me to work for them, but they warned, "Look, we have to tell you something. You are never going to get client contact."

Many firms, and the placement officers who reflected their attitudes, believed that women's clustering in certain specialties or their absence from others was a result of women's preferences. A placement officer (a woman) at a major New York law school in the 1960s asserted "Most [women graduates] go into public service work such as Health, Education and Welfare. They like it." As a consequence, she said, she often suggested government work in these areas to young women seeking career advice. A male associate at Sullivan and Cromwell (whose wife worked at another large firm), reported only a few years ago:

When I was interviewed . . . I asked several firms if they hired women, and they said, "Yes, we do, but they seem to like Trusts and Estates work better than anything else and that's just not a growing part of our firm." I think that is probably true.

Yet views about what women do or do not want to do change over time, challenging the notion that there are basic female dispositions. In the 1960s study, nearly every woman interviewed asserted that women didn't like to go into the courtroom because it was male; it involved fighting; it was unnatural for women. These thoughts could still be heard in 1980, most graphically from a woman in the corporate sector of a large firm who said, "Men go in there ready to lock antlers; I don't like that." But another woman, speaking of the preference of women for litigation at her firm, Sullivan and Cromwell, explained to my surprise:

More women want litigation . . . and they have models. . . . Some of the biggest women lawyers are litigators. Ruth Ginsburg is a litigator, a constitutional litigator. Harriet [Rabb] is a litigator. . . . you know, the traditional excuse is women like to argue; that's why they go into litigation.

For every stereotype that was used in the past to rationalize women's segregation in the hidden legal specialties, another stereotype can be mustered to explain women's entry into the areas of law once forbidden to them. One successful attorney answered an interviewer's question with a firm gesture, "Why do women like litigation? . . . Maybe women are more aggressive and the more aggressive people tend to be in litigation." Another, responding to the thought that women shunned corporate work because financial matters bored them, declared:

No, they really like it. You know, they say women are good at details. And there is a lot of detail work in corporate. You don't have the stress of court, you are not under heavy time pressure. It often ends at five or six o'clock; you don't take it home with you unless you've got a public offering.

More people seem to attribute patterns to some essential quality of human nature, or "woman's nature," than to the coercion of "the system." But it is often easier to capitulate when the system is hard to beat; most individuals compromise, after all. There are also secondary gains from accommodation, as Erving Goffman has dramatically shown in his analysis of the adaptive behavior of people in stigmatized situations— the sick persons who need not work; the impaired persons who need not compete.[32]

Patterns of Practice

There were, and are, distinct advantages for women practicing in certain specialties. Of course, the "advantages" are relative to the legal system and to women's other roles. There is a certain congruity in the ways in which women have been structured into the legal profession and the ways they are structured into society.[33] Women are supposed to invest themselves primarily in the family rather than in work; they are supposed to help men in the climb up, not to compete against them.

When a woman entered an appropriately "feminine" law specialty she was reducing the strain generated by her involvement in a male profession, making life easier for both herself and the system. Further, she avoided the antagonism of male colleagues that would result from her presence in a "male specialty" of the profession. In a female specialty, women were at least conforming to the profession's definitions of where a woman lawyer ought to be. In the outside world, the female lawyer *per se* was an oddity. Within the profession, the female lawyer who did corporate finance was an oddity, but the woman who became a law librarian or a specialist in trusts and estates was believed to be engaging in an entirely appropriate activity. Other lawyers could live with that. Thus, role conflict in the profession was minimized by adherence to cultural definitions.

Classic sociological studies have implied that the sex-division of labor within professions is an expression of realistic adjustments to private and professional demands. An analysis of the specialties of women physicians, for example, contends that in certain medical specialties sex roles and professional roles were more compatible than in others. Some medical specialties were said to offer institutional settings for reducing role conflict, and female physicians accordingly tended to select fields of practice where the sex and professional roles are compatible: pediatrics, psychiatry, gynecology, and public health.[34]

The study suggested, for example, that public health work offered women stable hours of work, "liberation from [the] entrepreneurial role," and the possibility of part-time jobs. Further study of women professionals has shown that such stereotypical reasoning seems not to be valid for all specialties—even those the study cited—thought to be appropriately female.[35] The attributes did not suit the temperaments of all female practitioners nor fit their needs as professionals, mothers, and wives. For some women, stable hours are an asset; for others, they are too inflexible. One woman lawyer told an interviewer that regular time schedules were a good reason *not* to go into government legal work: "I can't take a job with the City of New York as an attorney; it's nine to five; there's no room to adjust my schedule."

The "relief" from building a practice, supposedly attractive to the female public health doctor, is apparently not experienced by many women attorneys who enjoy having their own practices, complete with the problems of getting and keeping clients. Many women lawyers stressed how private practice offered benefits to the working mother which far outweigh the entrepreneurial demands involved.

Certain specialties were and are known as "female specialties" because they deal with family law issues or interpersonal services in which women often are clients. In law, skilled service is associated with work on legal problems and is not seen to include the personal counseling (including psychological support) that lawyers are often called on to give clients. Lawyers come closer to the professional "ideal" if they avoid cases that demand extra-legal services. Thus, the profession tended to delegate cases of matrimonial and family law, legal aid work, custody cases, and, under certain conditions, estate work, to lower-ranking firms and individuals.

Women lawyers served these needs well. Like other lawyers with low-ranking minority statuses, they performed the "non-professional" work, freeing white male lawyers to become more professionalized. Firms of all sizes seemed to refer their para-legal "emotional problem" cases to their women lawyers, if they had them. For this reason, women have often found themselves specialists in matrimonial cases or in the probate field. Although probate cases may be lucrative and profit the firm ultimately, probate law entails a surfeit of administrative detail that the client does not see. A male partner may provide the masculine shoulder sought by the wealthy widow, but he will assign the details of her estate work to others.

Women's placement in the low-ranking specialties provide them with a work context that is less competitive with their home life and therefore creates less role strain. If women must fulfill the demands attached to family roles, it is believed that they should seek the occupational roles of lowest demand. Therefore, low rank is seen as more manageable for the woman with a family.

A lawyer commenting on this expressed a popular view:

One should not conclude . . . that women lawyers are necessarily being forced against their will into unattractive, unremunerative specialties. In many cases, I believe that they prefer a low pressure area, for in this way the woman lawyer can combine marriage and a career more successfully.

It is the less attractive specialties that allow a woman to work less hours, that free her from the tense competition of the legal profession, and also, unfortunately, provide less money. But studies show that women do not give money the importance assigned to it by men.

The low-rank solution to the many demands on women may be viewed in a wider context. William J. Goode and others have pointed out that men are loathe to grant women opportunities to challenge their power in the occupational world and in other spheres.[36] They suggest that male resistance stems from the two sexes' rivalry for power, not only in the occupational and professional sphere but within the family.

The legal system creates conditions in which women will have less motivation to compete with men or, once inside the profession, will have less opportunity to rank equally with men or outrank them.[37] If women were permitted to work freely toward achievement and success, the possibility of upsetting the traditional power system would increase, with consequences for the profession and the family. Thus, the selective recruitment of women into low-ranking specialties in the law —and in other occupations as well—can be viewed as one of the mechanisms maintaining harmony between family and occupational institutions, and as a device for preserving the stratification system. Keeping women in their place and out of men's places helps keep intact the stratification systems of both the professions and the society.

Government Practice

BY CHOOSING government practice, lawyers move into a professional current out of the mainstream. More than a third of all women lawyers are in government work; they choose it disproportionately more than men. "Women like it," I was told by law school placement officers in 1965, but the picture turned out to be more complicated than that.

Most American lawyers work in the private sector; about a quarter of them are salaried and over half are "self-employed workers." About 20 percent work for the government. Although women lawyers in the 1950s and 1960s were proportionately represented to a greater degree than men (28 percent as contrasted with about 14 percent of male lawyers), by 1970 there was an upswing that even further widened the gap between men's and women's careers in government (37 percent, as contrasted with 19 percent of the men). This upswing probably reflected a concerted effort on the part of government to recruit women. Although the 1980 census has not yet been analyzed, two recent nationwide surveys of law school graduates show that women continue to be over-represented in government work, although to a lesser extent than before. In 1975, 24 percent of women graduates went to work for government, and in 1977, 22 percent of the women did, compared with 18 percent and 16.7 percent of the men, respectively.

Law school placement officers once encouraged their women students to seek government employment. They believed that women preferred it, and they knew that women would have a better chance of finding employment with the government than in the private sector. Government provides a "haven of universalism," as Robert K. Merton has pointed out,[1] and is "the institutionalized conscience of the people." Thus we might expect government to be in the vanguard where ideology about equality and practice otherwise diverge. Government's use of competitive examinations for jobs, for example, is an expression of

Government Practice

the commitment to merit as the criteria for job selection rather than the irrelevant characteristics of national origin, race, or sex.

Government local practice has other advantages, both ideological and practical. Most work may be thought of as attending to the public good. Government work also provides security, and one need not look for business. Much of the work is manageable during a normal work week, although some of the most interesting is not. But money is not among the incentives; comparable private practitioners invariably make higher salaries than those who work for government.

Many of the women lawyers interviewed who worked for government (excluding those in elected and appointed posts) agreed that they did so for the reasons expressed by an attorney in the New York City Corporation Counsel's office:

I don't think I ever thought of practicing privately. The security of this sort of work was very important, I think. I came from a very modest background . . . my family had no particular connections. This has been a shielded and protected situation.

Women in law interviewed in the 1960s tended to cluster in "protected settings."[2] In seeking government jobs, women have sought to avoid discrimination, as did minority groups such as blacks* or Jews before the 1950s.[3] Women facing discrimination in other male-dominated professions also went to work for the government disproportionately more than men (see table 7.1).

Although a large proportion of white women choose government

TABLE 7.1

Percent of Professional Workers in Selected Occupations in Government Service by Sex, 1960 and 1970

	Lawyers		Dentists		Physicians and Surgeons		Engineers	
	Male	Female	Male	Female	Male	Female	Male	Female
1960	14.0	28.3	.03	10.0	14.0	30.0	17.0	32.0
1970	18.3	36.8	3.7	7.0	15.2	26.4	16.5	25.2

SOURCES: Figures for the 1960 set taken from ULSL Bureau of Census, *1960 Subject Reports: Occupational Characteristics* PC 1 (2)-7A, 1963, p. 277. 1970 Data from U. S. Bureau of Census, *1970 Subject Reports: Occupational Characteristics*, PC (2)-7A, 1973, p. 705.

*For example, in 1960, 20.1 percent of black lawyers went into government work, as contrasted with 14 percent of whites; 24.7 percent of black physicians contrasted with 14.8 percent of whites. (1960 Subject Reports. *Occupational Characteristics*, Final Report PC(2)-7A, 1963. U.S. Bureau of the Census, p. 284.)

work, an even larger proportion of women of minority status, or those who graduate from low-ranking schools, have tended to seek government jobs.[4] In 1966, ten of the twenty-one women graduates of Columbia University Law School went to work with the government or in government associated agencies (a higher proportion than the men). During the 1960s, university affiliated law schools placed 20 to 30 percent of their women graduates in government service, but up to 90 percent of the women graduates of the lower-ranking part-time or independent law schools have gone into government service.[5]

About a third of the national sample of women lawyers surveyed by James White in the mid-1960s were working for the government,[6] compared with 16 percent of the men at the time that survey was conducted. In addition, a comparison of respondents' initial jobs and those held at the time of the study showed that a substantial number of women had sought government jobs after working as private-sector attorneys, while an equivalent proportion of male lawyers had left their initial jobs with government to work in the private sector. The two sexes' differing experiences in the work place seem to reinforce their initial perceptions of career opportunity in the government and private spheres.

A number of top-ranking male law school graduates commented in interviews that men sometimes take government legal jobs for different reasons than women. Many men use their initial government experience as training for the private sector—to gain knowledge of the way the government handles certain types of cases so that they may better *oppose* them later as members of private firms representing corporate and other clients.

Women often choose government service because of the expectation that they ought to do "good works," a typical traditional rationale for women's work. This is sometimes coupled with their expectation of becoming a behind-the-scenes influence. Judge Shirley Fingerhood of New York exemplified this as she related to me her choice of a first job:

The career which I had somewhere in the back of my head, although not very clear, involved solving the economic problems of the United States, or some career working in government planning areas. The first job I got was in Washington working for a man named Felix Cohn, who was the original Indian Affairs attorney—the first person to really become involved in Indian problems and work at projects designed to give them back land and give them money for land that had been appropriated.

Government Practice

Of course, some men gravitate to government service because of idealism, and because of swings in the legal marketplace. Carol Arber, a lawyer who recently joined the New York State Attorney General's office told me that many lawyers who identify themselves as politically left "are moving over" because the number of jobs in legal services are being reduced, and they can continue to do work they believe in such as consumer protection and prosecuting businesses which are dumping toxic wastes illegally.

But if women gravitated to government work because they could find employment and expected fair treatment, they did not necessarily find it a haven of universalism. They did not have an equal chance at the good jobs or promotions. Few women obtained important legal posts in government, few did trial work, and discrimination affected their careers in the public sector as it did in the private sector. Top-level legal jobs in state, county, and local agencies have generally been held by men. Within the legislative branches of government, women did legal research and drafted legislative proposals. Their median salaries were lower than those of men ($8,737 per year compared to men's $13,439 in 1970).[7]

But women lawyers found government careers considerably improved in the last decade. Interviews with women in city government in the late 1960s and the new breed of women lawyers who became district attorneys and lawyers for the Department of Justice in the late 1970s (jobs formerly unavailable to women) showed that the experiences of the second wave were in marked contrast to the first. Unlike the corporation counsel who looked to government work as a secure haven, young women in the 1970s viewed government work as good experience for legal careers in other sectors. They included prosecuting attorneys in both the New York and U.S. attorney's offices who were eager to engage in trial work.

There have been many "firsts" for women assistant district attorneys appointed in New York and elsewhere in the country in recent years. "Woman Lawyer Scores a First in the Bronx," read a headline in the *New York Post* on October 9, 1969: "The first woman attorney to try a case in the history of the Bronx District Attorney's office has won a verdict of guilty," reported the article, a newsworthy event despite the district attorney's comment that "I see nothing unique in this. Helen Johnson is a good lawyer."

Many lawyers are eager for prosecutors' posts because in them the neophyte lawyer has a chance to work on cases independently. It might take years to get this kind of experience in a private law office. Reciprocally, the offices can be choosy about the law school graduates they hire. The Justice Department, for example, tends to hire from prestige law schools, such as Harvard, Yale, Columbia, Georgetown, and George Washington Universities.[8]

Women's recent advancement into non-traditional government law slots is due to a concerted effort in government to recruit women and minorities, and some government offices took this mandate very seriously. District Attorney Robert Morgenthau in New York City has drastically increased the number of women assistant district attorneys over the past ten years. Of 255 assistant district attorneys, about a third (83) are women.* Dramatic increases have also been reported in the U.S. Justice Department. In 1976 there were 14,312 attorneys employed by the federal government, of whom 11.6 percent, or 1,663, were women. These figures have soared in the Civil and Civil Rights divisions, for example, from 9.8 and 7.8 percent respectively, to 30 percent of all lawyers in 1980. In the office of the U.S. attorney, they rose from 3.7 percent (only 33 women) to 17.3 percent (305 women). By 1980, 31.5 percent of the lawyers hired were women.[9]

The U.S. attorney is the chief trial lawyer and legal representative of the attorney general, prosecuting violators of federal civil and criminal statutes ranging from bank robbery to sophisticated securities frauds, white collar crimes and crimes by public officials, environmental litigation, and enforcement of drug laws. Offices range in size from 2 (Guam) to 161 (Washington, D.C.), with most employing between 5 and 15 attorneys. There are a total of 1,798 authorized attorney positions. There are 305 women, 17.3 percent of this group, including 194 in New York. The district attorney is in a similar position and handles similar cases at the city level.

Attorneys in both agencies prosecute criminals, and in the course of their work they must deal with police, defense lawyers, and judges and jurors. In the past, women were believed to be unsuited or incompetent to handle either the pre-trial negotiation, which may involve detectives and narcotics offenders, or the work of actually bringing an indictment and conviction in the courtroom.

The traditional description of women lawyers—the domestic restrictions on their professional time, their presumed lack of assertiveness, and the reluctance of those they work with (the police in particular) to

---

*They were clustered, however, in the lower grades (11–14) and there were only three women at the executive level (3.9%). Women's average ranks in all departments were lower than for men. This may be due, in part, however, to their recent recruitment and therefore lower seniority.

accept their authority—has been shown to be untrue. The dramatic change in perception of women as suitable prosecutors had less to do with a gradual change in attitudes in the attorneys' offices than in the hiring directives which were enforced by the government. Once women were appointed and it was found that the feared problems did not arise, women more easily became part of the staff. One section chief who was interviewed said that when he came to the office there was one woman prosecutor; at the time of the interview in 1979 there were seventeen or eighteen. His office often attempted to provide solutions for the problems women faced in managing the demands of jobs and family.

One of the most successful solutions was one arranged in the U.S. Attorney's Office for the Eastern District of New York, in which two young mothers were on the staff. The chief of the section created a six-fifths slot, to be shared by the two women, each working three-fifths time. They were to receive all benefits, even separate offices—"corner ones with windows," he stressed. Some women in the office were aided by assignment to appeals work because it can be scheduled well in advance. They found the solution agreeable, even with occasional breakdowns and trial periods when they, too, worked night and day. Unlike much of Wall Street practice and certain kinds of small practices, the time demands were manageable because generally they were predictable.

Practice in the federal courts, in contrast to many city courts, makes a law career more amenable for women attorneys. One woman had moved from legal services, where cases were heard in the lower criminal and family courts. "There," she said, "we were treated badly, and it was a struggle even to be recognized as having a right to be there." In the federal courts, women have the advantage of working in an environment where judges and other attorneys are polite and do not think it proper to embarrass them. In addition, as everywhere in the professions, the caliber of the client and of the case reflects on the attorneys involved. Because women have been structured in the legal system to have a disproportionate share of low-status clients and cases, their reputations have suffered. This is one reason why women who have gone to Wall Street from other legal practices, or to the federal courts from the city and state courts, feel positive about the way they are treated.

Women also feel an attraction to government work because they can rationalize their professional activity as public service. An assistant U.S. attorney gloried in a bit of sleuthing in her work for the government:

In the consumer affairs office there was a law reform kind of approach trying to stop illegal practices as a whole, instead of just dealing with individual complaints. One of the last projects I worked on there was investigating abortion laws in conjunction with people in the city health department. We sent people around to various clinics against whom complaints had been lodged because of fradulent pregnancy tests. We used male urine at each of these places. One place told our investigator that she was definitely pregnant, and scheduled an abortion for her immediately.

Another place, under the guise of being an abortion counseling service, did a test [on the male urine] and then tried to convince the investigator that she really should have the baby.

Anyway, that was very interesting. We did that type of thing and had a press conference. Afterwards we turned over the findings on the doctor who had said the woman was pregnant to the attorney general and they brought a case against him to the regents and finally his license was suspended.

A woman lawyer in the New York district attorney's office contrasted her own goals with those of her male colleagues. She said of the men: "I think a lot of them are there to get their trial experience and use it as a stepping stone of some sort." "To what?" I asked.

Private practice, politics, whatever. But I expect that the trial experience that I have here will help me in another area, but that area will be in the public sector.

Men come here because they know if they go to a Wall Street firm they'll spend ten years carrying someone's briefcase before they'd get to do a trial of any sort, and twenty years before they are doing major litigation, and most of that will be civil litigation which involves lots and lots of documents and years and years of motions and the other things before you get down to the courtroom part of the case, and that's the part I like.

But they get out of here before long so that they can get into private practice because that's where they can make the most money.

Not all male lawyers behave in this way, but they tend to have shorter government careers,* and the norms of the profession are that lawyers worth their salt will move on from government work. Married women are even less eager than single women to consider their next move, probably because they do not feel pressured financially. (They tend to be married to lawyers who are regarded as primary breadwinners.) But they are affected by the norms that define them as persons who have "overstayed," becoming less transferable because they are too special-

---

*The median age of women lawyers in government work is 44.1 years, while that of their male colleagues is 38.4 years. From U.S. Department of Commerce, Bureau of the Census: 1970 Census of Population Subject Reports, pp. 11–30, table 3, "Age of Employed Government Workers by Occupation and Sex." Age of women is for lawyers and judges. Age of men is only for lawyers.

Government Practice

ized or too old. Most women working for the government still have time to work out this problem because they are recent recruits and relatively young.

To some extent, government reproduces the structure of legal work in the private sector. Specialists are ranked and rated and there is competition for the favored places. If women didn't make it to the top in the past because of prejudice, they nevertheless had a better chance in government than outside. Today that still holds true. Women have even risen to the rank of assistant attorney general of the United States. But the government has not done much better than the private sector in placing women at the top. Here too, the returns are not all in, and we will have to await the career progress of the new generation of women lawyers.

8

Poverty Law and

the Public Interest

REPRESENTING the poor and disadvantaged is one of the major areas of "women's work" in the law. It is a realm in which women have found work in the past and in which they still tend to cluster. It is a field women like to work in and feel comfortable in, and, in the common view, it is one women should work in. One reason offered by lawyers of both sexes is that if women are to be in a male occupation, they should do something womanly within it: namely, help people.

Yet what is known today as poverty law or law in the public interest did not represent a major opportunity for lawyers of either sex until relatively recently.* The Legal Aid Society of New York,† one agency concerned with defending the rights of the poor and active in employing women, was set up in 1876 with only one lawyer. Then the idea of providing free or even low cost assistance to the poor was a novel one. The notion was prevalent then, as it still is more than a century later, that there were the "deserving poor and the undeserving poor."[1] The Legal Aid Society grew very slowly. It was set up to give free legal advice to "worthy" German immigrants,[2] but there was not much enthusiasm about extending free services to a wider spectrum of clients. Furthermore, the organized bar resisted providing free legal assistance, fearing it might undermine the livelihood of

*Three quarters of the lawyers in public interest jobs in 1972 graduated from law school in 1965 or later, compared with one third of the bar as a whole. Howard Erlanger, "Lawyers and Neighborhood Legal Services: Social Background and the Impetus for Reform," in *Lawyers and the Pursuit of Legal Rights*, ed. Joel F. Handler, Ellen Jane Hollingsworth, and Howard Erlanger, (New York: Academic Press, 1978), p. 136.

†Although there are differences between Legal Services and Legal Aid, we shall consider them together because of the similarity of their purposes.

its own members for whom the poor constituted a pool of potential clients.

Efforts to increase the availability of lawyers for indigent defendants only began about the time of World War I, but the scope was small.* In 1963, however, the Supreme Court under Earl Warren held that poor people facing felony charges were unconditionally entitled to have defense attorneys provided by the government prosecuting them. Eight years later, the Court ruled that a poor person facing any prison term had the right to defense counsel at the government's expense. Both these rulings meant far greater government involvement in providing and paying for criminal defense lawyers.[3]

The result was a dramatic increase in public defenders as agencies of state or local governments. The number of defender offices increased from 5 in 1917 to 28 in 1949, and 163 in 1973.[4] Government funding also made it possible to further professionalize legal aid and public defender offices. Legal aid previously had consisted of private organizations that raised their own funds and depended heavily on volunteer lawyers— lawyers whose services were offered by their firms or independent volunteers, such as women who did not engage in active practice. For paid staff, legal aid and public defender salaries were not competitive with work in law firms.

Poverty law programs came of age in the mid-1970s, having been born in the Office of Economic Opportunity in 1965 and later transferred to the Community Services Administration under the auspices of the independent, non-profit Legal Services Corporation, created by an act of Congress in 1974.[5] The act extended the right to defense for an indigent person to civil affairs as well as criminal acts.

Today, legal services programs, which depend on Congress for appropriations, operate in 1,100 communities throughout the fifty states.† The development of legal services opened opportunities for women lawyers who, like their male counterparts, were excited about the chance to help bring about the "New Society" envisioned in the 1960s.‡ Ralph

*A post–World War II survey showed that in 1947 the New York Legal Aid Society had 9 full-time criminal lawyers while the Los Angeles County Public Defender had 21. Each of these lawyers handled thousands of cases a year. As recently as 1960, little had changed. In those two cities fewer than 100 lawyers were responsible for tens of thousands of poor people charged with crimes every year. By 1980, the New York society had a legal staff of 700. Robert Herman, Eric Single, and John Boston, *Counsel for the Poor: Criminal Defense in Urban America* (Lexington, Mass.: Lexington Books, 1977).

†Between 1966 and 1975, the Office of Economic Opportunity (OEO) awarded grants totaling approximately $4.5 million to operate "judicare" projects in which private attorneys were paid for defending poor clients.

‡A study of Legal Services lawyers done by Howard Erlanger in 1967 found that 13 percent of the legal services lawyers were women, (1 percent was black women), while

Nader's "Raiders" were the models for young lawyers who worked tirelessly for subsistence wages in righting wrongs endured by consumers, the poor, the old, the young, and women—in short, any group without power.

Certainly, idealism accounts for the choice and commitment of many of the ten thousand attorneys involved in legal services programs, public defenders offices, and legal aid societies. These attorneys generally have had to face the difficulties of living on low salaries and dealing with time pressures, large caseloads, and, in certain areas, the growing routinization of work. Because of these developments public interest law groups have drawn on the growing pool of women attorneys.

Because providing free legal services for indigent clients traditionally was viewed as a feminine concern, the Legal Aid Society in New York City attracted and recruited a relatively high number of women lawyers even before the public interest-minded 1960s. Fifteen years ago, women constituted a quarter of the staff. Today, they make up one third of its approximately seven hundred lawyers.* (The society is the largest law office in New York City and the second largest in the country—surpassed only by the U. S. Department of Justice.)[6] This proportion far exceeds the proportion of women in the profession as a whole.

The Legal Aid Society, a para-welfare agency, commonly receives a high proportion of "family cases": desertion, matrimonial,† and juvenile problems, as well as criminal cases. Prominent women attorneys such as Judge Anna Kross were active in its formation, and the concerns of the society were initially linked with those of the women's rights movement. The society has gone the road of all such volunteer organizations, toward greater professionalization of its staff. The large proportion of women lawyers on its staff probably reflects the early history of the organization and its subsequent labeling as a legal milieu suitable for women.

The rationale that legal aid work was an appropriate professional extension of women's family roles rested on a stereotype that was incon-

---

only 3 percent of all lawyers were women. Erlanger, "Lawyers and Neighborhood Legal Services," p. 138.

*Women accounted for 35 positions on a staff of 160, computed from the *90th Annual Report of the Legal Aid Society for the Year 1965,* January 31, 1966. The 1980 figure, secured by phone from the Manhattan office, was 30 out of 125 management attorneys were women, and 156 out of 530 of the legal staff were women attorneys. The 700 includes 400 volunteer lawyers, many of whom work part-time and whose services are contributed by large firms.

†Up to 40 percent of the cases handled by the Legal Aid Society are matrimonial cases, according to Mary B. Tarcher, assistant attorney-in-chief of the society (reported at a meeting of the Association of Women Lawyers, January 1966).

sistent with many aspects of the work legal aid lawyers had to do. The "family" matters that often come up in legal and (and legal services) work are not familiar to the average middle-class lawyer—female or male. Although many of the offenders the agency helps are juveniles, their alleged crimes are often violent and brutal. A woman lawyer engaged in legal aid work in New York commented that the emotional toll on women lawyers in her office was severe because of the nature of the cases they handled. She cited a recent case in which four pre-teenage boys (between seven and eleven years of age) were charged as second offenders in the rape of a five-year-old girl. The boys confessed to the lawyer—without apparent shame—that they had dragged the girl and her young brother to the basement of their apartment house, held a heated hacksaw blade to the brother's throat, and told the sister they would kill the boy if she did not submit to sex and keep silent. The girl did confess the incident some months later and the Legal Aid Society was called upon to defend the boys. The expectation that women should bring a special understanding, compassion, or knowledge of society to this kind of case is, of course, questionable.

Apart from those who practice at the lowest stratum of the criminal bar, most male lawyers face cases involving nothing more brutal than tax, corporate, or real estate problems; at most they will deal with the gentlemanly crimes of embezzlement or tax manipulation. The lawyer in legal aid work has a far greater chance of being assigned to cases of child brutality, rape, murder, incest, assault, and torture—the "family" problems of the New York and other urban courts. Of course, they also get divorce cases, which may be more familiar to the attorney in private practice.* One reason that women have been accepted by the public interest bar is that the types of cases they handle there are not relished by most lawyers. Not only do they draw small incomes, they are seen as performing social work rather than legal work. Helping the poor is not the first step on the ladder to success. This contrasts strongly with the apprenticeship system in medicine, where fledgling doctors working as interns and residents provide services for social welfare clients and at the same time obtain valuable training for later practice.

The intellectual frustrations of poverty law also make it an unlikely continuous career for many lawyers. As Stephen Wexler noted, this kind of practice "is not stimulating. Much of it is dull and routine

*Similarly, women lawyers who ascend to the bench are often assigned to the family court and must mete out justice in precisely the same kinds of cases. These "family" matters are considered appropriate to the talents and emotions of women. See also Cynthia Fuchs Epstein, *Women's Place: Options and Limits in Professional Careers* (Berkeley: University of California Press, 1970).

. . . . What is exciting takes very little 'brains.' A lot of challenging intellectual work is, however, done in appellate poverty law, but appellate practice has only a limited number of openings."[7]

An even harsher appraisal was given by Jerold S. Auerbach in his legal muckraking work *Unequal Justice:*

Legal Aid was described by its chief attorney in its New York Office for many years as "deadening, routine work, which would kill any sensible, ambitious man in two months." Only the chief in each branch escaped professional tedium, for the others the pay was miserably low, and most of the work "can be done by an office boy."[8]

Because poverty law is not justified as training, pays poorly, and brings one into association with clients whom society looks down upon, it is unattractive to lawyers who aspire to prestigious and remunerative careers. Further, the bulk of public interest law is boring yet emotionally taxing. One lawyer noted the high "burn-out" rate among those engaged in it, and a study by Howard Erlanger shows most lawyers working in it eventually leave.[9]

In the past, women were in a poor bargaining position for more prestigious work. Legal aid and other poverty work represented an option that permitted them to practice law and to act on behalf of "worthy" causes. Auerbach has noted that "considering the nature of its constituency and the structure of the profession, legal aid societies offered temporary employment to minority group lawyers whose access to desirable sectors of private practice were strewn with obstacles."[10]

During the late 1960s and the 1970s, many issues that legal aid groups dealt with became politicized. Crimes committed by poor people— black, Hispanic, or white—were seen as political gestures; so were the cases of battered wives and tenants exploited by greedy landlords. Young men and women chose to go into legal aid work as an expression of political commitment. They resisted identification with corporations and business law: in fact, so many good law graduates elected to work in legal aid that there was competition for jobs. A woman attorney complained:

When I graduated from law school the Legal Aid Society wouldn't hire me. I hadn't gone to a prestige law school and hadn't been on law review. I went to them first. That was at a point in 1970 when they could get Harvard Law Review because that was the height of the pro bono movement among lawyers.

But the bright aggressive young lawyers were frustrated in the achievement of their goals and sought remedies to the problems that were keeping them from providing adequate attention to their clients—namely, their own overwork and low pay.

In the 1960s and 1970s, women, like other minorities in the profession, began to assert their claims of entitlement and to realize that some of the cost of low status and low pay fell on their clients. Perhaps when conditions became better—in this case with the introduction of government funding—formerly complacent groups were encouraged to hope for change and to fight to upgrade salary and working conditions. In spite of the government funding, which, for example, doubled the salaries of legal aid lawyers in 1967,[11] there is a history of programs cut back or eliminated or used as political footballs.

In an expression of growing militancy, more than four hundred lawyers from the Association of Legal Aid Attorneys (a two-and-a-half-year-old organization) went on strike in 1973 under the leadership of its woman president, Karen Faraguna. The lawyers demanded higher salaries as well as the establishment of limits on their caseloads so they would have more time to investigate, to research the law, to prepare the cases, and to stay with them from start to finish. Caseloads were running from 75 to 300 per attorney and the resulting poor quality of work was evident.[12] A study comparing legal aid attorneys and private lawyers retained on behalf of indigent clients quotes some defendents' opinions of their legal aid lawyers, one of whom said:

If I had a serious case, I'd hire a [private] lawyer, but only because Legal Aid has less experience. . . . Most Legal Aid lawyers are young and inexperienced.[13]

Clients also complained that many legal aid attorneys didn't show interest even though they were aware the attorney's heavy caseloads made it difficult. If legal aid lawyers were low on the prestige scale, their clients, too, thought less of them than of private lawyers.[14]

One legal aid lawyer told of how some of her clients would make obscene sucking noises at her from the detention pens they were locked in awaiting trial. She said she would respond by shouting, "I'm the only lawyer you're going to get so you'd better show some respect!" Another woman who had abandoned legal aid for private practice noted that even judges were disrespectful. "They would never call you by name," she said, "or call you 'counselor'; they would just shout 'Legal Aid' as if you weren't a professional person." The Hermann study reported that legal aid clients "think they [the lawyers] lack influence with the court," and complain "they were not aggressive," partially because of the demeaning behavior of judges.[15]

The support and sympathy given legal aid lawyers in New York during their strike was not in response to their low salaries alone.[16] (They started at $12,000 a year and by their twelfth year were making $21,450; starting salaries for Wall Street lawyers were at the time $18,000.) It was also due to the visible degradation of clients whom they could not effectively represent. Dramatic events such as the prison riots at Attica, New York, in 1971 and in other prisons alerted the liberal community to the abhorrent circumstances under which justice was being meted out to the disadvantaged. When legal aid lawyers went on strike, the New York Civil Liberties Union, the Puerto Rican Legal Defense and Education Fund, the National Lawyers' Guild, and the National Conference of Black Lawyers all issued statements of support. The political tone of the strike and the assertion that, more than a fight for money, it was a fight for the rights of the poor were expressed by the constitutional lawyer Arthur Kinoy, Professor of Law at Rutgers University, when he shouted to supporters in New York's Foley Square, "What you're doing is more important for constitutional liberty than a dozen important Supreme Court decisions."[17]

Turnover was so high among legal aid attorneys before the strike that in only eight months Karen Faraguna, the strike leader, had become senior lawyer in the court to which she had been assigned. Faraguna, one of the many women who had changed careers in the 1960s, had been a schoolteacher before attending St. John's University Law School in New York. She was like many women lawyers who preceded her in choosing social interest law, but unlike most, she was ready to exercise militancy in her profession on behalf of her cause.

The lawyers won wage parity with their courtroom opponents, the assistant district attorneys. Other gains were reforms in the lawyer's day-to-day dealings with their clients, recognition of the rights of individual lawyers to establish their own caseloads, and agreement in principle to allow lawyers to handle one case from start to finish.

The problems were hardly solved, however. In 1978 legal aid lawyers and lawyers from federally funded programs around the country met to form a national union again to address problems of lawyer turnover and large caseloads. Legal aid lawyers continued to use the threat of strike to improve conditions and pay, but much of the drama had been lost. Militancy cannot continue at a high pitch for long. The constraints built into the legal and social system make it fairly impossible for this type of movement to be attractive and effective, and turnover takes its toll. Among legal aid attorneys, 50 percent are under thirty years of age, and 83 percent have served less than five years.[18]

Some public interest groups have secure financial bases and others have virtually none at all. Many are funded by foundations (the Ford Foundation has funded thirteen public interest law firms), others by private contributors (the American Civil Liberties Union, for example). Some have received funding from the government, though federal funds are due to be cut by President Reagan.*

Gary Bellow of Harvard Law School has calculated that the more than $100 million currently available under federally financed legal services programs represents an allocation of one lawyer for every 13,000 people with incomes below the federally defined poverty standard. In contrast, there is one lawyer for every 900 persons in the United States, and it is a rare business enterprise of any size that does not have legal advice and representation on hand. Over 85 percent of the country is legally "uncovered," lacking funds to hire an attorney.[19] And yet the legal aid bar cannot expand to meet this need because of inadequate funding and because it cannot compete with the private sector in attracting and keeping lawyers.

Women, however, constitute a pool of lawyers willing to take on poverty law at a financial sacrifice because, as in the past, they feel they can afford it. Yet, some women hope there will be lasting professional opportunities in the public interest sector and that, even if low paid, it will acquire the status of a professional commitment. This was the assessment of one woman who was a volunteer lawyer for VISTA:

Money is a symbol of being accepted as a professional rather than something I need to have; I wish to hell I had it—for its symbolic value of being regarded well. I do not like the fact that I'm giving support to the volunteer principle. Here, I feel better about it because many lawyers in Legal Services, male and female, start out just as I did. I'm waiting to get into the program on a regular

*The Reagan administration has proposed to eliminate funding entirely for the Legal Services Corporation, beginning in 1982. Under the Administration's proposal the states could fund legal aid for the poor from the proposed consolidated social services block grants if they chose to do so. If the federal funding is eliminated by the Reagan administration, services to the poor and to women (who are a large percentage of the poor) will be in serious trouble. More than two thirds of legal services clients are women. Margaret M. Heckler, speech before House of Representatives, April 2, 1981, *Congressional Record*, vol. 127, no. 54, Thursday, April 2, 1981; reprinted by Women's Research and Educational Institute for the Congresswomen's Caucus, *Impact on Women of the Administration's Proposed Budget*, April 1981, p. 3.

basis. In the VISTA program, the Federal Government is subsidizing volunteer-ism and in a sense we are being made patsies. But on the other hand, if we want to do what we want to do, we have to be patsies for a while.

The VISTA lawyer didn't think exploitation was especially a women's problem, but a look at the sex division of personnel in some legal services offices showed that a high proportion of women seemed every-where to result from an organization's restrictive pay policies. In fact, there seemed to be new women's ghettos forming in some.

In the Brooklyn Legal Services office in 1977 about half of the staff lawyers were women. In the matrimonial division, where almost all the lawyers were women, one lawyer summed up her feelings about her work:

I'm doing things which have a meaning for me, and I'm helping people I want to help: poor people, and women instead of rich men who are exploiting those poor people and women. I love my job. I'm very happy here. I wouldn't want to work anywhere else or do anything else.

Although women cluster in the public interest sector of the legal profession, they do not seem to be represented in the upper levels of the hierarchy (for even if the sector is less hierarchical than private firms, it certainly has a hierarchy). In 1977 to 1978, when women com-prised about 27 percent of the staff of legal aid, they were about 17 percent of management. Erlanger, in his study of legal services lawyers, also found that women were more likely to be staff than to hold adminis-trative jobs.[20] The woman head of one unit looked around and reported:

There are no women in top management jobs. There's one woman project director. There is a woman administrator, but what she does is do the dirty work for the attorney, the big boss. She's not a policy-maker.

But some women have parlayed their legal aid experience into high-er-status posts, sometimes on the bench. "Actually, public interest work is a good route to the judiciary," explained a lawyer in the appeals division of the public defender's office in Boston. Musing over her own possibilities, she pointed out that as a graduate of Harvard Law School, a member of its law review, and a former clerk for a prestigious judge, she qualified for a judgeship. Yet, unlike her male counterparts, most of whom had gone on to large corporate firms, as the wife of a prosper-ous lawyer she could afford the relatively low salary ($47,000) of a judge, and it would be a step up from public defender senior attorney. Fur-thermore, as the mother of two small children she found the promise of regular hours attractive. "It's easier for a woman to be a judge than a partner," she said.

Although few would view the judiciary as "woman's work," it has been true that the competition for the bench in the lower courts may be easier for women because, first, there is pressure to appoint women, and second, highly qualified men find it difficult to take the kind of salary cut the bench entails at their level of professional development.

However, most women practicing public interest law do not seem bothered by lack of opportunity for advancement. Poverty law and public interest practice combine a number of factors which suit many of them. In addition to a sense of mission, interviews cited the regular hours of work and the egalitarian culture of the offices, which lack the competitiveness characteristic of many law firms and also tend to be less hierarchical. In the words of a Brooklyn Legal Services lawyer:

This office always had a lot of women. It was that way before I got here. That's true of Legal Services in general. Because, historically they didn't discriminate. Women could afford the lower salaries. . . . And they had no choice, because there weren't other jobs open. And just maybe there's a kind of friendly, non-hierarchical atmosphere or less hierarchical atmosphere here than in other places and women felt more at home.

There is a strong sense of working together toward common goals. "In legal services the people work together. There isn't even much of a division between the students and the head lawyer," was the comment offered by many women in the division. And one attorney added: "In an office like mine we deal with people and their problems. If we can't make the world better, at least we can make it more bearable for the clients we deal with."

Legal aid and the federal legal services programs have been among the most successful programs in delivering assistance to the poor and the aged (of whom a large proportion are women), and to children of these families. In some part, this is because the lawyers in these organizations have been willing to work enthusiastically for little money. Even those who have eventually turned to other things have had their impact. But a larger impact has been made by those who have stayed on—especially the women lawyers who, by virtue of their situation and their orientation, have brought excellence, energy, and dedication to this important work.

Feminist Law Firms and Feminist Law Practice

HE PRACTICE of feminist law is one important reason why it "matters" that women entered the law profession. It is doubtful that the kinds of strides which women made toward equality in the past two decades could have been achieved without the development of this sector of law. Feminist women lawyers can be found in all kinds of legal practice. By feminist lawyers, I mean those who identify with the women's movement and share a set of beliefs about women's rights to equality in private and public life, and who are themselves dedicated to working toward these goals.

To be a feminist lawyer is not the same as being a lawyer who is a feminist. There are women who work in large corporate practices who consider themselves feminists, but who do not see their primary function as that of advocates for women's rights; they may even defend corporate clients in sex discrimination suits brought by women employees. Many other women lawyers consider themselves feminists but are solo practitioners engaged in the general practice of law much as male attorneys are. There are also women lawyers who do legal work that is congruent with feminist principles, such as those who work in poverty law offices, whose practice would not be identified as feminist.

Feminist lawyers can be located along a political spectrum from center to far left. Those of the political center have more typically practiced through foundation-funded projects and universities, although some projects have been institutionalized into centers for feminist kinds of public interest law. The others can be characterized as "movement" lawyers—"new left" rather than women's movement. These too have depended on foundations and fund-raising activities to

130

support their activities. Movement lawyers feel they are outside the "establishment," but as a condition of their trade, all lawyers work within the legal establishment. Lawyers on the far left, however, may be more active in dissident political causes.

Most typically feminist lawyers are women, but there are a small number of men whose names stand out for their work on women's rights. George Cooper and Howard Rubin, who worked at the Columbia University Women's Rights Project, are two of these. So was David Copus, who successfully litigated against discriminating employers as an EEOC attorney until 1977, when he joined a private law firm.

Feminist law firms fall into several groups. There are the feminist firms that are law partnerships devoted to serving women clients and doing work for the public interest of women as a class. There are the single private practitioners engaged in the feminist practice of law. And there are nonprofit legal organizations devoted to women's rights, some of which work with university law schools.

These kinds of feminist law practices originated in the 1970s and have been important in establishing women's rights in a number of different ways. All have had successes in winning cases and establishing legal precedents that have aided the cause of women's equality. Their style of practice has stressed idealism and service. They have generally cooperated with each other, and they constitute a network, a referral system. They tend to know each other socially and to identify with each other. Some have had serious problems of survival, both individually and collectively, from the economic pressures faced by all small law firms and nonprofit legal organizations that are not government supported. But the problems of survival also come from social factors—from ambivalence and tensions caused by the incongruity of women's cultural and professional roles, by contradictions between an ideology of service and the necessities of law practice in the United States today, and by the conflict between the traditional norms of the legal profession and new conceptions of professional behavior.

Movement Law—Feminist Style

Women who came into law via the radical social reform of the 1960s did not come into the profession by only traditional routes. They were in sympathy with the revolts in the universities; they shared radical left politics, and they sought to apply their values to the restructuring of law as a step toward restructuring the larger society.

For those who were not absorbed into traditional practice it was a time of experimentation. Some young lawyers formed law collectives and law communes. These were intended to replace the usual partnership, with its hierarchy of partners on top and supporting associates and staff below, with a structure in which there would be equality of all without regard to job. In these firms secretaries were not to be regarded as subordinate to lawyers and were to be paid on the same basis. Furthermore, in the new communes lawyers hoped to serve the poor and oppressed, and to defend clients who were viewed as suffering from political antagonism—Vietnam war protesters and defectors, for example. The young people in these communes had ties to radical lawyers of previous generations. The New York Law Commune, for example, had ties to the National Lawyers Guild, an organization of older, politically left lawyers noted for taking up the cudgel for social reform and social justice.

Although there was much talk of social experiments at the start, not many got off the ground; they ran the risk of organizational and personal problems. The Law Commune disbanded in the late 1960s. The notions of equality on which it was based were not as easy to achieve as the members hoped. But more striking was the fact that the women lawyers in the commune became disenchanted with the roles they played within the group. Exposed to the new insights and perceptions of the feminist movement, the women lawyers judged their treatment by the male members to be exploitative and sexist, an ironic situation in view of the equalitarian ideals of the commune.

Some of the women lawyers in that group started private practices and others became founding members of feminist law firms composed entirely of women. (One of the first women's law firms in New York was started in the old quarters of the former law commune.) Most of the radical women maintained their affiliation with the Lawyers Guild and held to their ideological positions with regard to the defense of the economically underprivileged and politically oppressed.

What sort of people are feminist movement lawyers? Kristen Booth Glen, recently elected a judge of the New York City Civil Court, is a prototype—a dramatic prototype—of a movement lawyer in private practice. Her most recent legal triumphs include "saving" a classical music station, WNCN, from becoming a more commercial rock station. Glen not only took on the legal aspects of the case, but organized a "listener's movement": her passion for music matched her passion for the idea that "the airwaves belong to the people."[1] Glen, a beautiful tough-minded woman, has taught on women and the law at New York University and set up a Women's Law Clinic designed to bring students in touch with legal problems specific to women. She calls herself a Marxist and a lawyer—a "legal worker" whose aim is to help women to "engage in struggles that will let them have power over their lives." Glen has done work in private practice and for the ACLU on the issues of child custody, draft refusal, involuntary sterilization, prisoners' rights, and health care.

Emily Jane Goodman is another radical lawyer who has done work on behalf of prisoners, tenants, and women, and has written on those subjects. Slight and softspoken, Goodman defended prisoners who rioted at Attica Prison in a move to bring about better conditions in New York prisons. She wrote a book on tenants' rights, and with feminist psychologist Phyllis Chesler, wrote *Women, Money and Power.* * Goodman's professional and social circle includes another movement lawyer, Florence Kennedy, an ascerbic black lawyer who has essentially retired from the law to go on the lecture circuit, where she speaks for women's rights.

These lawyers are in networks that overlap and mesh with each other —with liberal and radical male lawyers and judges, with women in the law schools, with women's movement leaders, and with lawyers in politics. Former congresswoman Bella Abzug has, for example, tapped this network (and they have rallied behind her), as did Elizabeth Holtzman in her New York senatorial campaign. They also connect with feminist lawyers throughout the country who meet each year at Women and the Law Conferences.

Movement Men and Movement Women

As in other movements where emotions run high and behavior is accountable to principle, men have been both worse and better in their attitudes toward women lawyers, and history has played a role in creating confidence or cynicism on the part of women attorneys.

*This is a book on "the basic dilemma" of "how women can gain enough money and power to literally change the world, without being corrupted, co-opted, and incorporated on the way by the many value systems we must change" (New York: William Morrow, 1976).

Disenchantment with the early days of movement law and personal encounters with exploitation left some women lawyers bitter and angry. Furthermore, emotional involvements between some movement men and women lawyers have played a part in partnerships forming and dissolving. There is a different reward system in movement law than in traditional law. Winning brings prestige and respect in both legal systems, but in one the principal prize is money and in the other it is acclaim and love, which creates problems because negotiations may follow sex lines about who should grant and who should receive them.

Feminist lawyers have had complaints against some of the old and even younger male stars of movement law. Although they were men dedicated to justice, many had not been touched before by the women's movement, and even in the early 1970s they were not quite sure what it meant. Like other men, they were uncomfortable about how to deal with women lawyers and were suspicious of their competence. And perhaps because women lawyers expected them to be "better" than mainstream lawyers, some found them worse:

Men of the left? No better, they are worse. I don't know what it is. I puzzle about it. One is feeling that they're stars—that ego. In some ways they have adopted an extreme societal view of women as inferiors. It makes you wonder about their other alleged views.

Women's Rights? Because it's the left it gets more lip service but it's not just lip service that it needs. And it's never been made a priority. Probably because of all the unresolved conflicts the men have.

The male stars of movement law, however, exhibited behavior that ranged from egalitarian to sexist, and working with them—as generations of women have learned—could be at the same time inspiring and undermining. In the 1960s and earlier, their women apprentices faced dilemmas that arose from the fact that the women were selected because they were smart and attractive and hopefully deferential. Thus their sex status opened doors and sometimes gave them a real insider's view of certain aspects of the profession. On the other hand, they were not supposed to compete with or upstage their male masters. In some ways this is a manifestation of the classic master-apprentice relationship which Merton and Barber describe in their analysis of relationships that give rise to ambivalence.[2] In this instance, the resistance of the masters was compounded by the fact that they didn't think the women would or ought to develop into the next generation of masters, and they took it as a personal rejection when the women wanted to strike out independently.*

One angry woman vividly described certain of the "sexist" qualities of movement lawyers in an account of her early career at a famous civil rights law firm:

My clerkship was coming to an end and the only place in the world I want to work was Firm X, and I called up [A] and said "I want to work for you," and he said, "Come over," and I had an interview. It was later revealed that he hired me because he wanted to fuck me and everybody else in the firm knew that, but I didn't realize that at the time. I was there for three years and thought it was wonderful because there is almost no better lawyer in the world than [A]. But my days there were numbered because I didn't behave like the classic little girls who had worked there, particularly vis à vis [B] [the second partner]. Well, [B] had always had women who kissed his ass, and I guess blew him in the shower, or something, and carried his briefcase and did the work and kowtowed to him and called him "Glorious [B]" and that was even true of the young men who were there . . . well, I found that hard to do. But our final confrontation was because of a case. We had a discussion about whether to take it and [B] said it was a shit case, terrible. I said, "no, it is really exciting, I want to do it," and took it on. We had a lot of arguments about how it should be done, but finally I did the work on it and it went to the 7th circuit and I was getting ready to argue it—[B] had nothing to do with it for something like nine months—and I heard his secretary making plane reservations to Chicago, and there was this memo, "Get the X file." So I asked him and said, "What do you mean, get the X file," and he said "I'm arguing it." I said, "I'm not giving you the file, you had nothing to do with the case, you never wanted it, you were wrong about it, and I developed the theory on this case." We argued, I went to Chicago, and I won the case, but he was angry.

However the celebrated movement men fluctuated, meticulous about equality one day and less scrupulous another, they were working through the dilemmas they faced in thinking they were "for" women's rights and were behaving properly and then finding that women saw their behavior as sexist. Many, like the law school professors, were learning.

A well-known rights advocate, now in middle age, told us that his second marriage, to a feminist attorney, had changed him; he had learned not to think about his own work first, he was arranging his schedule so that she had the opportunity to conduct trials in cases which

*William J. Goode discusses this phenomenon on a more global scale as he points to the resistance of men to the women's liberation movement as an outgrowth of their loss of centrality; that is, they are disturbed when women do not make them their focus of concern. "Why Men Resist," *Dissent,* 27 (Spring 1980): 181–93.

they worked on together, and he was trying to advance her reputation (as his first wife had done for him).

A few older movement lawyers stand out because they trained and inspired women. Perhaps the most illustrious is Arthur Kinoy, professor of law at Rutgers University. Kinoy is a truely charismatic figure to his supporters and disciples. A former student described him as "probably one of the first women's advocates." In her book *Lawyering*, Helene Schwartz tells of fighting to become an active member of the "Chicago Eight"* appellate case after "the women's movement had been on my back because we never had a woman in a decision-making position on this case."[3] Schwartz drafted a jury motion which she expected would mean she would conduct the hearing before Judge Hoffman. This had been suggested by Kinoy, but one of the men warned that "the clients have no confidence in Helene"; he said, "I won't agree to Helene's conducting this hearing." Schwartz recalled, "I was a woman whose skills and time were being contributed to the 'cause' and I too was going to get skunked." Schwartz won the right to conduct the hearing with the support of Kinoy, who said,

I don't think this question over who should conduct the hearing would ever have been raised if a woman hadn't been involved. If they are senior lawyers with as much experience as Helene has, the men who write the papers are always the ones who appear in court.†

Kinoy was one of the small but growing group of lawyers, both men and women, radical and liberal, who were responsible for helping in the development of feminist law groups in the late 1960s and early 1970s. The groups often were linked by personal relationships or ideas to the other social movements of the times, especially the civil rights movement and the radical political explosion of the mid-1960s.

Among the first was the Center for Constitutional Rights. Founded by Kinoy, William Kuntzler, Morton Stavis, and Benjamin Smith, the

*The eight were Bobby Seale, David Dellinger, Rennie Davis, Tom Hayden, Abbie Hoffman, Jerry Rubin, John Froines, and Lee Weiner.

†But the strangely selective consciousness "movement lawyers" sometimes bring to women's issues was illustrated at a testimonial dinner on February 11, 1977, honoring Kinoy on the fortieth anniversary of the Lawyers Guild. Most of the lawyers who planned the dinner were women and most of those who made the speeches lauding him were women. They had worked with him on issues of civil rights, labor's right to organize, defense of the victims of McCarthyism, and other causes. They had also done important work on abortion and women's rights. Yet Kinoy's address to the gathering that evening never mentioned women's rights or the role of women lawyers in the Lawyers Guild. Furthermore, none of those attending the dinner, the women lawyers or men who regarded themselves as feminists, seemed to notice or mind the omission.

center was a law firm in New York City dedicated to handling cases involving constitutional issues. It was formed out of the southern civil rights movement and was devoted first to civil rights issues and later to defending antiwar activists. These four lawyers, with recent law graduates and law students, some of whom now run the center, took on cases that were important to school desegregation and the protection of Vietnam war dissenters. In 1969 the center filed an important challenge to restrictive abortion laws, and the women's rights component of its activity began. In 1970 center lawyers and other women lawyers conducted public depositions in which women told of the dangers and problems they endured because of unwanted pregnancies, creating a model for suits showing how denial of abortion violated their most fundamental rights. Over the years many lawyers have trained at the center and gone on to other legal work. Four women attorneys and one male attorney form the core of the center today and continue to work on legal issues relating to abortion, rape, and rights of minorities.

About the same time that the Center for Constitutional Rights was created, two other law groups were organized specifically around women's rights issues. They were the Women's Rights Project of the American Civil Liberties Union (ACLU) and the Equal Rights Advocates of San Francisco. Both are nonprofit organizations, supported by foundation funding and private contributions. They take on women's individual and class action suits, seeking to establish legal precedents in interpretation of the Constitution or of civil rights legislation. They also engage in the training of lawyers and law students in the litigation of sex discrimination and other rights cases, and in educating women and women's organizations about their rights and the remedies available if their rights are violated.

The Women's Rights Project was established by the ACLU in 1971 to work toward establishing sex equality under the law through litigation. It fought through to the Supreme Court some of the most important cases providing guarantees of equality for women and men. According to Ruth Bader Ginsburg, the first director of the project, it was started when ACLU attorney Marvin Karpatkin spotted a case in the Idaho State courts which he predicted would be a turning point in women's rights law. He approached Ginsburg and she agreed to write a brief with some other ACLU lawyers. The case, *Reed* v. *Reed,* went to the Supreme Court and became a milestone—the first time in its history that the Court had invalidated a state statute on the grounds of sex discrimination. That decision was basic precedent for all the cases that followed.*

Work on the case led to discussion about creating a special project on women's rights within ACLU. The idea was backed by the impressive array of talent on the legal staff, and three women on the ACLU Board pressed for the project. All three were lawyers: Faith Seidenberg, who was active in the National Organization for Women (NOW); Pauli Murray, a black civil rights lawyer (later to become the first woman Episcopal priest): and Dorothy Kenyon, a municipal judge during the La-Guardia administration in New York City and a leading advocate of liberal causes. At the end of 1971, the ACLU Board was asked to give the project priority and to commit its own resources for support. It agreed to supply offices and services and $50,000 to finance the project's initial operations (since then, it has received foundation funding).

At the time of the project's founding, it was the only women's legal organization in existence. The NOW Legal Defense Fund and the Women's Equity Action League (WEAL) came later and played limited legal roles because they did not have the legal staff necessary to follow through on cases. In Ginsburg's view, the project's success is due to its financial security and its integration in a national organization with state affiliates which can connect issues with the resources for dealing with them.

Ginsburg became counsel to the project, and ultimately stood before the nine black-robed justices six times to successfully argue the cases that created a new era of equality for women. It was the same Ruth Bader Ginsburg who, with training at Harvard and Columbia Law Schools and law review credentials, had not been able to find a job upon graduation. Today, after a distinguished career as a law professor at Rutgers and Columbia, she is a judge of the U.S. Court of Appeals in Washington, D.C.

On the West Coast, a somewhat different model of feminist legal enterprise was conceived when the Equal Rights Advocates (ERA) was started in San Francisco in 1974. This was to be a "teaching law firm." Four lawyers who had been in practice together, along with professors at the Stanford University Law School, sought to train students in litigation on sex discrimination issues. Supported by grants, primarily from the Carnegie Foundation, they worked with Barbara Babcock, the first

*The Justices ruled in *Reed* v. *Reed* (404 U.S. 71) that when a women and a man were otherwise equally qualified to administer an estate the male could not be given arbitrary preference. The Idaho state law had called for the automatic appointment of a male executor. The statute was declared unconstitutional under an equal protection clause of the 14th amendment.

woman law professor at Stanford, and Barbara Armstrong, the first women law professor at Berkeley. They also have contributed talent to the law schools: two founding members, Wendy Williams and Mary Dunlop, are professors at Georgetown and Stanford, respectively. The ERA continues to do work in the rights and constitutional field, although they no longer have funds for a formal teaching program.

They are, for example, establishing the rights of women to non-traditional blue-collar work, such as truck driving, and sales positions (for example, typewriter and computer sales). Their major concerns, according to ERA lawyer Judith Kurtz, are sexual harassment on the job and the general work atmosphere that discourages women from non-traditional employment. Another lawyer at ERA, Donna Hitchens, has developed a lesbian rights project that deals with child custody issues and women whose employment is in jeopardy because of their sexual preferences.

Other legal centers focusing on feminist law have come into existence around the country. Several have been supported by the Ford Foundation alone in Cleveland, Washington, New Haven, and elsewhere. Although these firms may receive court-awarded fees, they are supported by contributions and have never sought to maintain themselves financially by practicing law. Their problems have been more typical of voluntary organizations than those in the private sector.

The Private Feminist Law Firm

In a sense, the life histories of the first women's law firms represent in microcosm the achievements and problems of all feminist women lawyers and to some extent all women lawyers.

Unlike their cousin organizations, the Women's Rights Project and the Equal Rights Advocates, the first women's law firms were structured as ordinary law firms but were extraordinary in that they were composed entirely of women lawyers. They were extraordinary also in that they were founded with an ideology dedicated to social goals rather than profit, though they were in business.*

---

*Of course, the legal profession is supposed to be service oriented, according to a Parsonian model, but it is clear that profit in and of itself is primary for most legal firms. See Talcott Parsons, "The Professions and Social Structure," [1938], in *Essays in Sociological Theory*, ed. Talcott Parsons, rev. ed. (Glencoe, Ill.: The Free Press, 1954) pp. 34–49.

Two women's firms were organized in 1973. Up until that time, it had been relatively rare for women to enter into practice with other women.[4]

Why hadn't women organized their own firms in response to discrimination in the way that Jewish law firms arose in response to the discriminatory policies practices against Jews? There were many fewer women than Jewish men in the legal profession, and some were isolated from other women lawyers. But like Jews, the women tended to live in large cities, met at Women's Bar Association meetings, and if other conditions had been right, conceivably could have joined together. Few women lawyers considered it, and those few felt they couldn't activate enough business to support a group practice.

Unlike the Jewish men who formed practices and had as clients the Jewish business and professional community, women could not anticipate tapping a sympathetic and prosperous client group who needed them. There was no business women's or professional women's community because women faced prejudice in these areas as well. Although being a member of a firm is usually more profitable than individual practice, building a firm can also require a greater outlay of capital, and women rarely could count on sufficient economic resources.

It took the women's movement to change conditions sufficiently for women to form firms. The movement inspired women to become law students and lawyers and emphasized the value of serving women as a client group. It was striking and newsworthy when the first private women's law firms were begun. After the first two were formed in New York City, several were started in other cities such as Boston, Washington, and San Francisco. This, however, did not prove to be the start of a significant trend, although there are some in other urban centers.

The women attorneys who founded or joined feminist law practices in the 1970s were motivated by a number of concerns and ambitions: first, they wished primarily to use law to achieve social justice. They wanted to serve women, and they hoped to remedy the poor treatment women clients often faced from male attorneys contemptuous of their ignorance or vulnerability. They believed that the legal profession should be demystified and that clients should be informed of all choices and urged to make their own decisions rather than "leaving it in the hands of the lawyer." They intended to establish egalitarian working communities that would treat secretaries and office staff with dignity and, in some cases, as "partners." They also wanted varied practices. They hoped there would be some balance between the cases bringing money and those bringing intellectual stimulation. They sought financial success, not only to survive but to be able to handle work for people who could not afford legal services and to further the causes in which they believed.

Some firms confined membership to women, partly as a response to the sometimes hostile male club atmosphere of the legal profession. Their members sought to remove themselves from the sexist attitudes and behavior they experienced working for and with male attorneys. "We're going to come in in the mornings and see all women peers instead of men, and no superiors at all," said Mary Kelly, a founding member, to a *New York Times* reporter.[5]

Many of these women had worked well with men in the past and were not antagonistic to them. But they had found comradeship and support in women's movement activity. Not "one great example of sisterhood," one lawyer said of her firm, "because each of us is aggressive, competitive, and so on," but "we work together far more than men do. And we do not have to put up with the normal bullshit that women have to put up with—the sexist comments, the putdowns. Who needs it?"

How They Began

The original women's law firms were based in New York City. The larger firm was founded as (in alphabetical order to insure equality) Bellamy, Blank, Goodman, Kelly, Ross, and Stanley—informally referred to as Bellamy, Blank, even though Carol Bellamy left the firm shortly after it was formed to become a New York State senator. It was given a $150,000 three-year grant from the Sachem Fund of New Haven, an arm of the Mellon family fortune, to start a program of test cases in the field of matrimonial law and on sex discrimination in mortgages, loans, and other credit procedures. The firm was formed as a general midtown practice, although it did not plan to do securities, negligence, or criminal law work. The six lawyers were equal partners. They were young; most were in their twenties and early thirties, and four of the six were married. At the time of formation none had children.

The "downtown sisterhood," Lefcourt, Kraft, and Libow (the order of names was established by drawing names) started out with three lawyers and a para-professional. All former members of the radical "new left" New York Law Commune,* they established themselves as a collective. They did not have foundation assistance, but they did have contracts with the Task Force for Justice of the Presbytery of New York to conduct an experiment in do-it-yourself divorce actions and to write a manual for such actions. They also contracted to do a study with 300 volunteer observers in the city's Landlord-Tenant Court.

Members of both firms were women committed to social reform. Their backgrounds were in union organizing, in the civil rights movement, in the feminist movement, and some had been active in radical politics. Seven of the nine lawyers who founded the two firms had been educated at the New York University Law School. This was the first important law school to increase substantially the proportion of women law students and was the school at which the first powerful women law students' association was formed. The founders of Bellamy, Blank originally worked on projects in the Women's Rights Committee at NYU and fought to open the lucrative and prestigious Root-Tilden fellowship to women. This experience confirmed the need for legal help for women in a number of areas that the student lawyers intended to work on when their legal careers were established.

By the time the first feminist law firms started, all had experience as lawyers in a wide range of practice. Several had worked in large Wall Street firms. One had worked at the Center for Constitutional Rights. Some of the others who entered the feminist law firms had worked in small practices with varying degrees of dissatisfaction about the kinds of cases they were required to handle and the sexist treatment they received from male colleagues and senior partners. This was true both for those who worked in ordinary firms and those who worked with radicals.

The women founded their firms with high hopes and the expectation that they would find structural solutions to the problems they had faced elsewhere. Both firms were to engage in a general practice, divided among business law, matrimonial cases (a typical speciality for women lawyers in the past), and sex discrimination suits which were developing under the Title VII provision of the Sex Discrimination Code of 1964. In fact, the month after it was formed, Bellamy, Blank was asked to represent two groups of women who came to them with sex discrimination complaints.

One of the issues that the two firms had to deal with was that of male

*Among their clients were the Panther 21, a key black activist group and Abbie Hoffmann of the Chicago Eight.

clients. Unlike other feminist lawyers who were going into practice at that time, both firms decided that they would defend men but never in a case that was in conflict with their feminist philosophy. One lawyer reported to the *New York Times* "We would make that decision case by case."[6] Since that time both firms have, in fact, defended male clients and at times have been criticized by other feminists for doing so.

The sex discrimination class action suits they worked on set precedents for women's equality, as they had hoped. As the lawyers for female employees of the National Broadcasting Company (NBC) and the New York newspaper *Newsday,* Bellamy, Blank paved the way for more equal treatment of women in the communications industry. They won a suit against the Manufacturers Hanover Trust Company, which charged discriminatory credit policies putting married women at a disadvantage in establishing credit. They also broke a long tradition of employment discrimination against women by the New York Racing Association, winning the case of women who wished to become pari-mutuel clerks. The two firms joined forces in the famous "maids" case against Columbia University which charged that women in the buildings and grounds department could only get jobs as maids but that men had access to a much wider set of job opportunities. Although the case was lost in court, attorney Carol Arber (Carol Libow changed from her married to her original name when she was divorced) said she felt that they had achieved a victory because the university altered its practice and made more jobs available to women after the case was brought and before it came to trial. Lefcourt, Kraft, and Arber also won a number of other cases including one against the Beechnut baby food company for sex discrimination which resulted in developing better job opportunities for women at the company. Their efforts in family cases also raised some interesting issues such as women's right to live according to an "alternative life style," as in the case of a woman whose husband attempted to obtain custody of their children on the basis of his allegation that she was an unfit mother because she lived in a commune.

Thus the two firms achieved their early goals of helping women clients individually and helping women as a class to break down the barriers against them by using the law to fight discrimination. They also became "role models" for women lawyers in firms elsewhere in the country who found it inspiring that women could go into law together, establish a firm based on feminist ideals, and be successful in their work.

Though the first two feminist firms were successful in their cases, they faced severe economic problems and problems of career phasing of their members,* which led to their dissolution about five years after they were formed. Bellamy, Blank dissolved in 1977.[7] Lefcourt, Kraft, and Arber dissolved as a law firm in 1978.†

Some women lawyers viewed the dissolution of the two firms despairingly, although recognizing the contributions the firms had made. As the first women's firms, they had received a good deal of media attention, and their demise had an impact on the public. Members of both firms stressed that their financial problems were primary; Carol Arber made the point that they may have been worse in New York than other places.

Problems of Feminist Law Practices

There are problems that all women lawyers face, and those that are peculiar to feminist practices. There are also problems that all new firms encounter when they start out. These are social and economic. But "social" and "economic" are arbitrary designations because the two kinds of factors are interactive. Time and money are different problems, for example, but when time is money it is difficult to separate the two.

The problems faced by the feminist firms can be seen as sets of dilemmas. All professionals observe norms, for example, that stress "professional" or "distant" behavior and which at the same time stress "personal" or "intimate" behavior. Thus all professionals, as Robert K. Merton and Elinor Barber have described, must deal with the ambivalence created by the activation of two sets of legitimate norms.‡ Kai

---

*Carol Bellamy, who has since become City Council President in New York, left several years before in order to pursue a political career. Nancy E. Stanley went to work with the Civil Rights Division of the Department of Justice in Washington. Diane Blank left to practice law with her husband for two years before moving on to Gordon and Schectman in late 1979 to do litigation, (mainly in the area of labor law). Janice Goodman started her own private practice, focusing on sex discrimination cases. Jerma Rone, who joined the firm in 1975, became an associate in the litigation department of a large corporate firm. Mary Kelly had left some years before to start a practice in Westchester in order to spend more time with her young children.

†Carol Arber is now with the New York State Attorney General's office. Veronika Kraft, Carol Lefcourt, and a fourth member, Ann Teicher, are in private practice.

‡Merton and Barber, in "Sociological Ambivalence," *Sociological Ambivalence and Other Essays*, ed. Robert K. Merton (New York: The Free Press, 1976) p. 18, showed that Parsons's view that norms in various institutional realms would cluster around "polar" extremes in the pattern variables did not describe the complexity in those settings. See Talcott Parsons, *The Social Systems* (Glencoe, Ill.: The Free Press, 1951) pp. 46–51, 101–12. They demonstrated that norms could be activated on the polar extremes at the same time.

Feminist Law Firms and Feminist Law Practice

Erikson suggested there are axes of variation, contradictory tendencies, to be found in social life.[8] He found both dependence and independence among Appalachian coal miners, for example. Antagonistic norms, or contradictions in the axes of variation, confronted women lawyers in feminist practice in good measure. These were some of the unavoidable dilemmas:

- The need to make money and the desire to avoid money considerations in the choice of cases

- The wish to serve the special needs of women clients though women often brought matrimonial and other problems the firms did not wish to handle

- The wish to conduct business in a non-authoritarian manner though this produced objectionable reactions in clients

- The wish to make their partnerships entirely equal though they found it difficult to reconcile their sense of justice with inequities of participation

Economic Problems

Cash Flow. The feminist firms had to deal with the problem of cash flow as do all new firms. It is the nature of the law that there is a time lag between the handling of the case and full payment, either from the client or from judgments handed out by the courts. Although the foundation grants given to Bellamy, Blank provided some kind of cushion for the members of that firm, it was not enough to provide even basic incomes for the lawyers or to pay for their overhead expenses.

Bellamy, Blank took women clients who could not afford legal fees, refused cases that clashed with its ideals, and handled Title VII cases which took an inordinately long time to settle. Janice Goodman, for example, reported that suits against the Manufacturers Hanover Bank, the New York Racing Association, and the National Broadcasting Company suit took four years to reach completion. Although the firm received a $150,000 counsel fee, another lawyer commented that it was a low fee considering the work that went into the case.

Law firms usually have some high income-producing cases to offset those that will produce little. Aside from government or publicly funded law offices, most firms that take on public interest cases or poor clients also work for large corporations or rich persons. Although a few social interest "heroes" are said to take on mafia-type clients at high fees to pay for the defense of victims of racial discrimination on death row, usually the priorities are not quite so clear. In any case, the feminist firms had few rich cases to finance the others.

Women's law firms in New York and elsewhere in the country often find that the bulk of their cases tended to be matrimonial. In fact, the "downtown sisterhood" found that they became almost entirely a matrimonial firm. This is how one of the lawyers explained it:

We assumed we'd make money the way most lawyers do. We figured we would have a general practice, some commercial stuff, some matrimonial. And we figured also we could make some money from the Title VII cases because of counsel fees. What we didn't know was that each one takes several years, and you get awards only if they are successful, so we had problems with that expectation. The other problem that turned out to be our major stumbling block is that we got mainly matrimonial stuff. It bothered me for a long while (but it doesn't bother me now) that people only thought of us that way. It happened to us because we're women. There's no other explanation for it. That's sad.

What was wrong with matrimonial cases? "Divorce is to the practice of law as proctology is to medicine," was the caustic description of Professor Michael Wheeler of the New England School of Law in Boston.[9] Matrimonial law not only ranks at the bottom of the prestige hierarchy, it rarely pays generous fees. Matrimonial lawyers, as described by Hubert O'Gorman,[10] are not considered to be engaged in real law because of the highly charged emotional context in which they practice and the amount of time they must devote to human rather than legal problems. Even highly paid divorce lawyers seldom are highly regarded in the profession because they are not considered to be doing important work.

Few women starting divorce actions against wealthy husbands have sought the services of women in feminist firms or women lawyers of any label. Most clients of this type probably would suspect the competence of women professionals. One woman attorney told the story this way:

People would call me, people I went to law school with, people I dealt with professionally, and they would say, "Ha, ha, ha, I referred a case to you, but I don't think it will come." And I would say, "Oh?" "Yes, when I gave her your name, she said, 'Oh, I think I would rather have a man.'"

And most women rejected me because they wanted somebody they could lean on or somebody they thought would be tough, or whatever. Until about seven or eight years ago, 90 percent of the cases I had were men, because women would get a list of three names, two men and mine, and they would choose the man. I'm sure men did that, too, but I ended up with more men than women as clients.

Certainly, in the case of the big divorce actions, women attorneys do not have the reputations for engaging in the dramatic, outrageous,* and highly lucrative techniques "big bombers" such as Louis Nizer or Raoul Felder are known for.[11]

Feminist attorneys are ambivalent, too, about handling such cases. Although they want to protect women's interests, they sometimes express disapproval of "bombing men." One feminist lawyer described her firm's reaction to fighting a divorce action for a wealthy woman:

In our eagerness to support all women we worked ourselves ragged on that case. But then we found out that what we were really fighting for was not injustice for women, but for a swimming pool and a gardner. And that did not seem to us where feminists ought to be placing their efforts.

It proved to be difficult to attract business aside from the modest matrimonial cases that had always been relatively easier for women attorneys to get. Though most did not want such cases, the new women lawyers, like older women, found themselves channeled into matrimonial work because the referral system tends to reward success in a specialty with more of the same kind of case.

No women clients brought them corporate ventures or even substantial trusts and estates work. Even the establishment of the First Woman's Bank in New York did not change the climate. The bank did, however, retain a well-known Wall Street attorney, Rita Hauser, who brought the bank to her firm of Stroock and Stroock and Lavan. A partner in one of the women's law firms did not take exception to the bank's choice, explaining that the women's firms were still too unfamiliar with the legal work involved. The question was, how could they find a course that would give them such experience.

There are not many big businesses controlled by women that would seek women's legal counsel in the way that men have always sought the

---

*Among these practices are the use of private detectives to reveal embarrassing activities to defame spouses being sued for divorce by their clients. Being considered too moral or pure is a problem women also face in party politics. Cynthia Fuchs Epstein, "Women and Power: The Roles of Women in Politics in the United States," in *Access to Power*, ed. Cynthia Fuchs Epstein and Rose Laub Coser (London: Allen and Unwin, 1981), pp. 134, 137.

counsel of men. The relationships between big business and big law have always been close. They are based not only on the complementarity of resources of these institutions but also on the ties that men in corporations, finance, and law have to each other through family, school, or political connections.

Many of the lawyers who joined women's firms did so because they represented ideological positions that were contrary to the practices and values inherent in the business-legal world. Those women who were not opposed to such values were more likely to find a place for themselves in the large corporate firms. Furthermore, if women brought business matters into their firms, feminist or not, it would probably be the business of male friends. In fact, although only a small amount of business was brought to the feminist firms on referral from male associates in law practices to which the women had belonged prior to beginning their feminist practices, more business was generated from these sources than from males whose practices centered on social justice issues. The women's former colleagues in large firms referred cases that were not worth their firms' time but that were financially attractive to small firms.

Lawyers often get business through contacts made at professional associations. The New York feminist lawyers were members of legal associations, both those in the mainstream, such as the Association of the Bar of the City of New York and the New York County Lawyers Association, and the more radical New York Lawyers Guild. Although participation did give them some visibility, they did not bring in business in the early years.

Getting Business from Feminist Clients

The major source of business for the feminist firms was in clients who had a degree of consciousness about their rights in the occupational sphere. It took a high degree of consciousness to feel secure in retaining the services of a women's law firm, which was not only new and without reputation, but in which all the lawyers were young.

The feminist firms were a natural source of legal help for women seeking redress in sex discrimination suits, however. In this field, few lawyers had any expertise except those in the feminist firms, and a measure of ideological commitment was necessary to take on cases that might be alienating to the larger business community. The Women's Rights Project of the ACLU and the Women's Rights Project of Columbia Law School were taking on selected test cases and could scarcely handle the number of complaints that were generated.

But women—even feminists—who had ordinary problems involving civil actions, real estate, and business law did not bring them to the women's firms. A lawyer in one of the feminist firms confided bitterly that women could have helped feminist firms prosper if they had supported them. On balance, she said, not enough women put their money where their mouths were and some expected free or low-priced service because they were dealing with other feminist sisters. Such clients often cost more than they paid.

Many feminist attorneys reported encountering problem attitudes about money among their women clients, feminist or not. Women clients are often uneducated about fees for professional service or advice. They are unfamiliar with the financial conditions of the legal client-professional relationship, and those who do not have incomes often cannot pay for services. Very simply, they did not know how to play the appropriate client role.

Ambivalence About Making Money

Feminist lawyers of the late 1960s and early 1970s were decidedly ambivalent about going into business and making money. Most radical feminists, as well as moderate feminists of various persuasions, opposed women's assumption of what they called "male" values. Ideological lines were drawn in those years, and persist in altered form, between the women who felt it was legitimate and appropriate to work for high status and money and those who believed it was morally wrong and ideologically incorrect.

These women were different from the older lawyers studied in the 1960s, who believed it was "unfeminine" for women to seek success in male terms. Radical feminists of the 1970s believed that women had been prevented from achieving success before and that this had the unanticipated benefit of shielding them from the corruption of the "rat race." They argued that women who internalized establishment values were capitalizing on an illegitimate opportunity structure.

For feminist attorneys faced with the prospect of women clients who did not pay and their own ambivalence about charging fees, business problems were compounded. Some feminist attorneys harshly judged others who charged the going rate for consulting. Many feminist attorneys do charge modest rates compared to other lawyers, or charge on a sliding scale. Members of the [Washington] D.C. Feminist Law Collective and other lawyers interviewed in Texas and California felt they should charge fees that women in trouble could afford so long as they could survive doing so.

At the time lawyers at the Center for Constitutional Rights in New York were interviewed, an attorney noted that only one male lawyer was on staff with four women attorneys, and she hoped there would be a better sex balance. In 1976, attorneys for the center drew salaries of $12,500, the most salient reason why the center couldn't hold male attorneys. One feminist lawyer plaintively and angrily pointed out that "only" rich male lawyers, those with family legacies, could indulge their taste for social reform (they are the counterparts of the women lawyers who work at low-paying public interest jobs while their lawyer husbands work in corporate practice at high salaries).

Ideology versus Economic Imperatives

Some feminist firms faced problems arising from the ways in which they sought to break down hierarchical differentiation and eliminate differences in rank between professional and administrative staff. One firm started by paying everyone equally and attempted to create a participatory democracy in which everyone in the firm would have a say in decision making. Yet when secretaries drew the same salaries as lawyers, the lawyers developed resentment because they worked longer hours and carried the mental burdens of their cases home, while the secretaries left at the end of the normal work day.

Another system tried was an equal distribution of partnership shares. For the ex-law commune members, this followed in the tradition of their former firm. The system worked in some firms, but there were problems when partners put in different efforts and had different levels of experience. In the usual firm, partners who have experience train those who do not and receive proportionately more money. Similarly, those who bring in a great deal of business usually have larger shares than those who bring in little (although some firms have instituted equal shares for partners).

A former partner in one feminist firm described the stress resulting from equal shares for all:

Well, actually, it was partly because we had this Soviet-like insistence upon equality. You know, the old movement thing, instead of recognizing that people were really at different phases which meant that their income production abilities were very different, which they clearly were, you know, we had this damned insistence about people being equal right down the alphabet; and it created tremendous pressures. Because the people who were way above production were being held back. The people way below were terribly guilty . . . we definitely had that problem. It was a serious one. It was a strain throughout that we had to deal with.

Life Cycle: The Compounded Problems of Shared Age Statuses

Youth is an asset for those starting out in new ventures, but it has proved to be deleterious for some women in practice together and for one feminist firm in particular. Because most feminist lawyers are relatively young and without substantial financial support, they have limited resources on which they can depend to manage careers and families.

In one firm, the partners had entered their thirties together—up against the biological clock, as sociologist Norma Wikler has characterized it—and decided to have children about the same time, creating a problem that would not have come up in a firm with a mixture of ages (or sexes). In this case, the partners limited their lawyering to begin mothering when their children were born. A lawyer who had watched their practice over the years commented cynically:

Well, now they're basically working part-time, and it's like a funny novel . . . . I mean, you just have to laugh because there they're doing what some men would have predicted, "Well, they'll just have babies and drop out."

The issue of part-time work has been a thorny one for all women in professional life, and they have solved it differently depending on their definitions of their professional and private roles and the choices available to them. In one feminist firm, where only a few women were mothers or planned to have children, a two-month leave plan was set up, and as one partner put it, "you could use your two-month leave to have a baby or go to Europe." One lawyer on the staff did have a child, using her two-month leave, and resumed full-time practice. But another lawyer wished to work part-time while her child was young and was turned down because the other partners didn't think the newly formed firm could manage it. The partner who wished to work part-time left the firm feeling it had been unnecessarily inflexible. In another small feminist firm, all three women partners had children within a short time and all insisted on working part-time. Some lawyers in other firms chose to continue full-time legal work and hired housekeepers to take care of their children (as did many women who worked in large firms).

But there was an associated set of dilemmas for many feminist lawyers, who were ideological about the rearing of children. They wanted to work part-time because they felt they should have a special relationship with their children and should not ask other women to rear them. They also hoped that practicing in feminist firms would provide that option. The women in Wall Street firms who were mothers usually did not feel they had choices and so altered their family demands to meet the work demands. Some left Wall Street, but those who stayed usually expected to work full-time.

The feminist lawyers believed strongly that a feminist firm ought to be able to resolve the problems facing women elsewhere in the professional world because they were committed to solving them. Yet they were small, beginning law offices and really couldn't afford to be flexible in this matter. Perhaps had there been other institutional supports on the outside world, such as child care centers, or financial support for firms devoting a certain amount of time for social interest work, experiments in flexibility could have been worked out.

Women Lawyers and Women Clients: Learning New Roles

All professionals face role strain in their client relationships, and institutional mechanisms come into play to help reduce the strain.[12] For example, the pressures on the busy professional are directly observable, or at least well known, and the public learns that as clients they must tailor their expectations to fit the schedule of the professional. Crowded doctors' waiting rooms are testaments to society's judgment of whose time is worth the most. Clients of lawyers quickly learn that they will be billed by the hour for discussion of their problems. When clients do not know how to be "good," according to the demands of the professional, they face sanctions such as disapproval or abrupt service. But, in addition, there are norm enforcement agents—the receptionists, secretaries, and administrators who enforce regulations and inform the untutored how to live up to the rules or keep the "uneducated" away from the professional.

Women professionals and women clients, somewhat outside the system, are often imperfectly socialized and, in addition, may (in the view of the establishment) cantankerously reject the professional relationship with its crisp, specific, universalistic, and affectively neutral (to use Talcott Parsons's language) norms. When this happens, feminist professionals cannot rely on the usual mechanisms for the reduction of strain and thus are exposed to a great deal of it. Feminist women professionals, for example, believe they *should* be more open and less imposing to the client or patient than men. Women clients also think they are less imposing than the men and also expect women lawyers to be more open.

This should provide for a good fit of expectations. But it is one thing to believe it is socially good not to be imposing, and quite another to find that everyone with your credentials is imposing except for you. The imposing doctor relies on his aura to provide a barricade against the assaults on his time and his self, while the unimposing doctor is left naked. The imposing manner is often the foundation for respect, and respect is often coupled with distance. Male professionals can usually rely on respect, announcing their professional status with white coat or the grey suit, but women must do much more to announce their status and win acceptance as a professional. Because their competence has always been suspect, the onus has been on women to make their professional status visible and to demand that its rights and privileges be honored.[13] Women professionals who have been troubled by men's use of barriers as unnecessary mystification, and who reject their use, must battle prejudices against women in professional settings without the defenses and armor perfected by men through the ages.

Thus, feminist lawyers face contradictions in their roles which create ambivalences for them and for their clients. Certainly, many of these problems come from the fact that these relationships are still new and

153

there has been too little time for expectations to become shared. Furthermore, the boundaries of what is treatable by the practitioner have not yet been established in the minds of the clients. A patient knows that if he or she has a stomach pain and a toothache, the physician will do something about the stomach and the dentist will treat the tooth. Women clients in distress, like other new client groups such as blacks or consumers,[14] want the lawyer to handle the "whole" problem. One aspect of this was described by a feminist movement lawyer:

I have problems representing women clients. Most of my experience with them has been with discrimination cases, and any time you start dealing with discrimination plaintiffs, they can be hard to work with. My experience is that sometimes there is a history of injustices that they feel have happened all the way back and it's finally got to this point where they're going to set it straight and so they don't focus on the particular problem; . . . they want to right the whole picture and that gets to be a problem—when their head gets set to think that everybody's always doing them wrong.

Women lawyers are often torn between sympathy and a view that the client's expectations are irrational; they are subject to a special set of pressures from women clients in a feminist setting that they do not believe men experience.*

Women clients want, and expect, more of the woman lawyer than they do of a man—they want the lawyer to demonstrate more than a professional interest and to give more of herself to the client. The clients believe that the woman lawyer will be more sympathetic and understanding than a man and will spend a great deal of time with them. The lawyers think women clients feel it is appropriate to take more of their time than they would of a man. That is, they believe the male lawyer's time is worth more because he is a real professional in business. Women clients have been known to express shock on finding that their "talks" with women attorneys are professional interactions for which fees are owed. Their wish to establish a personal relationship with their attorney creates unreasonable expectations of attention to their needs and problems.

This situation creates ambivalence on the part of feminist attorneys. On the one hand, their aim is to help women and provide them with the support that men do not typically supply; on the other hand, they have the same problems of time and budget allocations as any other

*Carole Joffe points out this is also true for abortion counselors in her forthcoming book *The Family Planners: Some Contradictions of the Helping Professions* (Berkeley: University of California Press).

professional and cannot respond to all the pressures put on them by their clients. Even the most understanding feminist lawyers sometimes are sorely taxed.

Authority is another issue. Women are not alone in wanting a strong shoulder to lean on in times of trouble. But many feminists believe that male lawyers often take advantage of the dependency of women by making decisions for them and promising more than can be delivered. One woman lawyer felt that the feminist attorney, dedicated to being open about the client's options and honest about her limitations, is punished because the client would rather the lawyer demonstrated decisiveness and competence.

Like the women in corporate settings described by Rosabeth Moss Kanter,[15] women clients respond out of fear of the unknown, vulnerability about their futures, and what they perceived as assults on their self-image.

One woman attorney commented compassionately about the confusion and helplessness exhibited by many of her women clients:

There is a whole problem about representing women. Most of my clients are women, whether it's matrimonial or otherwise. So many women who come into a law office have either gone crazy or are barely coping or are so traumatized by what's happening, that you're dealing with an impossible situation. . . . Because of lack of experience, women don't seem to understand what's going on, don't know how to use the information you're giving them. You have to translate things into such simple terms, and even then they don't get it. For example, if I said to you, this is something we're going to negotiate. You would understand what that meant. Nine out of ten women who come into my office say what do you mean by "negotiate."

She added that such behavior was not limited to women; it was the product of powerlessness and helplessness and was shown by men as well.

A month ago somebody called who I have known casually for several years. He called me at home, apparently having tried me at the office. He was in tears. His wife had come home from a vacation and had decided she wanted a divorce and he should be out, yesterday! That's as bad a [situation] as any woman's. It happened to be a reversal situation; she was rich and he was not; she had all the power and he didn't. It was really more a question of power than wealth. She had a large family behind her. . . . He called me six times a day. Fortunately, it was over within two weeks. Everything signed, sealed, and delivered. If he had not been a friend and if it had gone on any longer, he would have been as large a pain in the ass as any woman.

But feminist attorneys feel it is important to teach women to face up to their responsibilities as clients and their rights as clients. One attorney who addressed the fee question said:

I tell women now, when I see them, that whether they decide to hire me or someone else, they simply must equip themselves with money in some fashion, even if it means borrowing from a relative, because otherwise they start out as a total victim and their lawyer treats them as that.

For the client who has the strength to work on her legal problem with a feminist attorney, there can be unanticipated satisfactions in the client-professional relationship. One lawyer, stressing the differences between her style of practice and the traditional legal practice, described the way she worked:

I am totally straight with clients. I tell them everything. I have done a lot of work and teach about health rights and hate it when a doctor won't tell me something; I think people have an absolute right to know everything. For the clients who have it all together, they love it and just say "It is wonderful; it is a breath of fresh air." But I am careful.

And feminist lawyers can find a community of interests with their women clients, one that facilitates the professional goal. Another lawyer commented movingly about the matrimonial cases she handles:

Sometimes I find it useful to share personal experiences with the client, not always, but I've gotten to the point where I can spot the difference. What happens is the personal experience is a source of a lot of humor and laughing. That is important, because when people come to me, I say, "We are going to go through a hard time together; don't kid yourself. I am not going to mislead you in any way about the amount of pain; it is going to be awful. But there will be some laughs too, and times when you will feel absolutely terrific and free. And you should feel comfortable with me and you should be able to understand what I can do for you during this period of time, and what I can't."

Relations with Staff

Some employees of feminist firms—secretaries and receptionists—came to their jobs with images of women as professionals which were different from their images of male professionals. One lawyer commented that her staff felt her office would be a nice place and they wouldn't have to work hard. They objected to the work pressures of a busy legal office and worked at a slower pace than was common in male firms.

Some may have been acting on messages picked up from the women lawyers, who often were uncomfortable about asking employees to work faster or put in extra time. Committed to being humane in their dealings with staff, the women lawyers rarely assigned tasks by giving orders; they asked if a secretary would "mind" doing something. This style is characteristic of women in their interactions with men[16] and of subordinates interacting with superordinates. Yet even women who hold positions of authority, such as those on Wall Street, often use the style. Some believe it is a more gracious way to treat those working for them. Others avoid the appearance of giving orders because they fear subordinates will resent taking orders from a woman (a commonly believed and partly substantiated view). Other lawyers simply had not had experience as "bosses" and were uncomfortable giving orders.

It is also difficult to recruit young women secretaries if there are no men in the office. Secretaries expect the possibility of a social life on the job and do not think they will find it in a feminist firm.

Dilemmas of the Reward Structure

Feminist lawyers cannot hope to become rich at their work. Certainly they have psychic satisfactions when they win important cases, especially those engaged in constitutional litigation or in suits in areas where they feel they have an impact. Some are satisfied by doing day-to-day work that they feel helps other women. However, for those who find themselves in a practice that begins to look like the ordinary work of a matrimonial lawyer there are fewer satisfactions.

While women are gratified by the rewards expected by men who do social interest law, they differ in some ways. For one thing, male stars can count on adoration, often translated into sex and love, from spellbound movement supporters and groupies. This kind of gratification is not commonly available to women, or at least it has not been so in the past. While the male movement lawyer's prowess creates charisma, which is sexually attractive to women, women lawyers find that many men are put off by a "take charge" woman. It is not that men do not find them appealing, just that men are defined as attractive because of their achievements while women are defined as attractive in spite of their achievements. Furthermore, women lawyers have found that their male colleagues are so used to deference from women that those who are not deferential seem aloof and unattractive.

Thus, women do not get the applause from men that men get from women. Oddly enough, women are paid with love in other spheres. Women who help men get ahead often are rewarded inappropriately with kisses, hugs, chocolates, and flowers, when they ought to be rewarded with money and promotions.[17] Perhaps as long as society's norms support women's helping behavior, they will get rewarded only for that, and not for achievement.

But women also seem to be satisfied with less in the way of material rewards. This may be due in part to socialization, but it must also be due to the fact that women know not to expect as much because they won't get it. Thus women are willing to work for less money even if they get less deference, while few men stay in movement law much past their youth because they begin to want financial rewards and they sometimes feel there are few glamorous movement cases left to fight.

It is probably harder for women than for men to leave movement law and do law for profit. Women do seem to be somewhat more committed to ideology and social interest. The university is one way out, and nonprofit public interest law is another. Perhaps most important is the fact that women's issues are still dramatic; and though most men in public interest law seem to find them less significant than the rights of American Indians or Puerto Rican nationalists or prisoners on death row, that leaves the work, the frustrations, the sacrifices, and the satisfactions to the women.

Social Assets of the Firms

One can gasp at the recital of problems faced by feminist lawyers or wonder at their achievements given the obstacles they face. New organizations are always beset with problems; and when they are pitted against established institutions the problems are made worse. But in spite of these difficulties, feminist lawyers have had support and have demonstrated strength in many ways.

First, they have drawn commitment to their work and legitimation from the feminist movement and from the larger society, which has endorsed the improvement of women's situation through legislative means. People may oppose them on particular cases, but their commitment to eliminating sex discrimination is now conceded to be part of the nation's civil rights tradition.

Second, they have performed outstandingly. Some of their legal accomplishments have been noted in this book, but beyond the renowned cases observers are struck by their skills in the courtroom and in negotiation. There are men who are noted for their florid styles and courtroom dramatics in social interest law, but who are considered poor and unprepared lawyers. The feminist law community is respected, almost without exception, and that reputation has bred confidence.

Third, they have had visibility. Lawyers for the feminist firms point out that judges recognize them, because as women they are still unique and thus stand out more. ("But," said one, "there are judges who never treat any lawyers with respect, and we are treated no better," the ultimate test of becoming integrated.)

Finally, but not conclusively, they are practicing in a more enlightened environment, in which more and more women are joining them in professional activity in law and in other fields, and in which men and women generally are accepting the presence and competence of the woman lawyer.

Because feminist attorneys are known to be helping the cause of all women, there are women's networks they can rely on. They find women on Wall Street, women secretaries, and women court clerks helpful, and their experience thus contradicts the popular view that women are antagonistic to each other. One lawyer pointed out that even the secretaries of attorneys they oppose often try to be helpful, making sure their calls get through and because they know feminist attorneys have slim economic resources, sending their own messengers to pick up and deliver materials. Such behavior demonstrates that women in the profession are "bonding"—that their sex status has become a basis for solidarity.* A feminist attorney described what the "grapevine" meant to her:

One thing has had a dynamic, positive effect. It's the grapevine. I can call a number of women in Wall Street firms or midtown firms and as a woman lawyer say, "Hey, I've got this problem, do you know anything about it?" Of course, this happens through other channels too, such as people you meet through the women's law conferences. In one case, a witness, a professor, told me her

---

*Arlene Kaplan Daniels has shown this kind of networking across professions in "Development of Feminist Networks in the Professions," *Annals of the New York Academy of Science,* vol. 323 (1979): 215–227.

daughter was a lawyer specializing in tax law. In fact, she had just finished writing an article on a subject very important to me which had to do with the tax implications of the back pay of lawyers. So I was immediately able to call up this woman, who had heard of me, and though we had never met she sent me drafts of her material. There are people that we now feel free to call, even if we haven't met them, tell them who we are, and ask for help. Most people recognize us and help.

A formal feminist network has come into being, based on the National Conference on Women and the Law, which holds an annual meeting that brings together feminist lawyers from around the country. The organization has grown from 50 to 2,000 participants in the past decade. Lawyers in firms such as ERA in San Francisco and the Women's Law Collective in Washington, D.C., describe the organization as important to them as a way to share experiences and problems. The conference's first meetings largely drew student activists, but they now are attended primarily by experienced women lawyers.

Feminist law practices have accomplished much, but it is clear they have limited possibilities for survival as private firms. Not all women's law firms are the same, nor have all followed in the path of the two New York City firms that dissolved. The law groups devoted to women's rights which did not attempt to be self-supporting have endured, albeit in changed form, and it is not impossible that private women's law firms will be created and succeed, surmounting the problems of the past, or simply having the good fortune to avoid them.

The first feminist firms, whether they endured or not, left an inheritance. They turned public attention to women's legal issues and won cases that gave women throughout the country greater opportunity and freedom to live and work more equally. They became role models for other firms and young women lawyers and demonstrated to the legal profession that women had the capacity to work as professionals in a world long dominated by men; to win in a competitive arena even when the opposing side had more resources than they; to work cooperatively in spite of the folk wisdom that women do not work well together; and to accomplish their goals of doing business humanely. There is still a network of individual lawyers who support each other personally and professionally, and who, on occasion, join forces in pursuit of common social goals.

The firms fought not only cases but stereotypes. The cases were easier to win than the stereotypes were to deal with. The stereotypes, for example, led potential clients into believing their special competence was the relatively low prestige, low paying speciality of matrimonial law.

The dissolution of some of the private feminist firms and the problems of some nonprofit groups are not due only to the dilemmas outlined but to the contradictory context in which they were operating. Feminist law arose at a time when humanitarian ideology was regarded as inspiring and when young people were self-sacrificing in their devotion to their ideals. Ideology carried young people forward for some time, but inevitably, keeping commitment alive in the face of ongoing personal deprivation is difficult without support from the society and one's peer group. Although American society was more egalitarian in the 1970s than ever before, the mood of the country was changing. Young men who had devoted themselves to public interest work turned to other kinds of practices to make money for the families they were planning or supporting. The women tended to stay with it longer than the men because of their commitment to the women's movement and because their financial situation as individual breadwinners or spouses of income producers gave them more flexibility of choice.

What of the independent private feminist practitioners? Like men who worked in public interest firms in the 1960s, supported by the government in one way or another,[18] they expanded into social interest work of various kinds. Of the women interviewed for this study, one joined the Human Rights Commission of New York; another became a judge; one combines law with writing and journalism; and several teach in law schools. Most have diversified their work, in that they do not solely concentrate on women's rights issues. Indeed, some have turned from these issues because they feel the current mood of the courts is unfavorable to the cause of women's rights. Others have attacked these problems with even more tenacity. Surprisingly, few have sought political careers, although they vigorously support feminist candidates who do run for office.

The impact of the feminist lawyers has radiated in other ways. Eileen Moran has pointed out how activist women, who could not win nominations or elections themselves because they are regarded as too radical by political clubs, nevertheless paved the way for the election of women who had not been involved in the feminist struggle.[19] Women lawyers who fought for women's issues may have been unsuccessful in creating viable practices, but they nevertheless made it possible for others to do feminist law. Like the politicians, the militant lawyers made women lawyers of more modest demeanor seem acceptable to their profession.

161

10

Small Private Practices and Husband-Wife Law Partnerships

ASMALL PARTNERSHIP of family members or the neighborhood law office inhabited by a single wise old lawyer who sits before his roll-top desk are probably more familiar images to Americans than the large law factories that have become the typical environment for legal work. Setting up one's own office, though increasingly difficult in this age of specialization and complexity, is still attractive to many lawyers who wish to live according to the model of the autonomous free professional.

Women, like men, have found this choice attractive or have selected independent practice because the alternatives were limited. And women, like the men in such practices, face a variety of problems as well as advantages, although some are accentuated because of their sex. Women in independent practices or in small firms face many of the same problems as those in feminist firms in getting business, although ideology is less of an issue in their choice of cases and clients. Women in family practices face still another set of experiences.

Lawyers who choose to enter private practice or small partnerships are entrepreneurs. Unlike lawyers in large corporate firms, the success of the single practitioner rests on his or her ability to maintain a social and professional network that will produce a steady stream of referrals. While Wall Street lawyers can rely on their cases being assigned through the firm, single practitioners must solicit business. This involves

Small Practices—Husband and Wife Partnerships

being an active member of professional associations, speaking at public meetings (both professional and non-professional), and most importantly, making contact with people who are likely to need a lawyer's help.

Establishing competence is especially problematic for the small practitioner. She cannot depend on the legitimation given by affiliation with a large firm. The importance and difficulty of establishing and maintaining a referral network varies with the nature of the lawyer's practice. A lawyer in independent practice in Westchester described the referral system:

Well, there are Bar Association referrals services. They are not as reliable as word of mouth actually, because they are impersonal for one thing, and they don't refer you all the time, they keep a list and rotate. I find that word of mouth is the most efficient. If I wanted a more active and aggressive practice, then I would involve myself in either more groups in the county in terms of various kinds of activities to make it known that I was in practice and doing certain kinds of law and so on and so forth. Because I have limited time, I've sort of allowed referrals to come on whatever basis they do.

Of course, referrals usually develop from satisfied clients in an area of law where the lawyer has demonstrated competence. Thus the lawyer tends to get labeled. This is even more limiting if the lawyer only has access to a segment of the community. A suburban lawyer described how women attorneys tend to develop practices primarily in matrimonial and real estate law:

Matrimonial [law] becomes a big breadwinning kind of area for women. Probably because a lot of them don't have connections to the business community in an area. In White Plains there are all kinds of businesses . . . at least 300 corporations and everything from the dry cleaner to the local insurance man. All those people need legal advice, and most of them go to men. Women without connections to these business people fill up the coffers with matrimonial or real estate, you know, the closings and all that comes in.

Some specialties provide lawyers with a regular clientele. For example, the clients of commercial lawyers are routinely in need of legal assistance and likely to know others in the same position. Specialties like matrimonial law do not provide a steady clientele, yet here too, word of mouth referral provides bread-and-butter work for most women lawyers in small practices.

There are no reliable figures on the number of women lawyers in the United States specializing in matrimonial law, but it is widely believed that a disproportionately high percentage of women work in the field. Whether in government public interest law, private firms, or "law collectives," women do a lot of matrimonial work. In James J. White's national study of women lawyers in the late 1960s, 50 percent of those engaged in legal work said they performed domestic relations work.[1] The proportion of men reporting they did such work was far lower (38.6 percent). In Jerome Carlin's study of the New York Bar, only 1 percent of the respondents said that matrimonial law was the main concern of their practice;[2] men prefer not to handle matrimonial cases. Like psychiatry, which is regarded by many in medical practice as not dealing with "medical" problems, matrimonial cases are regarded as non-legal by many attorneys. Hubert O'Gorman points out, in his study of matrimonial lawyers, that many of them judge the "real" issues involved in such cases as "psychological," "social," and "medical,"[3] and feel that matrimonial law "isn't really law." Matrimonial cases are believed to require a minimum of legal skill and to put a premium on non-legal and "emotional" knowledge.[4] And because women are supposed to be attuned to such emotional problems, it is usually assumed that they can and should handle them. Women lawyers seemed to accept this definition as much as the men in my 1960s study, and many lawyers today continue to feel that women bring a special sensitivity to this specialty.

Primarily I am known as a specialist in matrimonial work and family court problems, the kind of work many lawyers look down their noses on. I do matrimonial work out of choice because I feel I know the field and particularly because I have the ability to reconcile people. I try to keep a family together. It's very important. I give the women advice . . . and the male clients too. For example, I'll tell the male client, "You courted her, now try it again." You'd be surprised at the number of times the clients listen to me and I patch up their marriages.

On the other hand, some women lawyers share their male colleagues' distaste for matrimonial law and avoid it if they can. One woman asserted:

I have handled matrimonial cases, but I prefer to keep away from them. It's emotionally disturbing . . . the things people suddenly have to say against one another, people who at one time lived only for one another, this disgusts me, really disgusts me. I don't handle them if I can possibly help it.

For both female and male lawyers, it is initial placement in the profession rather than gender-related interests that usually determines whether or not they will go into professionally devalued specialties such

Small Practices—Husband and Wife Partnerships

as matrimonial work. However, once exposed to specialties believed to be suitable for them, many women lawyers do find they like the work. But this becomes true for both men and women in all specialties. O'Gorman found that 94 percent of the men in his sample whose practices consisted primarily of matrimonial work preferred these cases.[5] On the other hand, many of the women who had chosen to do business law expressed a distaste for matrimonial work for the same reasons as the men in O'Gorman's study.

Women in small practices also specialize in real estate law. Among the women interviewed for the 1960s study, ten out of sixty-five chose this field. These women maintained practices in their homes or were employed by medium-sized law firms. Real estate law is known to be routine and detailed work, and these women were mainly doing home-purchase "closings." High-level legal skills are not necessary, and there are few time pressures in the work because it does not involve court appearances or trial deadlines. The older women studied in the 1960s who were in real estate law did not express any special feelings about their work, except to echo the general view of their day that women were "better" and more thorough at routine tasks, and therefore particularly suited for that type of work.

But there are other reasons for women to handle real estate cases. Many real estate agents are women, and people feel comfortable with a woman handling business that has to do with their homes. Also, because women play an active role in home buying, women lawyers (and agents) are likely to hear about possible business. In fact, home buying is one of the few areas in which women are involved in large purchases, although the kind of real estate cases women handle typically concern family residences rather than large business holdings.

Women also engage in general practices handling a variety of civil law matters. They do tax work, negligence cases, and some commercial work. The occupational story of these women is not unlike that of men in small practice, except for the substantial problems that come from prejudices about women's competence.

The case of Harriet Gorin (not her real name) illustrates the point that when women have prospered in non-traditional careers (as elsewhere), "chance"—idiosyncratic experiences—have often helped them along. Harriet Gorin started out in a traditional path for a lawyer: while still in law school she married a man who had a criminal practice, which she joined on her graduation. Three months after she passed the bar, her husband was accused of involvement in a criminal case, and she was called upon to defend him. Her success in this case and in another large case that occurred while he was under indictment earned her an excellent reputation, so that when her husband was indicted for a second time and found guilty (she did not defend him this time), she had already taken over his practice and expanded it substantially. She not only had an intensive apprenticeship, but inherited a practice before the age of thirty.

The difference between this woman and others in small practices stemmed from the nature of her clientele—Mafioso and other criminals accused of such crimes as trafficking in drugs. These clients are in continuing need of legal services and are willing and able to pay high fees. Moreover, the endorsement of one satisfied chieftain is at least as good as that of a corporation. This women found herself in the unusual position of having a large and authoritative network composed of people likely to need her services on a regular basis and able to pay substantial fees.

While this success story is unusual, it illustrates another factor important for women in small practices—the extent to which these women depend on personal relationships with men. It was her husband's practice that had launched Harriet Gorin's career. Another lawyer I interviewed, who concentrated on trusts and estates, depended on her lover for referrals. Husbands and fathers who have reputations in a community and are willing to introduce their wives and daughters have assured many women's start in law practice. For women who are outside men's networks, such male sponsors have served as alternative channels to business opportunities. Women who do not have such relationships are at a distinct disadvantage.

Small practices provide advantages as well as difficulties for women working in them. Some give women the chance to try a variety of work and not to get stereotyped. In a large firm, men may exclude female associates from certain kinds of work. In small practices, the chronic lack of personnel to do urgent jobs can give women chances to work in different types of law. One lawyer described this phenomenon:

I shared offices with somebody who was away a good deal of the time, and when he was away, I would handle things for him and when I was away he would handle things for me. I wasn't away very much and he was away a lot and he had a lot of business and was quite casual about it. He didn't only refer things to me but I know two other women who got started by his referrals. He was always referring things out.

There is also the opportunity to set one's own hours. This can backfire if the lawyer is overworked, but several lawyers said they had formed their own practices so they could work part-time. A lawyer who left a large firm when she had children opened an office in her home, worked part-time, and employed a baby-sitter two days a week. She said,

I know then that she comes, and I can either stay here and see clients, or I can do work upstairs, or in the office I set up on the sun porch, or I can go to the library in White Plains. I can go see clients or go to court. Wednesday is one of the days because that's the court date up here for adoptions.

For some women lawyers with small children, having an office in the home gave them the opportunity to keep an eye on their children and keep a hand in law, while also keeping expenses down. Not as many younger women interviewed in the late 1970s preferred this solution as the older women interviewed in the 1960s, but there were a few who worked out the family-work problem in the way reported by this lawyer:

With three small children I aim to maintain a low-profile active practice. I expect I'll do that for the next six to eight years if necessary. Thus I can keep my hand in so that I don't do what other women have done, which is to drop out for ten years or so. At the same time I will not set up an office outside my home where I must go every day, maintain office services, a secretary, and so on.

Small practices located in neighborhoods make the lawyer more visible to possible clients. A small practitioner said she got business from "friends, and friends of friends." "I also get business from my beauty-parlor operator," she told me, as did others. A recent law graduate who did not want to practice in a large firm told us that the most frequent advice she received was to go back to her home town: "People would give me business because they knew me and my family and because I was a local girl."

Law Partners and Marital Partners

Probably the most important kind of small practice for women has been the husband-wife partnership.[6] The married couple in a professional partnership is a phenomenon associated with a former age and seldom expected in contemporary society. Of course, customers of small shops and neighborhood restaurants still encounter husbands and wives working together, but it is not common for husbands and wives to enter into business or professional work as a team. Indeed, informal and formal nepotism rules in large organizations have often prohibited it.

Although husband-wife partnerships are an extension of family businesses common before the industrial age, it is only recently that women have had the training and education necessary to become equals in professional partnerships. D. Kelly Weisberg has pointed out that some of the early advocates for women's entry into law schools were husbands who wished to practice law with their wives.[7] Through the years, family partnerships have been protected settings, providing places for women lawyers in otherwise blocked opportunity structures.

I studied twelve husband-wife partnerships in the late 1960s, followed some of them later on, and encountered several new ones in the late 1970s. No one knows how many of these partnerships exist, but they probably are dwindling, together with small practices of all kinds. Furthermore, opportunities for women to practice law in large firms or offices are opening all the time. A dozen years ago many women went into practice with their husbands because they couldn't get jobs elsewhere, or believed so. Partnerships with husbands were a safe harbor in a hostile professional world. Other male lawyers might be suspicious of a woman lawyer, doubt her capacities and her femininity, but the lawyer to whom she was married did not.

The partnerships studied in the mid-1960s were not characterized by equality either ideologically or in practice, yet they probably incorporated more egalitarianism than most other marriages of their generation. Yet husband-wife partnerships tended to adapt the sex-division of labor within their firms to that of the profession. The men usually did the "outside work," client contact and courtroom appearances, while the women typically did the less visible work, such as writing briefs and researching cases. They also tended to do the office management and "housekeeping" of the firm.

Neither partner saw these arrangements as problematic or as sexist. In the mid-1960s, it was still unusual for women to handle litigation or argue cases in court—it was believed they did not want to do those things and were not good at them. In essence, wife-lawyers accepted the prevailing male view; they were pleased that their husbands went to court and, if the husbands were trial lawyers, were pleased to see them "shine" and to vicariously share their glory. This was not atypical for many lawyer marriages.

Wife-lawyers usually defined themselves as their "husband's helpmates" and saw their professional roles as part and parcel of their family roles. They used this definition to explain their presence in a "masculine" occupation and, at the same time, to prove they were still feminine. Because they defined their activity as helping their husbands, it was socially acceptable.

Many of the women interviewed in the mid-1960s who were in practice with a few other lawyers or a husband or father reported that they were performing the practice's non-legal administrative tasks—hiring secretaries, running the office, and keeping the work calendar. These "housekeeping" responsibilities did not develop the lawyer's skill or contacts and often condemned her to obscurity. In addition to the housekeeping tasks, women in partnerships with their husbands practiced the same kind of law as other women lawyers of the time; matrimonial cases, probate work, and real estate. This level of participation prevented women from competing with their husbands. Although theoretically husbands and wives drew equal shares, in fact, women tended to earn less money, underscoring their subordinate status. Thus, the wife-attorneys usually failed to build independent professional reputations. They also literally felt wedded to their firms and unable to take advantage of opportune changes. In fact, many probably did not have any chance at all to improve their skills.

Nevertheless, women lawyers in such practices had real jobs in the law and many of their peers did not. They were subject to fewer role strains since the demands of their work were not in conflict with their domestic duties. Husbands knew it when their wives had a hard week at the office and, because it was a joint family enterprise, they could hardly complain. Furthermore, husbands could and did cover for their wives when children were sick or needed attention. Wife-partners could even put their careers on "hold" when their children were small and not worry about losing tenure. In partnership with their husbands, women could work shorter hours than other professionals and still maintain a steady relationship with the firm and their profession. These all were clear advantages to the women in these practices.

Although most wife-partners presented a rosy picture of their work situations fifteen years ago, and even those who were dissatisfied showed resignation more than resentment, other lawyers who knew the couples sometimes provided insights. Husbands could, if they wished, take advantage of their wives. Some wives complained to friends that their husbands tended to use them as secretaries. "He says its easier to give me the things to do than to explain them to a secretary," was a common observation. For most, the need to maintain good family relations was paramount, and there was no place else to go. One unusually dissatisfied lawyer-wife, however, turned the problem into an advantage, reported a friend. "She complained that it was always 'Darling, get me this from the file or that from the file,' until she was so disgusted she went out and got herself a judgeship."

Career competition between husband and wife can put a partnership under enormous strain. If competition on the part of the wife is intense and unmasked, it can become a central threat to both partnerships, professional and marital. Although high career aspiration is culturally approved for the husband, the career-ambitious wife faces disapproval. When she is not only in the same field as her husband, but engaged in practice with him, her hopes to rise may be intolerable to herself (especially if she feels thwarted in the attempt) as well as to her husband. Ambition and achievement on the part of the wife is tolerable and appreciated when the husband is equally excellent, or more so. Incompetence of either husband or wife is also a source of strain, and in the two cases studied in which it was reported—both attributed to the husband—the marriages had dissolved. Of course, charges of incompetence may really be an expression of personal animosity in marriages which are strained for reasons unrelated to work. Of two lawyers who said their husbands were not as competent as themselves, one went on to become a district attorney, an extraordinary accomplishment at that time, and later married a judge. The other attorney later married a man in an entirely different field and developed a successful independent practice.

Competitiveness is potentially so disruptive to the husband-wife partnership that arrangements develop to physically separate the partners or the spheres in which they work. The division of labor within the firm certainly reduces competition. The husband's hierarchic position as primary decision maker and the wife's more limited career aspirations and underlying "helpmate" ethos traditionally served to control competitiveness. Further, the structure of partnerships as nominal (if not actual) relationships of equality, served to reduce conflict.

It would be difficult to assess how much competition was reduced because women's work with their husbands was defined as "help" and not as the contribution of an equal. Although women in partnerships who were educated in the 1960s and 1970s felt strongly that women should not merely be considered helpers but true partners, a division of labor within the family firm usually served to reduce the potential for competition. But self-selection also played a role. Very competitive

Small Practices—Husband and Wife Partnerships

women rarely wished to practice with their husbands, and those who tried it soon turned to other kinds of practice.

It was hard to know what women's feelings were about the effects of their partnerships on their careers and their aims and hopes. A few of the women interviewed—those who aspired to judgeships, for example —indicated that if they were given a second chance they might not choose to work with their husbands. They believed the family partnership had restrained their career development. Most of the women in family partnership admitted to modifying their initial ambitions in law, typically setting their sights lower. (But this was true of many women lawyers of their generation, no matter what their type of practice.) It was noteworthy, however, that among the most successful women interviewed, most were married and not in partnerships with their husbands.

It is difficult, therefore, to determine whether the family partnership serves to restrict the woman any more than the profession at large. For those women who might have dropped entirely out of law work to raise families—as some did—the family partnership served to maintain identity as a lawyer and keep alive professional ambitions. Driving professional ambition has not been common among women in the United States, especially married women. Even where it does exist, there are strong cultural taboos against acknowledging it. And even men revise their ambitions and abandon dreams of what they might have been. Women who entered partnerships with husbands probably imposed limits on their career aspirations and decided it was a rational step. Although many had aspirations to law careers, they also aspired to become wives and mothers. Some would have completely forsaken their career ambitions to pursue the usual feminine roles if they had been forced to choose.

While the number of dual husband-wife practices has been diminishing, today it is not uncommon to find husbands and wives working in the same firm or partnership along with other lawyers. In these situations, the wife's role is not helpmate, but full partner. Women lawyers are very sensitive about establishing reputations now and are no longer willing to do work that may impede them. They resist being moved into work traditionally considered to be "women's specialties," nor do they wish to do invisible work. As a result, more and more women are handling litigation and meeting clients. Some lawyers in practice with their husbands have come into the partnership with established reputations.

Kathryn Marshall, an Illinois lawyer, commented on her own family partnership, formed after she had launched a successful career in a major U.S. attorney's office:

The problem of not having the option of inclination or opportunity to participate in the active litigation of the law firm has not been a problem in our firm. Perhaps this is the case because I came out of the very highly litigation-oriented United States Attorney's Office in Chicago. As a matter of fact, many clients came to me in the area of litigation because of my presumed expertise.[8]

Another woman reported that the pervasive sex division in specialties continued to shape her partnership, though she and her husband had quite different abilities:

I bring in more business clients than my husband does, although he handles more of that type of legal work for the firm than I do. By the same token, he probably brings in more family law problems than I do although I handle all of that type of work for our firm.

One couple who decided to go into practice together found themselves in traditional specialties. The wife, a lawyer in feminist public interest work, reported that:

"He had background in corporate work and I had background in Title VII. Neither of us had background in matrimonial work. So we both read everything there was to read on matrimonial work because we suspected that's where most of our clients would be coming from. I learned it by the practice, really."

"Does your husband handle matrimonial cases with you?"

"He hates it even more than I do, so I do most of it and he is trying to build up the corporate end of the practice."

Wives in the newer husband-wife partnerships are far more evaluative of their positions in the firm than their predecessors had been. An established lawyer who went into partnership with her husband in the late 1970s said she felt her "identity oozing away" when her husband's male clients made remarks such as "I see you have the little wife here." She could not be an "associate to my husband," she said, because "it sticks in my craw." It was mainly in this partnership that she realized "how much it meant to me to be a hot-shot lawyer."

Another couple evaluated their partnership as "being ideal for the family," because "it's our own outfit with our own hours and our own thing." In this partnership the wife took off a day a week to take courses that interested her. What about competition? "People always ask about competition," she said. "It's never an issue. If he does well I am delirious. The same is true for him."

Couples who had been in practice for a long time, and those followed up on ten years after the first study, had undergone changes. Some were ideological changes and some were life-cycle changes. The women partners were doing better in their professional lives than they had been previously. One had become much more active in bar association activities, had heightened her own reputation, and was bringing in business independently. The firm had grown, and she now had a number of younger attorneys working for her. She claimed to have been affected by the women's movement and seemed also to be assessing her marriage anew. Her children were now grown, giving her considerably more freedom.

An attorney in a labor relations firm had developed a speciality in sex-discrimination cases independently of her husband and another partner. She, too, had a heightened sense of career. Here again, grown children made it possible for her to have a more flexible schedule.

New and old partnerships thrive on similar assets, and husbands and wives each benefit from them. Kathryn Marshall illustrated this in a note about her husband:

I have had the distinct pleasure of hearing my husband discuss the awesome benefits of having his wife as a law partner. He tends to emphasize how his comfort level is increased because of the absolutely unquestioned loyalty that he knows is part of his relationship with his law partner. He also emphasizes his security in working with a known quantity and with someone whose abilities and capabilities he has no doubt about.[9]

Work in a family law firm may also permit women attorneys to take advantage of new opportunities in playing public roles, as in the case of Phyllis Schlafly, an active opponent of the Equal Rights Amendment, whose public speaking engagements take her far from home. Schlafly, who received a law degree in 1978 when she was fifty-three years old, went to work part-time in her husband's firm, Schlafly, Godfrey, and Fitzgerald, and continues to pursue her political activity.[10]

It is significant that husband-law partners today share the new ideology. When the women wish to come forward and take more visible responsibility in the firm, they usually have their husband's support in doing so. One prominent lawyer, well known for his dramatic courtroom style and for taking underdog cases, now practices with his young second wife. Friends of the couple noted that the wife started out in classic style, "carrying his briefcase." As time went on, however, the husband pressed his wife's career and gave her more prominent roles to play in court.

It is clear that norms have changed; women lawyers no longer want to do the firms' housekeeping or to be invisible, and lawyer-husbands accept this. Moreover, the benefits can be shared. Now that women's contributions have become more visible, men can more easily refer cases to their wives without incurring the disapproval of clients. Of course, the benefits and rewards in husband-wife partnerships ultimately depend on the personalities involved and the social context within which they move. Other family members, friends, and professional associates can have an important effect on the relationship, either strengthening it or undermining it. Still, couples can select supportive social networks and supportive work environments—two mitigating factors in the assessment of the constructive or destructive aspects of a relationship. From an outsider's perspective, the new husband-wife partnerships can be viewed as prototypical of a new and useful equality between marital partners and other professional partners.

Women in the Legal Establishment: Wall Street and the Large Corporate Firms

WALL STREET LAWYER—the words conjure up the image of a man in a banker's gray pin-striped suit. But today the Wall Street lawyer may well be a woman. A little more than a decade ago, a mere forty women were to be found in the wood-paneled offices, so much like gentlemen's clubs, of New York's largest and most prestigious law firms. In a change nothing short of revolutionary, the number of women on Wall Street has jumped, and today there are more than six hundred women in the three dozen select firms clustered along the elegant midtown sections of Park and Madison Avenues and in the canyons of Manhattan's financial district. These firms are known as the Legal Establishment, the "blue chip bar," or simply, if inaccurately, the "Wall Street" firms.

What happened in New York has also occurred across the nation in Wall Street–type firms, those with a hundred or more (one has 550) lawyers. The fifty largest firms in the United States—in New York, Chicago, Houston, Cleveland, Omaha, and San Francisco—numbered 1,297 women among a total of 6,034 lawyers, 21.5 percent of the total in 1980, most of them hired in the last few years.[1]

Before the 1940s there were virtually no women attorneys in New

York's Wall Street firms. But World War II created opportunities for women lawyers. Some had the remarkable experience of encountering a "sex-blind" hiring partner who was looking for talent. Some women were hired on temporary bases for special assignments and managed to stay on because they had become indispensable specialists in their areas. The war did not open significant numbers of doors, however. Erwin Smigel reported counting a total of only eighteen women lawyers in Wall Street firms in 1956, in his now classic book *The Wall Street Lawyer.*[2] And in 1968, I estimated that only forty women were working in Wall Street firms or had some Wall Street experience.

What makes the recent influx of women to the large firms particularly meaningful is that these firms constitute a network of legal institutions not matched anywhere in the world. Their clients are the largest corporations, commercial banks, and investment houses, and a few rich men and women. They derive a good deal of their power from their ability to "make" law in this country by influencing legislation and the way it is implemented, as well as by working on many precedent-setting cases. While the names of Wall Street lawyers are not generally known to the public, they regularly contribute legal talent to the highest levels of government and in turn draw upon Washington's leading political leaders, as in the case of Richard Nixon's recruitment by Mudge, Guthrie, and Alexander in 1960.

The fact that women are now working in these firms is particularly impressive. If prejudice against women was particularly strong in the legal profession, the Wall Street bar, its most powerful and prestigious sector, was the most impenetrable of all.

About five percent of the nation's law graduates end up at the big firms.[3] The top students are wooed by the firms each fall with interviews interspersed with first-class hotel suites, gourmet dinners, and tickets to Yankee Stadium and symphony concerts. The current style is somewhat more lavish than in the past, but putting a recruit through social paces was not only a lure but an attempt to assure that the young man (for that was what the firms were looking for) would fit in socially as well as intellectually. Now, of course, the top talent includes a large proportion of women, and they are also included in the courting process, by some firms eagerly and by others with some trepidation.

The ambiance of the Wall Street firm inspires respect and confidence. Descriptions of two trips to interview lawyers for this study may illustrate how pervading the Wall Street atmosphere is.

Women on Wall Street

I arrived to interview a lawyer at Rogers and Wells at 8:30 one morning, making my way to the fifty-third floor of the Pan Am Building on Park Avenue. Oriental carpets covered the floor, and stately wing chairs flanked sofas covered in striped silk. Fresh flowers were on the receptionist's desk. At 8:30 there was the bustle of work going on, but the people walking back and forth with papers were lawyers; the secretaries came in later, I was told. I had been asked to come during a quiet part of the day, but found that the person I was to interview had been called out of town on an urgent matter. Travel and early mornings and late evenings marked her work schedule.

At Sullivan and Cromwell the carpets were brown, but the walls were gold and the staircase winding. Here too, fresh flowers were arranged. Portraits of the founders looking serious and profound stared down. At six in the evening the receptionist was a young man with a flight bag on the floor and a ski jacket draped around the chair. A woman darted out of an office in slacks and a sweater, and another woman, black, dashed through the reception room. The brash young man I was to interview led me to an interior elevator that went down several floors to his office, containing a polished wood desk, china lamps, and leather chairs. Not a paper was in sight; the desk was clean, as are many on Wall Street. It seems fair to say the poorer the lawyer, the more cluttered the desk. The day was icy (a record-breaker in New York) but when the interview was over the young man put on a well-tailored herringbone tweed coat, no scarf, no hat; he was clearly in uniform, a dark suit over a striped shirt with the most modest round gold cufflinks. No one could fault him for not being well tailored. His brashness was evident only in the straying of a lock of hair, an intense honest air, a thoughtful opposition to majority opinion. It was he, not I, who noted that his brashness was its own kind of obsequiousness. He wasn't a yes-man, he argued in a way that showed his superiors he took them very seriously indeed and weighed everything they said. His comments were testaments to the care with which he listened to them and his disputes were carefully considered. Dissenting was his way of honoring them.

Some of the newer large midtown firms are decorated in a more modern glitter of glass and chrome, and their young lawyers reflect this ambiance in a somewhat more dashing mode of dress—the young men's hair a bit longer than their downtown colleagues, the shirts less often white. There are more to the differences between firms than meets the eye, but the look of their offices conveys something of their values, their ways of working, and the nature of their hierarchies.

Wall Street Lawyers: Their Profile

A large proportion of the members of the establishment bar come from the social elite of the eastern establishment. Yet background is usually only one factor in the social sifting determining the membership of the big firms. Merit has always been an important determinant of membership, and probably has become even more so. Merit here signifies a high grade-point average and membership on the law review at a few major national law schools, notably Harvard, Yale, and Columbia, although representation of schools in the large firms is now more diverse than in the past. (In 1957, 71 percent of partners in large firms graduated from these three schools; in 1962, the proportion went down to 58 percent.)[4] In other cities, graduates of regional schools of high quality are preferred, such as Stanford University for the large San Francisco firms.

Nowhere is the "old boy" network so characteristic of the formal and informal structure of an occupation as in the "establishment bar"; nowhere is tradition more important and the impact of background status so pertinent both to recruitment and to the style of doing work. These firms have constituted the quintessential upper-class male culture.*

A distinguishable segment of the New York blue-chip bar are the seven firms that constitute the "Jewish" establishment bar, which grew after World War II in response to old-line prejudice against Jewish attorneys. Although they are still characterized as "Jewish firms," these firms are today mixed in social and religious composition. Found mainly in midtown New York between Fifth and Park Avenues, they rival, or nearly do, the large WASP firms in size, resources, and power. Some are growing at a faster rate due to their specialization in "take-overs," work eschewed until recently by the WASP firms. Their clients, though increasingly Fortune-500 oriented, are traditionally weighted with large Jewish investment houses and businesses, particularly in the textile industry. The fields of entertainment, publishing, bankruptcy, labor relations, and tax work also characterize these practices.

Those in the Wall Street establishment readily distinguish the old-line "Protestant" or "Jewish" subcultures within it. The stereotypes held about each other by the two groups of firms have the character of all stereotypes, composed of germs of truth, as Robert K. Merton has ob-

---

*A sizeable percent of partners in large establishment firms are located in *The Social Register;* although it is going down in those firms, the percent is lower still in new, fast-growing firms. Erwin O. Smigel, "The Wall Street Lawyer Reconsidered," *New York,* August 18, 1969, pp. 36–41.

served,[5] and a certain amount of falsehood. The extent that these differences are really believed determines the extent of their consequences,[6] and there is a self-fulfilling prophecy at work that makes the establishment firms distinctive cultures with regard to recruitment and operating styles.[7]

From a position of virtual invisibility, women started to move into Wall Street significantly in the 1970s. In fact, it became a preferred place for them to seek and find employment.* The numbers of women on Wall Street are still changing rapidly. By 1977, women represented 12 percent of the legal staff in the thirty-two largest New York City firms (then having more than 95 attorneys each).[8] Of a total of 4,815 attorneys in these firms, 587 were women.

Legal staffs have two major divisions: partners and associates. Partners share the profits of the firm, though not always equally. Usually a few top partners take a larger share of the profits while the other partners share a smaller per capita pool of the profit. Associates are salaried, and they usually come into the law firm right out of law school. After a six- to ten-year apprenticeship, a small percentage of them are elevated to partnership status. Some leave because they dislike large firm practice, and some leave after being passed over in partnership decisions. A small percent become "permanent associates," lawyers who practice a particular area of law the firm needs but who, it is understood, will never become partners.

Of the 1,520 partners distributed among the large New York firms in 1977, 29 were women. This represented almost a tenfold increase since 1971. In 1956, Smigel found only one woman partner, and when I surveyed those same firms in 1968, that number had only increased to three. A survey made of women partners in Wall Street firms in 1979 showed an increase to 34.[9] By summer 1980 there were 41; at that time, of 3,987 partners in the top fifty law firms in the country, 85 were women.[10] This is a large increase in numbers, yet even today only about 2 percent of all partners are women. As of 1980, more than a quarter of the large firms (including some of the giants, such as Sullivan and Cromwell with 206 lawyers) have *no* women partners. Of the large firms, fifteen have 1 woman partner, six have 2, two firms have 3, and two have 4.

---

*The small law partnerships are where women law students are *not* going, according to Ron Ostroff, "Making Way for Women," *National Law Journal* 1, no. 21 (1979). A recent study of New York City women lawyers shows that 16 percent of them practice in corporate firms. Frances Della Cava and Madeline Engel, "Women Lawyers in New York City: Social Characteristics and Social Problems" (unpublished paper, Department of Sociology, Lehman College, City University of New York).

Of the thirty-four women partners on Wall Street in 1979, only three achieved partnership before 1970. Another nine joined them between 1971 and 1975, and another twenty-one were promoted between 1976 and 1979. Another seven were named between fall and spring of 1980, as I was assembling the latest statistics on partnerships for this study. Following the pattern reported by Smigel for male Wall Street lawyers, eighteen of the thirty-four women partners come from Harvard, Columbia, and Yale law schools, and another five from New York University Law School. Most of the other women partners went to other highly ranked law schools (only one had attended a "local" school).*

The large New York firms had 3,142 associates in 1977. Of these, 587 (19 percent) were women. In the twenty largest firms studied by Smigel in 1964, the number of women associates has grown dramatically. Smigel reported a total of 18 women working in the large firms in 1964. These same firms each employed an average of 4 women associates in 1971; by 1977 they averaged approximately 17 female associates each.[11]

Many firms reporting no women associates on their staffs in 1971 and earlier now have between ten and twenty, and some have as many as forty to fifty women working as associates. This represents rapid change. The step up in recruitment has meant that in some firms half of all the recruits hired in a given year might be women. Furthermore, the inclusion of women as partners represents a significant change in the status of women attorneys who come to Wall Street; it reflects a long-term commitment to their participation as peers at top levels.

Why Do Women Choose Big Firms?

Once women were considered more acceptable by the large firms and were actively recruited by some (under the gathering cloud of legal action by the Columbia University Women's Rights Project), the question arose of why those women who chose large firms did so. Women, after all, had been believed to be uninterested in business law, insuffi-

---

*Sixteen got undergraduate degrees from "Seven Sisters" schools, another nine from prestigious state universities or colleges, and eleven from city schools and other state universities.

ciently committed to the high pressure work ethic of the large firms, and hostile to the values implicit in much of corporate legal work.

Of course, the social setting for women in the legal profession during the 1970s was far different from what it had been in the 1960s and earlier. Women who joined the big firms in the early 1970s entered in cohorts that included other women; they often came from law schools where they had been from 10 percent to a third or more of their class. The question "Ought I to be doing what I am doing" was one that could be shared with other women colleagues, often ones with whom they had shared experiences in law school. This single element of added numbers created a strong sense of acceptance and legitimacy for women recruits, a feeling that unlike their predecessors they were not deviant and had less reason to be self-conscious or to fear discrimination. The new women who entered corporate work were also less bound by sex-typed motivations. They could join Wall Street firms, comfortable in sharing with their male peers a desire for high prestige and high monetary rewards, both goals that earlier generations of women found difficult to express.

Some women lawyers are "second deciders" in the sense that they chose law school after abortive careers as teachers, editors, or secretaries. And some are second deciders because, having chosen law with a motivation toward social action, they revised their goals to aim at "legal problem solving" rather than "social problem solving" and to seek success according to the standards of the professional establishment.

Many lawyers choose to start work in large firms because they are convinced these firms will give them superior training for any kind of practice they may later prefer. All recruits know that only a few will be chosen as partners; they weigh the cost-benefit equation, the possibility of success against the possibility of leaving the firm, having invested a major commitment of time and work. In fact, many young lawyers, both men and women, intend to spend only a few years in a big firm for the experience, expecting they will do something else later. They view it as postgraduate training to develop skills, but not as the beginning of a way of life. Some young associates do seek out other options and leave before they are evaluated in a serious way for partnership potential. But others who start out with ambivalent feelings decide they like the challenge of work in a large firm, grow accustomed to the high incomes, and come to see the challenge of climbing the firm's hierarchy as a test of their powers.

The Changing Context

Many attribute the increasing openness to women on the part of large firms to the enormous growth in size in the firms themselves. Size has not only created new places, but has had an impact in changing the firms' traditional ways.* Not only did the large firms look like gentlemen's clubs, with their paneling and Persian carpets, they were also conducted according to gentlemen's traditions. As the *New York Times* put it, "Gone are the days when partners sent handwritten notes to clients because they felt the telephone was too impersonal. Gone too are most of the male secretaries."[12] New associates are as likely to consult Lexis, a computerized legal information system, as to pore over dusty tomes, and even female secretaries' duties have been taken over by typists and word-processor operators.[13] Support services are also provided by a new semiprofessional class, the para-legals, or legal assistants, usually young women from elite colleges who do not have law degrees but who do routine kinds of legal work.

Many of the larger firms have also opened branches in other cities and even countries, so that even partners in the same firm may not know each other well. Partners and associates are likely to travel often to Boston, Los Angeles, and Milwaukee; and although many firms have long had branches in Paris and London, they are now opening new ones in Hong Kong, Tokyo, Rio de Janeiro, and Abu Dhabi. This also has opened up opportunities for lawyers with skills in languages and foreign cultures. There is also competition from the Wall Street–type firms in other cities—Washington, Chicago, San Francisco, and Houston.

The once tightly-knit community of lawyers whose social ties were reflected in their work relations has become diffused. Intimate knowledge of and concern with the lives of their associates became impossible as the size of firms grew, and increased size probably made the social dimensions of colleagueship less important. The need for a greater

*The enormous growth in the size of the establishment firms can be seen by looking at the top twenty law firms studied by Erwin Smigel in 1957, which had 50 to 100 lawyers. These have grown from 30 to 120 percent, which means some firms have doubled and more. Others not on that list have made similarly radical jumps, and some have become larger than those included in Smigel's sample. Thus the increase of women lawyers at Wall Street firms was accompanied by a general increase in personnel. This was probably an important factor in receptivity toward women since, although men were in competition with able women, there was an increase in the number of total places available. The change in size setting new structural conditions affecting social arrangements has been recently analyzed by Peter Blau, *Inequality and Heterogeneity: A Primitive Theory of Social Structure* (New York: The Free Press, 1977).

number of recruits, together with less concern about their backgrounds as long as they had the required training and brain power, means that the firms draw from an increasingly larger pool of recruits than in the past. This has occurred within the developing social context of egalitarianism backed by legal restraints on exclusion, but it has made for an ease of transition that might not have been created by the law alone.

Effect of Diversification

Law firms have not gone from clublike homogeneity to representative cross-sections of the society. But there is no doubt that the backgrounds of young lawyers now entering the larger firms are considerably more diverse than they once were and that the most talented of them can confidently look forward to partnerships.

For Jews, Catholics, and other men from ethnic minorities, the first breakthrough came during the Roosevelt years. The New Deal gave many Jewish and Catholic lawyers an opportunity to work in government, thus making valuable contacts that were later transferable to private corporate practice. These contacts led to a greater acceptance of minority group members as colleagues. Although many Jewish men chose to begin their own firms rather than join old-line firms, many did enter WASP firms in the 1940s and 1950s and began to achieve partnerships in these firms in the 1960s.[14] Updating his study of Wall Street lawyers in 1969, Erwin Smigel found that "discrimination of Jews in Wall Street is just about finished."[15]

Of course, many Jews and Catholics were readily able to fit the image of an establishment lawyer. They tended to be already assimilated enough into establishment culture as to be hardly distinguishable from Protestant men in appearance, tastes, educational background, and behavior. The visibility of women's undesirable trait—their non-male-ness—could not be diminished so easily, nor could the entrenched prejudices about their incapacities to join in the work of the male world.* The women's movement, which was highly visible by the late 1960s, played an important part in altering ideas about the legitimacy of women's work in high commitment careers and helped create oppor-

---

*Of course, black lawyers also remain distinguishable. In the nation's 50 largest firms, 20 of 4,251 partners are black (less than one half of 1 percent); of 6,408 associates, 151, or 2.4 percent, are black. "Data on Law Firms Raise Racial Issues," *New York Times*, July 29, 1981, p. A20.

tunities for women in the larger firms. Men became more open to the possibility that women could be seriously committed to highly demanding work. Women, in turn, were asserting their right to be evaluated on an equal basis with men.

Legal Changes

Appropriately enough, it was legal action that made the difference. Women took legal action against the law firms themselves. How it happened is something of a David and Goliath story. The foundation for the cases was laid in the fall of 1969, according to Diane Blank, a leading actor in the discrimination claim against Sullivan and Cromwell. Her firm was one of a number chosen by women students at New York University and Columbia Law Schools, with the participation of the Columbia Law School's Employment Rights Project, to break exclusionary practices.

It was Blank's second year of law school at New York University, and the big firms were interviewing on campus. The women students pooled experiences and surmised that the elite firms were giving them short shrift. In particular, Shearman and Sterling had only "law-review type" men lined up outside their interviewing room, and were not inviting women with the same qualifications.* Several of the women complained to the placement office, which called members of the firm and insisted that they talk with women too. The firm consented and conducted several interviews with women. But the women selected charged that the interviews were "outrageous." They reported that the hiring partners had implied that women were no good at litigation, that "a woman in litigation would be [only] O.K. because much of our litigation involves brief writing so she wouldn't have to appear in court." In addition, the women were told that the firms were looking for candidates with military experience because it was a good indication of responsibility.

After this experience, reported Blank, several NYU women met with law students at Columbia and pledged to report on their next year's experiences with interviewers. They did this, collecting evidence of a pervasive pattern of discrimination. Columbia had, in the meantime,

*Shearman and Sterling, a firm of 312 lawyers, named their first, and as yet only, woman partner in 1980.

set up its Employment Rights Project with EEOC funding and brought in Harriet Rabb to head it. She was joined by Professor George Cooper and attorney Howard J. Rubin. Complaints were filed against ten firms on behalf of all women law students in New York with the New York City Human Rights Commission, then headed by Eleanor Holmes Norton, herself a lawyer and interested in the issue.

Diane Blank's interview experience with Sullivan and Cromwell was one of the two chosen by the project to proceed on, and it formed the basis of her case against them. The other was Margaret Kohn's case against Rogers and Wells for discrimination on the basis of sex in hiring, promotion, assignment, and pay.* The Sullivan and Cromwell interviewer had admitted to Blank that the firm was biased; that, for example, women there were put into "blue sky work."† He discouraged her interest in the firm. "Some of my partners are prejudiced against women," he told her. Blank, who had worked with Cadwalader, Wickersham, and Taft as a summer intern, claimed she was genuinely interested in a Wall Street job. She had been editor-in-chief of the *Annual Survey of American Law* (not law review, but one of the specialty journals) and had an impressive two-page résumé. The interviewer did not even bother to look at it. Instead, he asked her about her lawyer-husband's career—what kind of work he did, his ambitions, how she expected to juggle home and career demands with such a busy husband. "If he'd looked at my résumé," Blank said, "he'd at least have been warned—it showed that I'd had a lot of experience in civil rights work."

Kohn's case against Rogers and Wells was tried first, and it was decided against them on the basic issue—that they had systematically discriminated against women in hirings. In 1976 Rogers and Wells agreed to a complex formula, including a guarantee that the firm would offer over 25 percent of its positions each year to female graduates.[16]

The case against Sullivan and Cromwell was then enlarged to challenge the firm's system of evaluating associates for promotion to partners. The handling of the case followed an interesting and ironic course. Sullivan and Cromwell hired Ephraim London, a venerable civil rights

---

*The name of the firm at the time was Royall, Koegel, and Wells.

†"Blue sky work" is a specialty concerned with keeping abreast of changes in the securities laws of each of the fifty states. Lawyers in blue sky work are not usually in direct contact with corporate clients, and this is one reason why it may be delegated to women. Smigel describes blue sky work as a job that most lawyers especially shun: "It is not considered a creative position. Much of it is tedious repetitive work of almost a clerical nature. The firms have difficulty in keeping young lawyers at this task." Associates who have been passed over as partners also are delegated to blue sky work. Smigel, *The Wall Street Lawyer,* p. 41. Today most blue sky work is delegated to paralegal assistants in the large Wall Street firms.

lawyer, to head their defense.* The case was assigned to Justice Constance Baker Motley, a black and the only woman among thirty-two judges on the federal bench in New York at the time, whose name had been selected at random from a drum. Sullivan and Cromwell asked her to excuse herself on the basis of her sex: since she was a woman, she could not decide the case on issues, they asserted. They also cited her civil rights background as grounds. She refused to withdraw, and they entered a writ of mandamus to get her removed. An appeals court turned down the request. Motley declared the case a class action suit.

Sullivan and Cromwell pleaded entrapment, suggesting that Diane Blank was not serious about working there. (She relates the experience of sitting in a room with three of their lawyers "glowering" at her for saying she thought she was qualified to join a firm of their caliber.) But Sullivan and Cromwell settled, after Harriet Rabb began to question them about their partners' earnings and other matters firms are loathe to make public.[17] The settlement was on terms similar to the Rogers and Wells case, and even included a provision not to hold firm-sponsored events in clubs that barred women.[18] The other firms that were cited in the initial complaint by the Columbia project (and whose anonymity was guaranteed in their settlements) adopted similar guidelines.

By now it was clear that the shoe was on the other foot. Firms were looking for women to hire and even promote, and they began to court and find women who were developing outstanding reputations elsewhere. The 1972 amendments to the Civil Rights Act changed the entire process of big firm recruitment. The anti-discrimination legislation made it illegal for firms to discriminate in hiring, and pressures on law school placement offices by students and dedicated faculty alerted them to the injustices of past practices. Beginning in the late 1960s, firms had to face challenges from women students who reported discrimination or sexism of hiring partners to law school placement offices. Because law school placement offices were under the gun, they now compelled law firms to interview women seriously or be subject to sanctions that included being barred from interviewing anyone.

A partner at a Wall Street firm who went to Boalt Hall (Berkeley Law School) in the late 1960s, remembered how the school allowed recruit-

*Not only was Ephraim London well known as a civil rights lawyer, he was the law partner of Helen Buttenwieser, who had been the first woman lawyer at Cravath, Swaine, and Moore and a "first woman" in many spheres: the first woman to become a bank director in Manhattan as a trustee for Title Guarantee and Trust Co. and the first woman board chairman of the Legal Aid Society in New York. Buttenwieser was also a director of the American Civil Liberties Union and was active in the NAACP Legal Defense Fund. Judy Klemesrud, "At 74, A Life-long Achiever Has a New Job," *New York Times,* November 26, 1979, p. B12.

ers who baldly said "we don't hire women" to interview on campus. She complained to the dean, who said: "Do you really want to hurt your male colleagues' job chances? If we bar firms who don't hire women, they'll lose out on job opportunities." The next year enrollment of women increased, and twenty went into the dean's office to complain. She described what happened next:

> The school wrote all firms recruiting on campus asking for their hiring practices toward women and minorities. One firm had written that they had no policy either for or against minorities, but their interviewer had a strong prejudice against women. The school asked that he not be sent but he came anyway. The students then engaged him in an all day debate in the school cafeteria. He wanted to interview—to have something to show his firm for the day. But we wouldn't let him. We kept changing shifts—men joined us too—to debate his prejudice against women. At the end of the day, he said, "O.K., we've had this little charade. Now when can I interview?" He was absolutely unrepentant. At this point the dean said, no, he couldn't interview at all, and wrote a letter to the man's firm explaining why he had not been allowed to interview. The whole matter of firms not hiring women got cleared up at Berkeley in that one year.

In some cases, women law students sued their placement offices; as a result of one such action, the federal Equal Employment Opportunities Commission ordered the University of Chicago Law School in 1972 to help women get jobs with law firms that recruited on campus. In a complaint filed two years before, women at Chicago had charged that firms granted only perfunctory interviews to women, explaining that "senior partners . . . won't stand for it."[19] There were regional differences in discrimination, and school placement offices have remained sensitive to the issues.*

With this kind of incentive, firms, or at least the sophisticated ones, moved swiftly to correct the tone and behavior of their representatives at interviews. Women who had complained of inappropriate questions or demeanor in employment interviews at Harvard, Yale, Columbia, or Berkeley were even accepted into the offending firms after they had entered formal complaints with the law school administration. An associate at an establishment firm recalled an unpleasant interview with the hiring partner:

> The man who did the interview said a lot of negative things, but then tried to backtrack because he realized he was being very rude. The way in which he

*Dean Albert M. Sachs and Dean Alfred Daniels, of the Harvard Law School, went to Cincinnati, for example, to explore why the seven firms in the city employed only 1 black and 17 women among 253 lawyers. Patrick Kennedy and Steven Miller, "Gauging Discrimination Behind Closed Doors," *Harvard Law Record,* November 5, 1976.

tried to mend the situation was by scheduling another talk [because] . . . he couldn't say no "to a pretty girl."

The hiring partner had gone to school with one of the applicant's teachers, who was working in Title VII litigation and who had filed a complaint against the firm. The partner's behavior was noted as part of the suit and brought him reprimand from his colleagues. The norms of the firm were such that his rudeness was regarded as a form of interpersonal incompetence. The woman involved now works for the firm, and she reports that many of the firm's partners have personally apologized to her for the interviewer's behavior.

While firms have different reputations for their treatment of women —many have been under threat of sex discrimination suits or have settled them only recently—all are conscious of the "woman question." Blatant discrimination has been done away with. Where prejudice continues to exist, its expression is subtle.

Once the doors opened on Wall Street, women started flocking through them. Some large firms today have many women; some have few. Some have no women partners and others have several. Some integrate women more fully than others; some are enthusiastic about their new recruits and others are merely tolerant. Firms have different traditions, different profiles, and different patterns of courtship of new recruits. They are different communities; some are casual about their boundaries and some guard them rigorously. Women and other minority recruits are either attracted or repelled by the differing styles and images of the individual firms. The resultant sorting of young attorneys in the legal marketplace reinforces the patterning of the firms and determines the places held within them by minority and other recruits. In the end, despite professions of principle, some welcome more women lawyers than others do—or more Jews or blacks, or more of the old boys that used to typify them all. A young woman (Smith College; Rutgers Law School) in a big midtown firm, whose husband (Princeton; Columbia Law School) works in a large downtown firm, described the sorting that took place in her own family:

John and I, one thing we talk a lot about is the difference in our firms. They are enormous. He is with *the* blue chip of the blue chip firms. And I'm with this young bunch of hustlers. It is funny and it suits our personalities, where each of us is.

Sometimes it takes a while to "feel out" a firm and to feel one's rightness of place within it. In a satiric story about a young Jewish woman associate in a "blue blood Wall Street firm," Cynthia Ozick gave this description:

> Immediately after law school, Puttermesser entered the firm of Midland, Reid & Cockleberry. It was a blueblood Wall Street firm, and Puttermesser, hired for her brains and ingratiating (read: immigrant-like) industry, was put into a back office to hunt up all-fours cases for the men up front. Though a Jew and a woman, she felt little discrimination: the back office was chiefly the repository of unmitigated drudgery and therefore of usable youth. Often enough it kept its lights burning until three in the morning. It was right that the Top Rung of law school should earn you the Bottom of the Ladder in the actual world of all fours. The wonderful thing was the fact of the ladder itself. And though she was the only woman, Puttermesser was not the only Jew. Three Jews a year joined the back precincts of Midland, Reid (four the year Puttermesser came, which meant they thought "woman" more than "Jew" at the sight of her). Three Jews a year left— not the same three. Lunchtime was difficult. Most of the young men went to one or two athletic clubs nearby to "work out"; Puttermesser ate from a paper bag at her desk, along with the other Jews, and this was strange; the young male Jews appeared to be as committed to the squash courts as the others. Alas, the athletic clubs would not have them, and this too was preternatural—the young Jews were indistinguishable from the others....
>
> The squash players, meanwhile, moved out of the back offices into the front offices. One or two of them were groomed—curried, fed sugar, led out by the muzzle—for partnership: were called out to lunch with thin and easeful clients, spent an afternoon in the dining room of one of the big sleek banks, and, in short, developed the creamy cheeks and bland habits of the always-comfortable.
>
> The Jews, by contrast, grew more anxious . . . became perfectionist and uncasual, quibbled bitterly, with stabbing forefingers, over principles, and all in all began to look and act less like superannuated college athletes and more like Jews. Then they left. They left of their own choice; no one shut them out.
>
> Puttermesser left too, weary of so much chivalry—the partners in particular were excessively gracious to her, and treated her like a fellow-aristocrat. . . .
>
> . . . For farewell she was taken out to a public restaurant—the clubs the partners belonged to (they explained) did not allow women—and apologized to.[20]

Recruitment to Wall Street

Some sociologists tend to focus on the sorting done by "self-selection"[21] —"They left of their own choice; no one shut them out." The woman who spoke of the differences between her firm and her husband's explained how it happened:

We got there because of P.R.—perspective interviewing. We tend to choose to go to one of these places based on tone, how you think and how you view the personality of the firm. . . . They pick you on the same basis, asking, "Is this person going to fit in?" Given two people who are equally well qualified, who have done very well at a very good law school and who seem to be fit for the job, they would certainly go with the person who seemed to fit in.

While many law firms are well aware of how they evaluate whether a candidate will "fit in," they are not always so aware that they too are being evaluated. Wall Street firms with few women often explain that they are ready to hire more women, but they say that when they make offers they have few takers. It is probably true, and for a number of reasons. One is the history of the firm and their past willingness to hire women (the forty-plus bachelor usually isn't considered a good marriage prospect). Reasonable women are wary of firms that have never hired women or have given them the run-around. Firms are perceived in certain ways, and the way they see themselves may not be the way others see them. Law firms, like individuals, engage in "presentations of self," as Erving Goffman has described the process.[22] Image is also conveyed by the interpretation of cues or symbols. Thus a firm might be perceived as conservative through the questions asked in an interview, the furniture in the office where it is conducted, the clothes the partners are wearing, whether they smile or are stern. Since views of what a firm's character is also come from a firm's history, and since many firms' histories are "bad" as far as women are concerned, the firms feel that history should be forgotten. But women candidates do not forget; they suspect that firms which excluded them in the past want only token women.

A young male associate at a large firm reported that the women he knew were attracted by the less stuffy, less "social register" firms, those that had a larger proportion of Italian Catholics or other ethnic outsiders. Their attitude was, "If the Italians can fit in, why can't I?"

Two New York City "Jewish firms" (Stroock and Stroock and Lavan, and Proskauer, Rose, Goetz, and Mendelsohn) seem to have been most open to making women partners. Stroock and Proskauer had four women partners by the spring of 1980.* Kaye, Scholer, in late August 1979 before its reorganization had three women partners; one WASP firm, Rogers and Wells had three also. These were the largest numbers

---

*Strook also named one of the first black partners in a major firm, Samuel C. Jackson (an Assistant Secretary of Housing and Urban Development in the Nixon and Ford administrations) in 1973. "Data on Law Firms Raise Racial Issues," *New York Times*, July 29, 1981, p. A20.

of women partners to be found in any large New York firms. Of the six large firms with two women partners, two are Jewish firms. In sum, the seven Jewish firms have seventeen different women partners, while the twenty-five leading establishment firms have twenty-four different women partners. All New York Jewish firms have at least one woman partner; nine leading non-Jewish firms have no women partners.

Paul Hoffman, in a book on the Wall Street firms, reported that the leading establishment firm, Cravath, Swaine, and Moore, was one of the first to hire women lawyers.[23] Helen L. Buttenwieser, of the Lehman family (her uncle, Herbert Lehman, was New York's governor and later a U.S. senator), now a senior partner in the midtown firm of London, Buttenwieser, and Chalif, served as an associate at Cravath in the 1930s. At the time, only one other Wall Street firm, Cadwalader, Wickersham, and Taft, had a woman lawyer. "Cravath got a woman . . . and a Jew . . . and got rid of her in one year," Helen Buttenwieser recalled without rancor.[24] Cravath also had one of the three women partners on Wall Street in the late 1960s, Christine Beshar. Today, however, when there are a total of forty-one women partners in Wall Street firms, Cravath still has only one female partner.

While the Jewish firms do have more women partners, they have so few more that it is difficult to be certain whether it reflects a real difference in attitudes. Women and men within these firms claim they are more receptive, but outsiders are more skeptical. It is probable, however, that although the difference in numbers is slight between WASP and Jewish firms, the Jewish firms have gone beyond tokenism —they have doubled the women partners on a percentage basis.

Women at some WASP firms claim that Jewish firms are "sweat shops." They are said to be "newer," less assured of income, more eager for new clients, more demanding of their associates' time, and therefore harder in terms of total time demands on the women working for them. Women at the Jewish firms often agree, but point out that they are "warmer"—reflecting Jewish ethnic styles of interaction—and therefore more resonant to women's interpersonal styles.

One Jewish associate mused:

I wanted to be in an integrated firm, not in a Jewish firm. I don't see that as a particular benefit although there's a particular emotional atmosphere and warmth . . . and a kvetchiness that I can identify with. And I can't identify with someone who's starchy about his emotions and wears saspenders and goes to the yacht club. It's not that I can't deal with it. I could deal with anything. But I don't know if I could ever be happy, or that they would ever understand me.

There is a question as to whether all women, or only Jewish women, respond to that style. One young woman associate who was not Jewish and who formerly had worked at a WASP firm found a Jewish firm more compatible because the members were supportive of new women recruits, trusted them with independent work, and simply had more women around.

I have never felt discriminated against—in high school or college. The first time it really hit was when I went to [L]. I met a friend from law school who was militant and always thought I was quiet and serene. I answered her questions about my job—"It was awful, they were a bunch of sexist pigs." She said, "My God, *you,* Anne, saying this, I can't believe it." Now I work hard at [X firm], much harder than at the other, but I'm treated like one of the guys and I trust them. Most of the partners are young—in their forties and fifties. They feel totally at ease with the number of women there. They make you think you have as good a chance for advancement and to make partner if you want that.

Another woman made a similar observation:

Before I accepted the job here I checked it out, and people said that I was going to give up my life and happiness and the whole bit. But I trusted this place instinctively and felt I would like the work I was doing, and I think I made the right choice. Rather work under a lot of pressure and have a lot of work and have interesting work than work at a place like [Y] (a non-Jewish Wall Street firm) . . . that is nine to five.

While all Wall Street firms have been growing, the Jewish firms have grown even faster. This fact alone could account for a certain greater receptivity to women as well. And, of course, there has been a greater commitment to social justice issues by lawyers in the Jewish firms because of their own past experiences with discrimination.

But even if atmosphere and style were the major issues it is important to remember that women are not alike in personal style and many who gravitate to old-line firms report the atmosphere suits them. Many find the courtly air of the traditional firms easier to take than the brusque, if warm, quality of the Jewish firms. The differences between the Jewish and Protestant firms may diminish in time; all Wall Street firms are tending to recruit from a similar pool of men and women and are moving toward shared values. Although no doubt there will always be some differences in ambiance among various firms, they may not affect the integration of women in quite the ways they have in the past.

The background of Jewish men and the diversity of their other statuses may contribute to an atmosphere more comfortable for women and more accepting to them. Fewer Jewish men than WASP men have the history of single-sex schooling from preparatory school through college. Jewish men also spent less time at sports when they were growing up and are less likely now to participate in the all-male environment of athletic clubs. These clubs are often Protestant only, or effectively so, even today. However, Jewish men are now gravitating to the squash courts and gymnasiums, newly integrated or newly built.*

Tokenism

I suggested that the Wall Street firms which had several women partners may have gone "beyond tokenism"—but it is a difficult subject to appraise. Tokenism exists when one or a few members of an undesirable group are accepted by an in-group in order to counter criticisms of discrimination without a genuine commitment to accept all qualified candidates from the undesired group. Rosabeth Moss Kanter and Judith Long Laws have both shown that not only does the "in-group" get off cheaply by admitting a few persons from groups not considered acceptable, but the token members have difficulty in performing their roles, whether professional or not, because they are under stress.[24] The token members suffer other consequences,[25] such as self-consciousness, over- and under-conformity,[26] and distancing. But the process by which the many confront the few is much more complex; in fact, contradictory processes may operate at the same time to both include and exclude the outsider.

Kanter argued (and it seems perfectly sensible) that in corporations the more women (or members of any undesirable group) come into a setting where they were regarded antagonistically, the more "natural" will be their behavior and the behavior of others toward them. The acceptance of women as their numbers have increased in law schools supports this view. And it is easier for women to function effectively in law firms where there are several women partners instead of one; and three may be significantly different than two. Why? An increase in the size of the group permits expression of a range of personality and talent and undermines the stereotypes that may emerge from one individual's unique behavior.

*At one firm, a woman partner reported that while there is a softball team, the men who play on it are casual (which is to say, not very good) players. Therefore, women players are welcome and don't pose a threat as they would in a "serious" sport situation.

But at the same time that forces move a system to open there are also forces that move it toward closure. These may be forces that seek to encapsulate and resist the newcomers, especially as their number grows. Like white cells surround offending matter, the dominant group may continue to regard women as something different and unacceptable, perhaps tolerated but not assimilated. The new entrants may be sabotaged as the majority group, protecting its community (here, the legal community), musters its forces to control its culture and its boundaries. When outsiders manage to establish themselves, strong but subtle forces may come into play to keep them from taking positions of command.

This has been shown to happen in "women's professions," where somehow the larger social system works to facilitate men's acquisition of power disproportionately to women.* Rose Laub Coser has shown how the same process occurs in other countries where large numbers of women have entered high-status professions, but men still cluster disproportionately at the top.[28] Whether the same pattern will emerge in America's changing professions is an open question. For no matter how dialectical the situation, it tends to resolution. At the moment there does not seem to be a shared perception on the part of the Wall Street "community"—either among male partners or women associates —as to how many women partners are an acceptable number in terms of both the women's competence and the preservation or modification of the firm's culture. These are issues still in "negotiation," sometimes in the courts, but more often, these days, among the women and men who work in the firms.

Across the Country

The clustering of women in the New York Wall Street firms is also reflected among the "Superlawyers," as Mark Green refers to them, of Washington, D.C.[28] At Wilmer, Cutler, and Pickering, a firm that doubled in size between 1970 and 1977, of 108 lawyers in 1977 there were 16 female associates and 1 woman partner (and 4 black associates). Ac-

---

*Epstein, *Woman's Place.* Recently, Gaye Tuchman has written a critical analysis of the assumption that the negative consequences of tokenism may be done away with by a mere increase in numbers. Gaye Tuchman, "The Interpenetration of Public and Domestic Spheres: A Review of Recent Sociological Research in United States," *donnawomanfemme* (Rome) in press.

cording to Mark Green, in *The Other Government,* founding member Lloyd Cutler, a Yale Law School *wunderkind,* had been hired at Cravath "despite the prevailing anti-Jewish sentiment of the large New York firms."[30] He went on "to fulfill his goal of creating a great Washington Law Firm, 'an all-around firm with people in it who could do anything and who would have the resources to do anything.' "

Green also described the climate at the rather more austere and WASP Covington and Burling. Women had worked there since at least the 1940s, and by 1970 Covington had 8 female associates, about 5 percent of the firm, but the first woman partner was not named until the mid-1970s. One recent female graduate of the firm said in an interview that women were "cubbyholed there, doing dribs and drabs of cases," and were given jobs specifically reserved for women—in trusts and estates, tax, and food-and-drug law. Covington's weekly luncheons were held at the Metropolitan Club, which prohibited women. When a woman associate complained, the partners replied that no alternative lunch site could be found to accommodate so many persons. Washington law firms are much more in the public eye than the New York corporate firms; Covington's "decision" became the subject of an item in the September 1973 *Washingtonian* titled "Oink, Oink" for the firm's unliberated view. Covington has since "improved" and now has 30 women associates and two women partners among its 185 lawyers.

Wall Street–type firms in other cities have drawn mixed reviews for their treatment of women. The same issues of image, whether conservative or liberal, exclusionary or not, have a bearing on the "woman question." Although Wilmer, Culter, and Pickering is a "liberal" firm and has seemed relatively receptive to women, the firms of other liberal lawyers active at the centers of power have seemed less so. In Texas, Fulbright and Jaworski, the third largest law firm in the country, was the target of a sex discrimination suit and was forced to settle the case in 1979, after two years of resistance.[31] (Leon Jaworski, the firm's politically liberal senior partner, was a former American Bar Association President and Watergate Special Prosecutor.) A reporter for the *American Lawyer,* Steven Brill, noted that one condition of the settlement was a court order sealing all documents in the case and gagging all those involved. The complaint had charged that the firm did not hire its first female attorney until 1956, and then only to handle miscellaneous small matters firms take on to accommodate clients, which male associates did not wish to do. The firm's first female associate was, according to the complaint, "instructed to do her own typing and was on at least one occasion requested to assist the typing pool." Another two women associates who had worked at the firm in the 1960s were also said to have been given work of low regard.* Fulbright and Jaworski now has 123 partners and 138 associates. One partner is a woman—she is in the Washington, D.C., office—and there are 28 women associates.[32] After the settlement the managing partner of Fulbright and Jaworski told Brill that the firm "intend[s] to make more women partners" and that "standards and perspectives on [partnership criteria] in the '60s are obviously not appropriate for the '70s."[33]

King and Spalding of Atlanta, a 102-lawyer firm, faced a sex discrimination suit in 1980; the firm had never admitted a woman to partnership in its 100-year history. Senior partners of the firm are former Attorney General Griffin Bell and Charles H. Kirbo, close friends and advisors of former President Carter.[34] In July 1980, they named their first woman partner, municipal finance attorney Ruth Garrett.

Discrimination within the Firms

Are the men of the legal profession waiting for the day when, having fulfilled their new obligations with token appointments of women, they will attempt to reinstitute a male world? Given the history of women's participation in the law and some women's fears about the fragility of their gains, it seems a fair question.

Most young women lawyers interviewed in the past few years report that resistance to their presence has disappeared to a large extent. A few, however, did report bad feelings, hostile comments, or obstructionism by men in their firms, though such cases today are usually limited to older male partners.

One woman graphically described the response of men in her firm:

I was having a discussion with one of the partners of my firm about hiring the women—they are really good about it. You know, as I said, half of the people they hired last year were women and half of the people they will hire next year will be women. So I mean they are certainly not discriminating, but some of the older partners are saying "God damn it, why are we having all these stupid women?" But the other partners were educating them. They were saying, "Well, by the way, do you know the percentage of women now in the law

*Of Houston's "Big Three" firms, Fulbright and Jaworski, Baker and Botts, and Vinson and Elkins, only Fulbright and Jaworski has a woman partner. Margolick and Gelberg, "Few Dents in Male Dominance," *National Law Journal* 2, no. 47 (August 4, 1980): 59.

schools?" One senior man said, "I don't know, 10 percent." "No, you are wrong, it is 45 percent and if we don't interview women, we are not going to get the best people."* The younger man's response was, in effect, "Go sit in a corner." And so it was. The firm has become really good about women. I don't think they were one of the forerunners, but once it became apparent that there were a lot of women out there in the market of jobs, they hopped on the wagon and they hired a lot of women.

Few women lawyers are obtuse about discrimination any longer, yet in recent Wall Street interviews accounts of discrimination came more often from the younger men than from the women. For example, one young man said he thought the large number of women recruited for summer jobs during their law school years were hired because it was an easy way for the law firms to look good in the statistical count of women. Certainly the Martindale Hubell legal directory— which does not include summer associates in its listings—reports far fewer (50 percent fewer) women than the self-reported data on the number of women given by the New York firms on their résumés to Columbia Law School. While this large discrepancy may be partially due to Martindale's lag in updating its directory, the way in which some firms have combined the numbers of female summer associates with those of full-time women attorneys suggests they are choosing—rationally enough —to present themselves in the most favorable light possible at the recruitment table and in public life.

Specialization

In the past, women who ventured into Wall Street firms were funneled into trust and estate and blue sky work. It was virtually unknown for them to do corporate work or litigation. Today, stereotyped women's work, trusts and estates in particular, is still attractive to some because one has more predictable hours and there are fewer crisis situations demanding late night and weekend work, a real asset to a woman lawyer with a family. But today many women believe they will have access to the full range of legal specialties and many make a point of not accepting assignments stereotyped as appropriate for them.

In one large firm, an associate reported that a majority of the women of his entering group asked for assignment in the litigation department,

*Both of these estimates are, of course, wrong.

one of Wall Street's most demanding specialties. Many are being granted their request. This is reflected even in the wide range of specialties of the women who have become partners. Of the forty-one women partners on Wall Street, six are in trusts and estates as compared to five in litigation and thirteen in corporate work. Various other kinds of specializations are also represented among the women partners, such as antitrust, labor, tax, interstate commerce, and real estate law.

When considering the forces affecting the status of women, a change in the configuration of the professional field itself may be as important a factor as an altered view of women's capacities. To look at litigation again, as growth in the size of the big firms have created places for women, the growth in the size of litigation departments may also have affected women's opportunities to work in them. More and more, large firms that once shunned the courtroom because it was thought to be not profitable and somewhat undignified are building up strong litigation departments.[35] In fact, litigation departments are the fastest growing sections of these firms.* Thus women's new ambitions and images of themselves match a growing development of opportunity.

Most women in law today want to do everything men do. In fact, some characterize themselves as "big talkers" and "dramatic" and aggressively seek the drama and recognition of the courtroom. They are quite the reverse of the other women lawyers interviewed in the 1960s who shied away from courtroom combat as being inconsistent with woman's nature generally and their own personalities in particular. Of course, few attorneys in large firms go into the courtroom, and only a small number of big cases go to trial. Even the people in the litigation department may not engage much in courtroom activity.

Yet some lawyers report a continued subtle differentiation between men's work and women's work, even as they have a better chance to enter untraditional departments. Some litigation lawyers report that they go into the courtroom less often than men of their class and rank and are disproportionately assigned the backroom paper work. Of the trial specialists, significantly fewer are women.

At the same time that women's role in the Wall Street firms is changing radically and rapidly, the traditional pressures persist and push them toward backroom tasks. One prominent women attorney in a large firm complained that for years she had been its leading brief

---

*Lawyers and legal scholars ascribe this unprecedented rise in litigation to two factors —an increased litigiousness among citizens and increased government involvement in regulating corporate and individual behavior.

writer and theoretician. She was complimented on her talents and took pride in her work. It was only late in her career that she realized she had been kept in the background and isolated in the firm. Although she ultimately became a partner, she was rarely exposed to clients. Furthermore, her income was lower than men of her rank. As she described her situation:

I never got to the point where a male litigator was. I had been undertrained because my peers had gone into court a great deal and I had gone into court just a little. Even when I asked for more litigation experience it wasn't assigned to me.

They never actually refused to give it to me. They just kept saying how valuable I was to the firm writing conceptual memos, analyzing problems, doing briefs, and things like that. They told me that it was logical to keep doing what I had been, because it was in the best interests of the firm. Through it all, I thought "how great I'm doing," because that's what they told me. It was only later that I realized that the men were advancing much further in their careers than I was. ·

On the other hand, women are becoming integrated in firms and are on less stereotyped career tracks. An important element in the integration of women is the fact that it pays firms to take their women seriously and to insist others do as well. After all, there are many of them, and they carry the name and reputation of the firm. The Wall Street firms can no longer afford to keep women deviant by casting them as outsiders and incompetents.* The policy makers in these firms, having acknowledged that women are now part of the pool of future colleagues, are best advised to learn to live with them.

Once a women has been hired by a large firm, she typically acquires the authority of the firm's name. She is not just a lawyer, but a "Sullivan and Cromwell" or a "Cravath" lawyer. Usually clients of these firms are large corporations or businesses that will only be represented by a prestige firm and have a long-term relationship with it. Sometimes the impact of the firm name is such that when a woman attorney joins the legal team assigned to a client no questions are asked.† The rank of the

*Technically the process is "altercasting," so named by Eugene Weinstein, who pointed out that role partners may play to one aspect of a person's role and thus force that person to reciprocate. Eugene A. Weinstein and Paul Deutschberger, "Some Dimensions of Altercasting," *Sociometry* 26 (1963): 454–66.

†Firms doing business with foreign corporations (for example, Japanese corporations), however, are extremely sensitive to cultural innuendos, and some attorneys have confided that women are excluded in such cases. Nevertheless, a partner in a major firm, who said his firm represents more Japanese clients than any other, makes it a point not to exclude women.

firm has a great deal to do with the acceptance of a woman lawyer. Women lawyers representing elite firms do not have their expertise called into question.

The entry of women into traditionally male specialties in corporate firms seem to be determined primarily by the firms' sex-typing practices, not the attitudes of clients. The following account by a woman lawyer with an outstanding reputation is noteworthy because it happened thirty years ago.

I gained some of my reputation by doing labor relations and litigation with the top litigation partner. In that firm they even let me go right into the courtroom, which was fantastic. Most of the firms that were forced to take women during the war, because they needed them so badly, had them all in the library. They didn't want anybody to know they had women. With me they really played very fair; they took me right where I could see, and they'd take me right into the courtroom and set me right at the counsel's table, and they never pretended I was a secretary.

The Process of Promotion

Will women's numbers and greater visibility insure their permanent place on Wall Street? The general view in the firms, among onlookers in the law schools, and among feminist attorneys has been "Let's wait and see what happens when it becomes time to promote women to partnership."

Since women associates who were coming into the large firms in the mid-1970s had been saying in interviews that they thought they were doing well, and were on track, we asked a male law school professor how he assessed their perception. He agreed with us that women thought things were "terrific," but was more cautious about the future for promotion.

. . . as a matter of fact it remains to be seen because the real test is not hiring at the associate level. For that job the only thing the firm has to go on is the record, and the women's records are as good as the men's, so they have no choice but to hire them. Especially now when they are defensive they hire them.

But when it comes to partnership, where the decision is made on the basis of very deep subjective evaluation, whether that person has the right chemistry to be made a partner, . . . and you can understand how these law firms think about a partner . . . you don't get "made" a partner," or even get "promoted" to partner. You get *elevated* to partnership. (You learn this in the discussion periods. Really, that is the very word they use. *Elevated.*) We haven't seen what is going to happen at that level, and there are the subjective biases of men, as men . . . because the firms are still dominated by older men with the whole built-in cultural issue of that generation. . . . it would be very surprising if that built-in mind-set didn't effect the decisions.

In New York the associate-to-partner ratio ranges from two-to-one at some firms to three-to-one at firms like Cravath.[36] In large firms in other cities the ratio is closer to one-to-one. A survey by Columbia Law School showed that 50 percent of the graduates who joined major firms outside New York City stayed on to become partners.[37] In New York City only 20 percent did so. That means that most associates do not become partners.

The fate of young lawyers who are passed over for partnership, having put in six to ten years in a firm and earning $50,000 a year, was diagnosed by the legal reporter for *Esquire* a couple of years ago.[38] Many of the young men, it was pointed out, get trapped by the high salaries, and although it was common to go from the firms to corporations, their high salaries are too much for the corporations to pay. For one thing, there is some resistance to employing them. Furthermore, many of the lawyers are so specialized their skills are not portable.

However, a recent *Wall Street Journal* article notes that in-house lawyers are gaining in numbers and the increase may provide more employment.[39] Yet the opportunities may be even better (proportionately) for women. A woman partner in a large firm told us that headhunters for corporations often call the firm looking for women lawyers. Although the law firms are in good shape with regard to the proportion of women hired, the corporations have not caught up.* Hiring women attorneys is one way, she pointed out, for them to fulfill their affirmative action requirements. Another benefit they get, is that although in-house lawyers often have executive status, it is on a different track than the corporate hierarchy, where women may be resisted. That is, corporations may feel that women will stay put in legal departments, while men might use them as a springboard to a job elsewhere in the corporation. In my earlier study, it appeared that women lawyers in corporations cared more about remaining a lawyer than doing business work far removed from law even though it might mean a step up.

Women lawyers also feel more comfortable in taking a reduced salary

*Women make up .25 percent of the general counsels of "Fortune 500" firms. (That is, five women.)

for a job which seems interesting to them or whose benefits can be calculated in other ways (less demanding in time, for example). Corporation executives, who would be unwilling to pay men less than they had been making before, expecting discontent on the part of the lawyer taking the pay cut, may be willing to offer a pay cut to a woman. Many women associates in large firms, pondering having families, were at least considering jobs with reduced time demands (such as those that corporations offer), especially when they had husbands whose incomes were high. In a sense, this ability to transfer from one legal domain to another might be considered a "secondary gain"[40] of holding status regarded as second rank in American society, but for some women, and some men observing their opportunity, it is perceived merely as an option. But many women associates on Wall Street do not want to exercise the option of moving to into a corporation. It remains to be seen what their chances are of joining the 20 percent who are elevated to partnerships, but it appears that the odds are better than before. With women now at 20 percent of the lawyers in large firms, the pool of women eligible for promotion is sizeable for the first time. But most women partners are between thirty-five and forty-two years of age, and since the average age of women in the larger firms is younger (most associates are recent recruits and have yet to put in their apprenticeship years), the relative proportion of their advancement will not be known for years.

Promotion in Wall Street firms is similar to the tenure system in academia. Young lawyers serve as associates for six to ten years while they await final appraisal, leading to partnership or departure from the firm. Only a few denied full membership may stay on as "permanent associates," and in the past these often were of minority or ethnic background, or women.

The promotion system is fairly well institutionalized and modeled after the plan put into practice around 1906 by Cravath, Swaine, and Moore. It was instituted as a way of insuring quality by devising a merit system that would select lawyers directly from law schools and then train them from within. The recruit was to have an excellent law school record, preferably to have served on the law review, and have "force of personality." The recruit would then go up through the ranks until he became a partner with life tenure; if he was not chosen for partnership, he left the firm. Nepotism was not to be practiced and, in fact, the sons of Wall Street partners were not hired by their fathers' firms, though they were helped to find employment in the firms of colleagues. Mr. Cravath, of course, hired only socially "correct" young men.[41] It is improbable that he ever entertained a notion of admitting women to his firm, any more than the senior partners of the other large firms who followed his example. In the past decade, when women did appear on the scene and became associates in the large firms, the view of almost everyone was that they would not become partners.

In the past, women did not expect partnership. Now many consider it a normal ambition. Like their male peers, they calculate their chances and work to meet not only the formal but the informal requirements. Like the men, they change their ambitions in the process, some becoming more ambitious for success, Wall Street style.

Different Routes to Partnership

Many of the women who have attained partnership on Wall Street follow a pattern typical of women's career sequencing in other high places—a pattern atypical for men, and a diversion from the "Cravath system." Although many women partners come from top colleges and law schools and compiled first-rank academic records, they tended to prove their legal talents outside the firms that ultimately granted them partnership, often engaging in activities that give them a special competence and immediate value to a firm. Several worked in high places in government—in the Securities and Exchange Commission and the Federal Trade Commission, in state banking departments, in the Office of Equal Economic Opportunity, and in various advisory positions to the White House and state houses. Several had worked abroad and had special language skills, two in Japanese, which made them especially attractive for work in international law. Several have credentials in other fields; one is a CPA and followed a "normal apprenticeship from a role as associate to partner in the typical way."

Women partners and associates today express high motivations for success. Some were motivated from the beginning of their careers, especially the younger ones. But it was striking to find repeatedly in interviews the extent to which women who came into large firms with no clear set of ambitions quickly internalized the goals of success. To the surprise of some of the women themselves, many who earlier wanted only to try a big firm found themselves increasingly drawn to working on the problems of large corporations for large rewards. It is theoretically important to find that early socialization can be offset by

later experience and by the changed realities of the opportunity structure.[42] One Wall Street lawyer expressed her feelings about getting
ahead:

> I feel I am as good as anyone there and maybe even better. I started out trying
> to liberate everybody. I didn't win. I've given up hours with my children; I've
> given up friends. Yes, I've enjoyed it; yes, I've made a lot of money. You don't
> want to be in a place where you won't succeed.

Concurrently, however, the women's movement was making it more
legitimate for women to display interest in such values as power and
success in the workplace. This was not the view of radical feminists, nor
of groups who are believed to hold views quite opposite—who believe
in traditional sex roles for women and who oppose their assumption of
"male values." The ideological lines drawn in the early 1970s between
women who feel it is legitimate and appropriate to work toward high
rewards and those who believe it is morally wrong and ideologically
incorrect persist today. The latter group is different from generations
past who believed it was unfeminine for women to seek success in
"male" terms. Rather they believe that women should avoid the corruption of the "rat race."

To some extent, women with these different ideological points of
view have clustered in different kinds of legal practice. Those who
accept mainstream definitions of success in both personal and material
terms have sought careers in the large firms. Those who disclaim material success have joined legal services, feminist legal practice, "movement law," and private practice.

There are, however, many exceptions. Ideology may have motivated
some women, but professional practice often has altered ideology.
Some women joined large firms because friends advised them to try
Wall Street or they felt it was prestigious to be asked to join. Some did
so because the best jobs in legal services or government practices were
limited. Some elected to earn high salaries to repay law school loans
before going into other kinds of practice, like this lawyer:

> I got this job because I wanted to make some money, you know, struggling along
> for all these years. I paid my way through law school. And then I got married
> and we were really struggling away, and I just wanted to make life a little bit
> easier for myself and also to see whether or not I would like to do corporate
> work on a full-time basis. And I did enjoy it. The good work I got was really
> interesting. Maybe I should have tried to get a job in constitutional law, but then
> it is very frustrating. It takes a lot of effort and you don't know if you have a
> job until the day before. Because it is always a question of funding.

It is not unusual for a neophyte lawyer to join a Wall Street firm only to hate it or find other kinds of practice more attractive and leave. Men and women both have done it. Although the work is often interesting and the pay spectacular, some object to the long hours and the routine aspects of the cases, or feel morally offended at repeatedly representing the interests of "big business."

In the 1970s, many women who had not thought previously about it found themselves thinking about how they might achieve partnership. They thought the chances were better than ever before, and with few role models available they began by discussing it among themselves or with husbands or friends. They were willing to pay high costs and structure their lives to minimize competition from their other roles. A number of the current group of young women lawyers on Wall Street clearly express their hopes and expectations of partnership, and they are organizing their lives and choosing their specialties to achieve it. One woman said it clearly:

I liked trusts and estates. I know that there is a stereotype about women going into it, but I figured out who was going where and it seemed the best opportunity. Tax was my second choice but they were pretty full there. [But] I finally chose litigation because it was the best place for a woman to make an impression. I learned in my summer here that although your clients are corporations the people who rule the corporations are sixty-five-year-old men who know everything about the corporation. But when they are in trouble they turn to a litigator and they are not experts in that. They have to listen to me—it doesn't matter that you are a twenty-five- or twenty-six-year-old woman. If you are in the corporate sphere the client knows as much if not more than the lawyer, and it is really a matter of trust and persuasiveness, hard enough for a twenty-five-year-old man to go out and be confident in a business relationship with a sixty-five-year-old corporate president, and even more difficult for a woman. And so many of their conversations take place in men's rooms, golf clubs, that I felt it would be impenetrable, and so I chose litigation.

Making "Rain"

Another dimension that is important in making partner is potential ability to bring in business. Those who can draw business—known as "rainmakers" in the profession—have always received special consideration. Law is an area in which women have not usually made much rain, nor do they express high hopes for improvement. Of course, client relations are themselves highly institutionalized. Corporations have never switched law firms in the way they might change advertising agencies, and few lawyers, even the wealthy and socially connected, can bring in a corporation account such as Xerox or General Motors. Because their female college friends do not usually go on to build business networks, women are less well placed structurally to pull in new business. Most women, and the men who share their limitations, try to establish themselves as "talents" who can better serve existing clients.* Yet today many women can count on male friends to refer business to them, and some women partners we spoke to felt a heavy responsibility to try to develop business contacts, especially in firms known to be more "nervous" than others about revenues. A lawyer in one of them pointed out:

In the midtown firms there is more of a sense that they have to be very sensitive to picking up business and getting new clients and things like that. My firm is funny because in the last ten years it has become incredibly rich and has all of these clients, but there is still that edge of insecurity. You don't dare turn a client down because he might go elsewhere.

One barrier to a large increase in the number of women partners, observed a woman recently "elevated," is that the forty-year-old men look at the women entering their firms and ask themselves who will bring in the business when the women are the forty-year-old partners and they are sixty and want to slow down. This probably will become more of an issue over time if the pool of women associates continues to grow.

Putting in Time

Large corporate firms come close to being what Lewis Coser has described as "greedy institutions." Like the Catholic Church, which demands total involvement by its priests, or utopian societies, which demand total loyalty, these firms "seek exclusive and undivided loyalty

*It is generally assumed that few women are likely to bring in business, and some places are more anxious about it than others. One senior partner in a large midtown firm told me that all that really mattered in his firm was how much money a potential partner would feed into the till. It was a simple financial calculation, he said. Obviously, it is one that includes a number of variables. But a low rating on one variable means one must over-perform in some other dimension.

... and attempt to encompass within their circle the whole personality." Like other such institutions, the firms "tend to rely on voluntary compliance . . . and . . . aim at maximizing assent to their styles of life by appearing highly desirable to the participants." And although this is less true today than in the past, "they exercise pressures on component individuals to weaken their ties . . . with other institutions or persons that might make claims with their own demands."[43]

Part of the myth and tradition of the Wall Street firms centers on the long working hours demanded of the lawyers working for them. Because the stakes are high (partnerships provide lifelong tenure and average incomes of over $200,000), young lawyers typically invest a large amount of time and energy at work. Stories abound about the "old days," when it was typical to work every night and weekends for months on end. Even a decade ago, interviews with women lawyers revealed the dilemmas many of them faced about putting in the same time as their male associates. This was particularly true because women practicing in the 1960s were still caught in the cultural expectation that if they were to give their "all" to any institution, that institution should be the family.*

Full-time work is not enough for a lawyer, especially for an ambitious younger lawyer. The norms of the legal profession equate excellence with hard work, measured in part by long hours. Smigel's study of Wall Street firms indicates that overtime work is not only common but expected, and that "going home is the wrong choice if an associate wants to stay with a law firm or to get ahead in one. So wrong in fact that it is not uncommon to hear that the New Haven Railroad cost a lawyer his job or his chance for a partnership."[44]

A young lawyer in a Wall Street firm noted that time demands pervade the entire schedule of the fledgling career lawyer. It is not enough that one gets the job done:

There's a terrific pressure to work hard and spend all of your extra time on legal things—such as evening classes, going to bar associations, which gives . . . very little time for anything else. And the guy that doesn't . . . and that spends time on . . . varied interests outside has sometimes a tough time of it. I don't see any change in the stereotype.

Expectations about time commitments in firms ranging from Wall Street on down began changing in the 1960s, as young men became

_____
*Lewis Coser and Rose Laub Coser point out that the family is another "greedy institution. "The Housewife and Her 'Greedy Family,' " in Lewis Coser, *Greedy Institutions: Patterns of Undivided Commitment* (New York: The Free Press, 1974).

more family-centered and wanted to spend time with their wives and children. Smigel noted the change in his interviews with Wall Street lawyers, and Louis Auchincloss, a Wall Street lawyer himself, has documented it in his short stories published in 1967. The change drew this complaint from an older lawyer in one of the Auchincloss stories:

Young men in the nineteen-sixties, however hard they may work . . . do not feel the zest for it that we knew. However brilliant they may be, work is work; their 'fun' is always home . . . just as their emotional needs are supplied by women, by wives.[45]

A male partner in a large firm commented, in an interview in the 1970s, that the time demands in large firms were not as all-consuming now as they had been.

When I came to the firm, it was considered unmanly to take a vacation. The question was, "Do you *need* a vacation?" People would brag, "I have not had a vacation for . . . " When I came, the story went around about Old Man Moore, I guess, who would come to the partners meeting and say, "What do you mean we need more associates, why, when I went home at three this morning, there was hardly anybody here."*

When I first started work [in the 1950s], the work day was from 9 A.M. to 10 P.M., with half an hour for lunch, maybe forty-five minutes. And an hour and quarter for dinner. And a half day on Saturday. Everybody put in that time. Most of the partners and all of the associates.

But things changed in the 1960s. You know, a man would say, "I was really planning to have dinner with my wife." And they weren't so devoted to Wall Street. They knew there were other kinds of legal work they could do.

Although it may be true that some young men are less "totally" invested in their Wall Street careers today, there is no doubt that the ambitious lawyer will put in long hours. This is the expectation both of the firm and the lawyer's family. In many high prestige firms, the senior partners (whose own apprenticeships were served in an era of total involvement in the law) still expect that promising young lawyers will be prepared to work late into the night, for weeks or even months, if necessary.

In the past, women faced three sets of problems in "putting in their time." First, those with family obligations felt stress at the kind of

*This incident is often cited among Wall Street lawyers and comes from Swaine's history of the Cravath firm. Hoyt Moore, one of the founding members of Cravath, Swaine, and Moore, was supposedly approached by a partner who complained that the associates were being pushed too hard. "That's silly," he replied, "no one is under pressure. There wasn't a light on when I left at two this morning." Cited in Paul Hoffman, *Lions in the Streets,* p. 7.

overtime commitment demanded of young lawyers. Second, women often were not given the same opportunity to show they were willing to put in time. And finally, because they had such slim chances to become partner, they were not motivated to put in long hours.

One young woman associate, a specialist in trusts and estates in an outstanding Wall Street firm, put it this way:

> I certainly have not worked as hard as the men worked . . . because I got the impression right away at the beginning that—in fact, I was told sort of off the cuff—they've never had a woman partner and they don't expect they ever will. So I took that at face value. . . . Since . . . I would never be a partner here . . . I have also taken advantage of that situation by not working night after night the way a lot of the men do in order to compete.

The appraisal of young lawyers' competence and commitment thus often eliminated women from consideration as partnership "material," especially in the high prestige specialties such as litigation, in which there was pressure to put in a "round-the-clock" effort. The specialties women were assigned to were not the most taxing ones because it was believed women would not put in the hours men did. At the same time, however, women were often condemned for their lack of ambition in not working as hard as the men. The circle was completed when this was offered as the principal reason they were denied partnership.

A Price-Waterhouse study found that the average Wall Street associate billed clients for 1,667 hours of work in 1976, but this is considered sluggish.* Associates at various firms say they are expected to bill 2,000 hours a year.[46] The "Protestant Ethic" may not only be alive in law, but spreading. A young woman in what she described as a "laid back" San Francisco corporate firm (120 lawyers) said that more and more lawyers are billing 2,000 hours there. Many lawyers bill far in excess of 2,000 hours and a woman lawyer I have kept in touch with who was billing 200 hours a month in 1976, in 1980 said she had been billing 300 hours a month for one three-month period on a particularly demanding case.

The hours of work demanded by Wall Street practice are both real and legendary. Some specialties require round-the-clock activity during certain periods; the necessity for long hours was explained by a woman partner:

---

*Billing applies only to the time a lawyer works on clients' cases. Time spent in firm administration, training, on committee work, vacations, and holiday leave are not included, and firms customarily bill for fewer hours than were actually worked. (Bernstein, "Wall Street Lawyers Thriving on Change," p. 109) Some firms permit pro bono work to be counted as billable hours. Others do not.

Litigation is an area that's full of surprises. Everything can be going along at an even rate, coming in at 9:30, leaving at 5:30; than all of a sudden one day your opponent will serve some kind of motion on you and you've got to stay downtown every night for two weeks. Or you'll be working on something and a new case will come in, and just all of a sudden you've got substantial overtime commitments. We had a case come in about a month or so ago. We represent [X oil company] and when the government decided to try to enjoin the acquisition of [X oil company] I guess they called us at about 3 o'clock one afternoon; and 3 o'clock the work began. They were trying to get a restraining order. A couple of people stayed here all night that night, trying to get together a brief. And I think they were working all night the following night as well. And there was a period of about two weeks when there was, if not around-the-clock work, there was work from nine in the morning until midnight.

Most women interviewed in New York firms said that they devoted the same amount of time to work as their male colleagues and some even claim to work harder in order to prove themselves.* One woman who had been in a firm for two and a half years remarked:

I have taken two and a half weeks of vacation, although I have been entitled to eight. And although no one has ever stopped me, I just have never been able to find the time. Part of that is bad planning, part is that I am in two of these cases, part is because the work load here is enormous. They will give you as much work and as much responsibility as you indicate to them you can bear. And, there are people who spend every night here, and every weekend here, and then there are people who come in very early, eight o'clock, and who work very hard, don't shoot the breeze during the day, take a short lunch break and get out by six or seven. . . . On a normal week, when things are not busy, I tend to come in at 9:30, I take about a half hour to an hour lunch. I do relatively little kibbitzing during the day, some days I do, some days I don't. And I am usually here until between 7 and 8. When really busy times come, in cycles, well, then I will be in here at 7 in the morning, no lunch, work till midnight or one, and that can go on for three or four weeks. When I am in the fourth week of such activity, I am cursing the firm, cursing the client, and cursing myself, cursing my husband, and then, you know, things get easier, and the bonus is coming, and I think, gosh, this is fine.

A law professor who is married to a Wall Street lawyer said that if law firms planned their work better, if they had more realistic views, and if they required more work during the day, they wouldn't as often have to work late at night. Some of the women lawyers interviewed believed that overtime work was not always necessary, and that young lawyers did it out of choice. Many lawyers do it in order to impress the partners, as an excuse to stay away from home, or because they prefer to linger

*See Chapter 15 for a discussion of women's need to prove themselves.

at lunch and make up the time in the evening. The mystique and the excitement of working long hours at night reinforces commitment and helps to make the firm their social life as well as their work life.

The attitude at different firms was reported by one young male associate:

A very good friend of mine worked at [firm X] where people wanted to stay late because they perceived that as a way of impressing people who cared, and therefore would not do any work in the morning so that they would have something to do in the evening.

At [firm Y] everybody always stays late. There is a tremendous esprit de corps. They say, "We stay late, we work our asses off, and we do a great job."

But at our place, every once in a while, your number comes up and you have to stay late and boy is it lonely in the library because there will only be two people there and they are all cursing. It's a different attitude. . . .

Some women, like some men, however, managed to differentiate between time-serving as an expression of legal machismo and the demands of a case. But "being there" confers a symbolic wallop that the aspiring find hard to ignore. The fact remains that, whatever the reasons, men who wish to climb the ladder of success often work late, and this looks good to the senior men in their firms. The workaholic puffery of the Wall Street style gave rise to the following tale:

Two associates at Cravath, Swaine and Moore, one of Manhattan's most prestigious law firms, were said to have bet about who could bill the most hours in a day. One worked around the clock, billed 24 hours, and felt assured of victory. His competitor, however, having flown to California in the course of the day, and worked on the plane, was able to bill 27.[47]

The big difference as the 1980s began was not so much that the big firms were permitting leisurely and flexible hours (they were not),* but that women were permitted to put in long hours and were doing so. This was true of associates and partners, whether single, married, or with children. A woman attorney reported:

*A good friend who had reported on the relaxed time demands in his firm behaved quite contradictorily just after our conversation. He came to visit in the country one weekend. Although expected early Saturday he had to drop into the office in the morning and found he could not leave until late in the afternoon. Arriving about four, he donned swim trunks to go to the beach with everyone else. But he became locked into business phone conversations until dinner time. He then left early Sunday morning to return to the office. He had "vacationed" two hours in the car each way to and from the country, and at dinner, but never did get to the beach.

When I went to work for [X] in the city I can remember going in the first week and I noticed there were a group of men, ranging in age from twenty-five to thirty-five, who worked every night until ten or eleven, and they would come in weekends, all day Saturday, and six hours on Sunday. These men have families, and I said I wasn't going to do that—no way. But it's kind of an insidious thing. As soon as I got assigned to cases I got interested in them and found myself doing the same thing.

Everyone agrees that women who expect to make it must play by men's rules. One woman partner, questioned as to whether women behave differently than men, replied:

These are the people who choose to be chosen, if they decide to go for partnership . . . the types "who roll very high dice." Women on Wall Street have adopted a male view of success. You have to work very hard in these jobs and qualifying for partnership requires having stamina and cool.

Some women have tried, without much success, to circumvent the system. One woman came into her firm with an agreement to work nine to five and have summers off. By the end of her first year, the partners could no longer "remember" it, and she was subsequently given an ultimatum: full-time participation or no future in the firm. And since she liked the work and wanted to be successful, she accepted their terms.

The hours worked by lawyers in large firms are highly visible. Lawyers keep a close record of their working time for the purposes of billing clients, but the firm's time diary also has the latent function of a control device. The senior partner who reviews the diary knows who worked overtime and on the weekend, and who limited himself or herself to a standard work week. Peers also know and they keep tabs on each other. One lawyer in a large firm commented:

I know that sometimes when I sign out when I've worked late that there are quite a few names on the list. Sometimes on Saturdays when I come in, there are a few people, but I really don't know whether they are in because they wanted to get away from home or whether they have something pressing.

Time diaries and the billing system are watchdogs that keep young lawyers overproducing.[48] It is precisely this over-production that is difficult for women in such firms who want to spend more time with their husbands or children.

How Are the Young Women Doing?

The Wall Street firms have ways of letting young lawyers know, both explicitly and subtly, how well they are doing. Everyone in the same firm starts at the same salary and most large firms tend to offer about the same amount to start. Two years ago, it seemed incredible that law school graduates were starting at $30,000 per year, but one firm started offering $37,000 in 1980 and the rest are expected to follow suit.[49] Until recently, bonuses and yearly increments informed associates how they were assessed. Bonuses varied from a few hundred dollars to a thousand or so, and yearly increments were $1,000 and up. (An associate can eventually earn more than $50,000 a year, but a partnership brings a jump to more than $100,000 with a possibility of making $250,000; name partners may draw half a million dollars.) Many firms have dropped bonuses in the past few years and reward associates by the size of their increments.

Most women associates interviewed in 1976 and 1977 had received positive messages about their work through these indicators. Indeed, several have become partners in the intervening time period. The general attitude of the women was that they viewed the opportunity structure as open to them. Their performance was not the only reason they felt hopeful.

Most women associates are just becoming experienced enough to be in line for partnership. Unlike women who made partner before them, the younger women are following a male path, promotion from within a "normal status sequence." Although they do not talk about it, they know their firms must be committed to promoting some proportion of them. (Some firms must, if only to comply with their sex-bias settlements.) Moreover, the firms do watch each other and compare how they are changing. When they were surveyed in preparation for an article[50] and asked the number of women partners they had, spokespersons seemed intent on emphasizing the progress they were making. The contact at one firm that had no women partners reported with embarrassment:

No, we still don't have a woman partner; our women just haven't been with us long enough. But we have a great group of women coming up in the next few years; we'll have women partners soon!

Yet there also appears to be a reinforcement and elaboration of the hierarchy in some firms, with more gradations emerging among partners and associates. Advancement may have a different meaning than before. For women and minority associates, there is a greater chance of becoming a partner, but promotion may be to a partnership bearing less power and influence and a proportionately smaller share of profit at the end of the year. There is some suspicion on the part of older women attorneys that this is the kind of partnership many young women are likely to get when their firms feel pressed to promote them. Although it is a step upward, it does not mean that women have "made it" in the same way as the men who are rising in the hierarchy. Nevertheless, partnership is sought after and no women interviewed in a large firm questioned the conditions of partnership which might be offered to them.

In essence, success-minded women are behaving as if, by complying with the rules set for men, they can win the same rewards. Yet there is still uncertainty about what their chances will be because they are women. They believe that being a woman is less salient than before, but at the same time they express apprehension about the future. One woman lawyer looked at it this way:

When I come up [for partnership] there will be ten of us coming up almost at the same time, and I'll bet you anything that it won't be like ten lawyers coming up. It will be *ten women*. And I don't know what the impact will be.

David Margolick, a reporter for the *National Law Journal,* studied Sullivan and Cromwell as archetypical of the problems women encounter in Wall Street firms, after hearing that women in the firm had expressed discontent about their chances for partnership. Margolick quoted a legal recruiter as saying that "several S & C women lawyers complained that their experience at the firm was good for roughly three years and then began to sour."[51] (One Sullivan and Cromwell woman, who spoke glowingly of the firm's attitude in an interview several years earlier, confirmed Margolick's report in a later re-interview.)

The nature of the complaints against Sullivan and Cromwell was different from those made before the firm settled the discrimination suit brought against it. Allegations now focused on the subtle problems that arise from women's perception that the firm is inhospitable rather than discriminatory. Margolick noted that some Sullivan and Cromwell women say there is no problem, but they may feel as one lawyer who responded to questions about problems of advancement in large firms

—"The only way I can function is to ignore precisely the questions that you've been asking me!"

Sullivan and Cromwell still had no women partners in 1980. A young woman associate who recently had left the firm to join another said that many of the women in her "class" were leaving because they felt they had no chance of promotion. Each woman in line to become partner thus far had been informed that she was unworthy or unassessable. An associate who had been there for five years was told in a statement reminiscent of the past that the firm's partners did not know enough about her work yet. She, in turn, felt that she had been assigned work that gave her no opportunity to show ability, since she had been buried in details on big cases. Women still have problems getting good assignments, she said, "and also get flak with the assignments they get."

Whatever happens they make you feel it was your fault if it goes wrong. . . . Many women at this firm have been going through a soul-searching analysis. They've talked together and as a result many have left; eleven from the corporation and litigation departments, and two more are leaving as soon as they have the opportunity.

When women leave, the associate said, the firm always suggests they leave for personal reasons rather than because there was something wrong at the firm: "They have an answer for everything and an explanation for nothing. The problem is that when they think of a model for a partner, it isn't a woman. No woman would possibly fit that mold."

What is the perception of the woman in such a firm? The first woman to become a partner in another firm summed it up:

The concept in a law firm is "my partner, my brother." There is great acceptance among partners for each others' views. But they haven't quite gotten to "sister" yet.

If women are categorizing the firms as male establishments, the woman are often categorized as non-partnership material because their attributes are not perceived correctly. One inside informant was told indirectly that she "smiled too much"—a sign that she was not seriously committed. Men in that particular firm dwell on the sex status of the woman and not her work, extrapolating, perhaps, from their own private lives. That, at least, was the thesis of some women, as summed up by one lawyer:

Solving the problem is mainly a question of time—ten, twenty, thirty years. Death will solve some problems. The older partners are unconscious of their biases. Look at the wives of the partners in the firm for a clue to consciousness. I know of only one who is a professional.

Ideas about what women can do and are doing, are often preconceived because of expectations linked with their female sex status.*

A troublesome case of partnership selection at a large Chicago firm recently led Margolick to canvas the women and senior partners at the firm for the *National Law Journal.* Jenner and Block had been considered a pacesetter in the country with regard to the hiring and promotion of female attorneys. Its recruitment of Jill Wine-Banks (formerly Volner), former counsel to the U.S. Army and a special Watergate prosecutor,† brought the number of women to five out of sixty-nine partners. Of the firm's seventy-eight associates, thirty were women, and in recent years women have constituted more than half of the new law graduates hired by the firm. Against this background, six men and five women were considered for partnership early in 1980. Of the eleven, six were elevated to partnership: five men and one woman. Founding partner Albert Jenner defended the selection, asserting that "It would be well for other firms to have the same attitude towards women that we have had." He suggested the women needed more experience: "The ladies not admitted," he said, "weren't ready yet."[52]

The women had another opinion, namely, that more stringent standards had been applied to the female associates. Male associates hinted, said reporter Margolick, that the decisions were sex-linked:

*Robert K. Merton has pointed out in "oral publication," i.e., unpublished lectures, that when people have statuses in a setting where they are infrequently seen, instead of the appropriate status being focused upon, the irrelevant one may be made salient, such as is in the case of women and black doctors or engineers. See also Everett Hughes, "The Dilemmas and Contradictions of Status," *American Journal of Sociology* 50 (1944): 353–59; and Cynthia Fuchs Epstein, *Woman's Place: Options and Limits in Professional Careers* (Berkeley: The University of California Press, 1970), p. 89. The Stanford University group of sociologists who have been doing research on these issues, (a review of a substantial body of work may be found in Joseph Berger, Susan J. Rosenholtz, and Morris Zelditch, "Status Organizing Processes," *Annual Review of Sociology* 6 [1980]) have also identified processes in which expectations of behavior are linked to status characteristics independent of task setting, and indeed, prestige-power dimensions were carried over from the outside, regardless of whether or not the status characteristic had any prior established association with goal of task of the group. They show that "expectation-states . . . are . . . created by prior beliefs about and evaluations of the characteristics possessed by members of a group who are strangers but differ in external status" (p. 4).

†A *People* magazine article on Volner led off: "Most Americans probably best remember the miniskirts she wore, but Jill Volner impressed Washington observers as the only woman trial attorney on the Special Watergate prosecutor's staff and a courtroom examiner of considerable skill. Among others, Rose Mary Woods and Jeb Magruder squirmed in the witness chair under Volner's incisive questioning." "Jill Volner of Watergate Fame Turns Down TV to Stick With the Law," July 7, 1975, pp. 10, 11.

"By nature most of the partners are more comfortable with male lawyers than with women lawyers." Another male associate was quoted as saying, "In effect, a woman would have to be a higher achiever than a man to receive the same level of acceptance." The firm had promoted nearly 100 percent of its senior associates in the past, and Jenner denied suggestions that the firm had adopted the "New York system," under which lesser lights were winnowed out in the first few years.

However, optimism about women's prospects reigns elsewhere. Kutak, Rock, and Huie of Omaha is a "top 50" firm with a large number of women partners and a climate that is one of the most open and liberal with regard to its women members.* Lindsey Miller-Lerman, recently appointed as one of the firm's six women partners, described the "dream" Wall Street type work situation she was offered as a lawyer with young children.† She was promised a three-day week for an indefinite period at a pro rata salary, and, unlike the New York associate whose firm forgot its agreement, Kutak delivered without penalizing her chances for promotion. Miller-Lerman, a graduate of Columbia Law School and a law review editor, had clerked for Judge Constance Baker Motley from 1973 to 1975, before going to Omaha with her physician-husband.‡ Asked what accounted for Kutak's enlightened attitude, she attributed it in part to senior partner Bob Kutak, "a 'work is joy' person who understands that there are also other important things in life," and to the fact that "the critical event in that firm is getting hired." "The assumption is made," she said, "that if you are good enough to get in the firm you are good enough to move up, and most people do in about five years." Miller-Lerman, who specializes in litigation, is not exempt from the pressures of trial work, and abandons her three-day-a-week schedule when a case goes to court. Another young mother in the firm is accommodated with an 11:00 A.M. to 5:00 P.M. work day. Kutak has shown how the system can bend without detriment to the person or the firm.

---

*Kutak, Rock, and Huie is also the only one of the top fifty firms in the country to have more than two black partners. "Data on Law Firms Raise Racial Issues," *New York Times*, July 29, 1981, p. A20.

†In a telephone conversation, May 1980.

‡This was not quite the traditional marriage where the wife follows the husband, subordinating her career to his. It was, in a sense, his turn; he had taken off time from work from 1973 to 1975, when their first child was born, so that she could devote herself to her clerkship.

Women's Networks

Women also have a certain amount of support from other women in their firms. In the early 1970s, women who went into the large firms often made it a point to get together and discuss their mutual problems. They had a "sort of group unity," one of their male colleagues observed. This lasted in some firms and petered out in others. Women did make friendships with each other as have men in the firms, and the friendships were made both by class and across the hierarchy. A woman associate in a large firm considered leaving for a job with more regular hours and found that one of the women partners in the firm had been "looking out for her" and urged her to reconsider. The older partner was not her "mentor" but someone accessible to have lunch with and talk about the informal side of firm life.

The next few years will regularize women's presence on Wall Street. How women will act and be regarded by their male colleagues we have yet to see. But several things are certain. In most firms, whether women "make it" or not will depend on whether they can adapt to the existing system. Although the legal profession is moving toward more flexibility in recruitment, change in the demanding style of work is slow. Furthermore, a discrepancy remains between women's personal styles and those expected of the successful Wall Street lawyer. "Women were not convincingly pompous," reported an assessor of change at Sullivan and Cromwell. "In general, it always seemed to me that as a class the women were much brighter than the men, that they could in fact grasp things much more quickly," noted another informant, "but the rest of the trappings don't come easily at all."[53]

12

Women Law Professors

IF in former decades women constituted a small proportion of law students, their percentage among law faculties was miniscule. Law professors are an honored elite in their profession and among the highest paid in the university. While they do not have incomes as large as partners in major firms, many continue to practice in their free time. For a lawyer of any ideological persuasion, becoming a professor can open the way to an honored, prestigious career that does not demand any compromise of principles and that offers the flexible schedules and other amenities of the university. Jobs as law professors have been plums, kept from the reach of most women. But women on law faculties

TABLE 12.1

Full-Time Tenure-Track Law Faculty Who Are Women, 1967–1979

Year	Percent of Tenure Track Law School Teachers Who Were Women	Number of Tenure Track Law School Teachers Who Were Women
1967	1.7	39
1968	1.9	48
1969*	. . .	. . .
1970	2.2	66
1971	2.9	91
1972	3.7	129
1973	4.7	176
1974	5.9	243
1975	6.9	296
1976	7.5	337
1977	8.6	391
1978	9.5	444
1979	10.5	516

SOURCE: These figures were taken from Association of American Law Schools, *Directory of Law Teachers* (various years). Cited in Donna Fossum, "Women Law Professors" *American Bar Foundation Research Journal* no. 4 (Fall 1980):906.
*No data were available for 1969.

have increased from 2.2 percent in 1970 to 10 percent in 1979, a dramatic change (see table 12.1).

Women were qualified for law school posts on the basis of some criteria, but discrimination prevented them from meeting others. A brilliant academic record at a top law school, service as an editor on a law review, and clerking for a judge of outstanding reputation, preferably a justice on the Supreme Court or another important federal court, is the typical route to becoming a law professor at a prestigious law school. Many women made good grades and served on law review, but the final credential was almost impossible for women to acquire; it wasn't until the 1970s that any number of qualified women could hope to become clerks for Supreme Court justices.[1]

There are no accurate statistics on women faculty in law schools prior to the late 1960s, but it is widely recognized that even among the few who had appointments, fewer still had visibility because they often were hidden away from the classroom and were not on tenure tracks to professional status. In 1967, women constituted only 1.7 percent of tenure-track teachers in law schools.[2] Data from the 1970–72 edition of the *Law Teachers Directory* compiled by Professor Ruth Bader Ginsburg for a text on sex discrimination[3] showed that there were 200 women associated with law faculties (which included non-tenure track appointments) in 1971–72, or approximately 4 percent of law faculties. But, as she pointed out, these figures concealed the fact that most jobs held by women were bottom-rung positions assigned faculty status. In the entire United States, only 33 full law professors, including one dean, were women. Of these, one was a director of research and nine served as librarians. That is, only 22 were "full professors" in every sense of the term. In the associate professor category 37 women were listed: of these, 12 were librarians and one was a clinical associate professor. Furthermore, 27 women listed as faculty members were registrars, placement and admissions officers, and librarians without faculty rank. There were 48 assistant professors.

A number of studies have shown that a library job was a common assignment for women on law school faculties over the years.[4] But the 1970s were to see a partial remedy of this pattern, although women face a persistent pattern of prejudice in academic careers.

Formal affirmative action programs regarding women first appeared in the law schools after 1967, when President Johnson issued Executive Order No. 11,375 that prohibited sex discrimination in employment and mandated all recipients of federal contracts to take affirmative steps to hire and promote women. In 1972, feminist advocates in the American

Bar Association won support for a resolution urging law schools to actively recruit and admit female students and to "make substantial efforts to recruit, hire, and promote women professors."[5] The resolution also urged law schools to deny placement help to employers who discriminated against female attorneys in hiring and promotion. These actions, as well as the efforts of women law students, accounted for the fact that between 1967 and 1975 women law teachers increased in number at an even faster rate than the number of women lawyers as a whole. According to ABA statistics, in 1975 7.9 percent of all faculty members were women, as compared to men. This was a considerable increase over 1967.* After 1975, the increase in the number of tenure-track women was much slower, reaching 10 percent between 1975 and 1978.[6]

But the gains made in the aggregate did not accurately reflect the overall picture of women in the law faculties. Sociologist and lawyer D. Kelly Weisberg analyzed the 1972 percentages, showing that the aggregated figures stemmed from increases at only a small segment of the nation's law schools.[7] Increases in the number of women teaching at New York University and a small group of liberal and experimental law schools had pushed up the overall percentage of women faculty. In particular, faculties in new schools such as Northeastern and Rutgers—both the Camden and the Newark campuses—were composed of 15 percent or more women; at Antioch, 26 percent; at the University of California at Davis, 25 percent; and at North Carolina Central University, 44 percent. None of these are considered prestigious national law schools. Weisberg noticed that the proportion of women law faculty in 1975/76 in twelve top law schools,† including the most prestigious, was below average—only 4.8 percent.

Furthermore, percentages can portray a picture that numbers contradict.‡ The national averages also conceal the fact that a number of

*While the percentage of full-time female faculty increased, part-time faculty actually decreased, from 7.6 percent to 6.1 percent between 1974/75 and 1975/76. D. Kelly Weisberg, "Women in Law School Teaching: Problems and Progress," *Journal of Legal Education* 30 (1979): 227.

In 1977, the average percentage of full-time women faculty for all law schools had risen to 9.8 percent and part-time faculty to 8.1 percent. "Number of Degrees Conferred in 1977–78," in American Bar Association Section of Legal Education and Admissions to the Bar, *A Review of Legal Education in the United States* (Fall 1978): 62.

†Boston University, Columbia, Georgetown, Harvard, Hastings, Stanford, University of California at Berkeley, University of Florida, University of Michigan, University of Texas, University of Virginia, and Yale.

‡For example, in the fall of 1978, the percentage of women lawyers at Antioch Law School was 13 percent, four out of twenty-three members of the teaching staff. At the University of California at Davis, 16 percent represented five out of thirty. Women did not constitute a sizeable proportion of any large teaching staff except at New York

law schools do not employ any women and, despite an increase in women faculty by 1979/80, half of all law schools employed two or fewer women.*

Yet even where numbers or percentages were small, they represented an increase over previous years (often an increase over zero). Among the top ten law schools in the country, only the University of Chicago had no women faculty in 1976, but in 1979 they remedied this —and others considerably increased their percentage of women law faculty, as table 12.2 shows. Many women joined law schools as the first women member of the regular law faculty. Some were proven lawyers and scholars already; some started at the rank of assistant professor. Although the increase was not dramatic at these schools, it did reveal a shift in trends. In 1974/75, about 25 percent of law schools (forty schools) employed no women and 28 percent (forty-four schools) employed one woman. In 1979/80, those percentages had been significantly cut to 3.6 and 17 percent, respectively.†

The academic credentials of male and female law school faculty members do not differ significantly. As I pointed out before, the important criteria for professorship are a degree from a top law school and academic honors.[8] Less than 15 percent (20) of all 160 law schools had graduated approximately 60 percent of all law teachers in 1975/76, as a study by Donna Fossum shows. Those who did not earn degrees from

University, where eleven out of sixty teachers, or 17 percent of the full-time faculty, were women. The loss or gain of a single woman at any of the schools could cause a radical increase or decrease in the percentage reported for that school. Thus, Antioch's percentage dropped from 26.3 percent to 15 percent when one woman left and its faculty grew from nineteen to twenty-three. Weisberg, "Women in Law School Teaching," p. 228.

*In 1975/76, of the 164 ABA accredited law schools, 18 percent employed no women on their full-time faculties, and 54 percent employed none on their part-time faculties. Moreover, the majority of accredited law schools that were reported to employ women employed only one. Thus, while in 1975/76 women comprised 7.9 percent of all law school faculties, the percentage of the accredited law schools employing no women or at the most one as full-time faculty members was 49 percent, and as part-time members 83 percent. Ibid., p. 230–33.

It is particularly discouraging that the fifteen law schools that hired a large percentage of women in 1975/76 did not continue to do so the following year. Over one third of these schools showed a sharp decrease in their employment of women, which suggests that many schools are not replacing women faculty with other women. Ibid., p. 233.

†See Donna Fossum, "Women Law Professors," *American Bar Foundation Research Journal* 1980, no. 4 (Fall 1980) Table 1.10. The AALS and *Review of Legal Education* (RLE) data for 1978/80 differ. The AALS reported that 15 percent of full-time faculty were women (820), while the RLE reported about 11 percent (429) women as full-time faculty members in 1979/80. According to the AALS study, the percentage of women faculty members has actually decreased from the peak. Elizabeth A. Ashburn and Elena N. Cohen, "The Integration of Women Into Law Faculties," American Bar Association Division of Public Service Activities, Section of Individual Rights and Responsibilities (November, 1980).

TABLE 12.2

Percentage of Women Who Are Full-Time Tenure-Track Faculty in the "Top Ten" Law Schools, 1974–79

	Percent in 1974/75	Percent in 1975/76	Percent in 1976/77	Percent in 1977/78	Percent in 1978/79	Percent in 1979/80	Increase in percent 1974–79
National Average	5.9	6.9	7.5	8.6	9.5	10.5	4.6
Law School:							
Berkely	5.2	3.7	6.3	5.7	4.2	7.8	2.6
Chicago	0	0	0	0	0	3.0	3.0
Columbia	4.2	4.0	3.7	3.9	5.5	5.8	1.6
Harvard	4.2	4.1	2.8	4.4	5.7	3.8	-.4
Michigan	0	1.9	3.6	3.9	4.1	4.0	4.0
NYU	6.8	11.1	9.2	11.4	17.6	15.4	8.6
Pennsylvania	5.1	2.3	2.7	4.9	7.3	4.9	-.2
Stanford	2.4	2.3	4.9	2.4	7.3	7.3	4.9
Virginia	3.3	3.5	1.8	1.8	1.8	3.3	0
Yale	3.9	4.4	4.4	5.9	2.6	7.1	3.2

NOTE: Based on data in the *Review of Legal Education* for the years 1974 through 1979/80, as compiled by D. Kelly Weisberg. None of the schools (with the exception of NYU and Georgetown) reported more than four women in 1974/75 or more than five women in 1979/80. "Top Ten" refers to those used in D. Kelly Weisberg, "Women in Law School Teaching: Problems and Progress," *Journal of Legal Education* 30 (1979). Cited in Elizabeth A. Ashburn and Elena N. Cohen, "Women's Integration into Law Teaching," American Bar Association, Section of Individual Rights and Responsibilities, August 1980.

these "producer schools" taught at less prestigious law schools through-out their careers.[9]

Women are sharply under-represented on the faculties of the pro-ducer law schools. In 1975/76, 2.5 percent of all women faculty (8 out of 319) and 5.9 percent of the men (193 out of 3,265) taught at one of the top five producer law schools, Harvard, Yale, Columbia, Michigan, and Chicago. In 1979/80, 2.8 percent of the women (14 out of 498) and 4.9 percent of the men (180 out of 2,648) taught at the top five schools.*

The women who succeed in gaining appointments to the nation's prestigious law schools are no less qualified than the men they teach with. When Fossum compared men and women, no major differences were found. Women were as likely as men to have received a law degree from one of the twenty producer schools and to have graduated with academic honors.[10] Slightly more of the men than the women were members of law review.†

The major difference between male and female faculty members is that men generally get tenure more quickly than women and thus occupy higher academic ranks. A woman's decision to apply for a job at a particular law school may be influenced by its anti-nepotism rules. Restrictions on geographic mobility because of a husband's job also make it difficult for women to exercise free choice. Fossum found that more women than men are hired by the school they graduated from, and this may be the source of some career restrictions.‡ Although women have academic credentials, prejudice and the restrictions that come from family obligations have meant that the routes women fol-

---

*In 1975/76, approximately 16 percent of the women (52 out of 319) and 20 percent of the men (669 out of 3,265) taught at the remaining producer schools. In 1979/80, 13.5 percent of the women (67 out of 498) and 18 percent of the men (658 out of 3,648) taught at these same schools. Similarly, RLE statistics for 1975/76 show that only 7.2 percent of women, compared to 11.5 percent of men teach at the top ten law schools. Ashburn and Cohen, "The Integration of Women Into Law Faculties," pp. 46–47. The remaining producer schools are (in order of the number of graduates in law teaching): N.Y.U., Georgetown, Texas, Virginia, Berkeley, Pennsylvania, Wisconsin, Northwestern, Stanford, Iowa, Illinois, Minnesota, Cornell, Duke, and George Washington. Donna Fossum, "Law Professors: A Profile of the Teaching Branch of the Legal Profession," Table 2, p. 507.

†AALS data on law teachers for 1979/80 showed that 47 percent of the men and 40 percent of the women were on law review; 17 percent of the men and 9 percent of the women were members of the order of the Coif (an honorary legal society. A class rank in the top ten percent of the graduating senior class is a requirement for eligibility). Weisberg, "Women in Law School Teaching."

‡Fossum found that academically inbred professors tended to spend a significantly longer time in the rank of assistant professor than non-inbred law professors. Inbred women associate professors had spent two years longer at that rank before becoming full professors than had inbred men in 1975/76. Fossum, "Women Law Professors," pp. 910–11.

lowed into law faculties and up the academic ladder were different from the typical male routes.

At the University of Chicago Law School in the mid-1950s, it seemed extraordinary that two outstanding faculty members were married to each other. It was generally known that Karl Llewellyn, one of the liveliest legal minds of the time and an architect of the unified commercial code, would never have left Columbia if his wife, Soia Mentschikoff, also a brilliant scholar, had been able to secure a post there. Chicago, long noted as an innovative school, hired the two. This move was unique not only because Chicago had selected a woman (she was the first they hired) but because it hired a couple. Other schools had anti-nepotism rules, usually informal, which prohibited couples from teaching in the same discipline or in the same school. A survey of 124 AALS accredited law schools indicated that 34 percent had such policies.[11] The time was not right for Chicago to set an example, and no breakthrough occurred even there. Soia Mentschikoff taught for eleven years as a "professional lecturer" before being named professor of law in 1962.[12] But by 1974 there were again no women on the Chicago faculty, for Professor Mentschikoff left that year to become dean of the Law School of the University of Miami (the second woman to head a law school)* after the death of Karl Llewellyn.†

Ironically, one "victim" of anti-nepotism rulings also became one of the few women deans of a law school in the country. In 1969 Judith Younger (former dean of the Syracuse University Law School) was denied tenure at New York University Law School where her husband was on the faculty. One of her colleagues told her confidentially the reason was that "there are just too many Youngers for this faculty." Another said, "What it is, is your sex. We have too many women already." "Two percent of the full-time faculty is female," noted Younger at the time.[13]

*To this day there have been only five women law school deans. They are: Jean Camper Cahn of Antioch, Judith Grant McKelvey of Golden Gate, Judith T. Younger of Syracuse (now professor at Cornell Law School), Soia Mentschikoff of the University of Miami, and Dorothy Nelson of the University of Southern California. Weisberg, "Women in Law School Teaching," p. 236.

†However, I did not get the impression that Mentschikoff served as a role model for the few women students at the school in the 1950s when I was a student there. Imposing in manner and stature, she seemed to them somehow larger than life. It might have been different for those who actually took her classes, but since her work was specialized, women could go through Chicago without knowing her firsthand. With no women's organizations to recruit women faculty members, contact hinged on happenstance and personal style. Menschikoff was not likely to initiate supportive relationships; indeed she was often quoted as saying women were responsible for their poor representation in law. There were no women faculty members at the University of Miami in 1978 according to the survey reported in "Law Schools on the Approved ABA 1978," *Review of Legal Education in the U.S.*, Fall 1978.

The use of anti-nepotism rulings as an excuse to limit women's participation on faculties is decreasing. Today twenty-two lawyer couples serve as faculty members in the same schools.[14] Indeed, for the first time, men are following their wives onto law school faculties.[15]

To return to the past, it was unlikely that a student attending law school in New York in the 1950s and early 1960s would ever have seen a woman professor. In my first study of women lawyers, each of two women faculty members, interviewed several months apart, informed me that she was the "only" woman law professor in New York State. Each of them had taken a route into law school teaching that differed from the typical male status sequence.*

Had she listened to her professors at Columbia University, Edith Fisch, an assistant professor at a small, local New York Law School, would never have become a faculty member. Late in her training (she was then in the law school doctoral program, having earned an LL.B. in 1948, an LL.M. in 1949, and a JSD in 1950 from the school), they told her that they would not have encouraged her had they known of her ambition to teach. Fisch was proud to be the "only woman law professor in New York" at that time, having been appointed at the age of thirty-nine in 1962, after having practiced law and published a number of textbooks. She held her post for three years.

Professor Fannie J. Klein represented—as a cohort of one—the "woman's route" into a major law school. The first woman appointed to the faculty at New York University Law School, she was named to professional rank in 1966 at the age of sixty. For ten years she had been teaching an introductory seminar, having come to the law school as the assistant director and librarian of the law school's Institute of Judicial Administration, started by the New Jersey Chief Justice Arthur Vanderbilt, her professor at New York University Law School. Klein lauded Vanderbilt for being egalitarian, but she said of her selection, "He needed help, but so many boys were in the Army he had to turn to a woman. I was ready, so it was a wonderful opportunity."[16] Klein and many other women of her generation felt they were privileged to have done as well as they had and did not measure themselves by the standard of success of their male peers. Asked if she was paid the same as other professors, Klein said she probably got the minimum but she did not care.

Another woman faculty member (with an LL.B. and a JD) at a small

*Robert K. Merton has defined the "status sequence" as "the succession of statuses occurring with sufficient frequency to be socially patterned." *Social Theory and Social Structure* (Glencoe, Ill.: The Free Press, rev. ed., 1957), p. 370.

New York law school in the 1960s started in a glorified secretarial job with "a Professor [K] that we had."

I worked with him in the Practice Court and helped him on books he was writing. He wrote [K] on corporations; [K] on insurance, and numerous case books. I was with him until he became ill and died. . . . Then I became secretary to the faculty, for the Practice Court. I made appointments for the faculty, and any time a professor wanted research done, or help on books, I did it. . . . Then I did free-lance work (while my child was small) and in 1950 they asked me to become librarian.

Women lawyers gravitated toward law librarianship, in part because discrimination is less pronounced in this area and in part because women have always been found in "backroom" areas of the profession.[17] For the past thirty years, women constituted 25 to 30 percent of the head librarians in American law schools.[18] In 1975/76, 30 percent of law librarians were women, and more than half of them (twenty-five) were also faculty members; 14 percent of the women full professors were also law librarians, and 10 percent of associate professors.[19]

Until very recently* the peripheral route to professorship was also followed by the only female full professor of law at Harvard University, Elisabeth A. Owens. She also worked in a special institute, heading its research program in international taxation. Owens was named to a tenure-track faculty post in 1962 after having been at Harvard Law School since 1955. Asked in a 1977 interview why she thought she had been appointed, she explained, "the passage of the Equal Rights Act in Washington."

When the mood of the country made law faculties uncomfortable about their lack of women, and especially once affirmative action guidelines were established and schools began to feel nervous about possible sex discrimination suits, many of them began searching for promotable women. "Columbia was facing a little problem they had with H.E.W. at that time," commented Professor Ruth Bader Ginsburg, looking back to the time of her appointment.† Often it was not difficult to find suitable women, because many of them, like Elisabeth Owens, had been doing important work in institutes or in research appointments without the prestige or security of tenured positions. A woman appointed in this way at an elite school commented:

*In 1980 there were two female full professors.
†The hiring agreement between Columbia and H.E.W. concluded: "The School of Law foresees the appointment of at least two women and one minority member to its full-time faculty over the next five years." Hiring Agreement, p. III-E-8, quoted by Bill Bonvillian in "Law School Escapes Lax HEW Guidelines," *Law School News* 7, no. 3 (November 10, 1972).

I was around a long time, and they knew I was harmless. You know, harmless in the sense of no threat to them. That is, they knew I wasn't going to be the wrong kind of person. They don't want anyone rocking the boat. In the present search, they feel very hesitant about the women. They don't know the young men they are considering, but somehow they find it more difficult to be able to choose a woman they predict will be good enough.

Professor Ginsburg went to one elite law school, Harvard, and graduated from another, Columbia. She tied for first place in her graduating class, made law review, clerked for a federal court judge, Edmund L. Palmieri, but unlike the men in her class, was unable to find a job with a law firm.[20] She had taught a few courses at Columbia between 1961 and 1963, and Columbia had helped her get a teaching job at Rutgers Law School. So when Columbia began to look for a woman of stature they knew exactly where to find one. Ginsburg was appointed as its first woman professor in 1972.

The same year, Columbia appointed its first woman dean, Harriet Rabb (who later made history in litigating a series of sex discrimination cases in the clinical program at Columbia). Rabb, also a Columbia Law School graduate, had been lecturing at the school and had also been practicing law for five years.*

In the past, non-elite institutions were no better than elite institutions in their hiring practices. Even when they had the opportunity to appoint extraordinarily qualified women who wanted to teach but who could not get jobs in elite institutions, they were resistant to do so. In recent years, however, non-elite schools wishing to improve in stature and grow in numbers have been hospitable to women. New York Law School, for example, had hired six women faculty members by the mid-1970s and thus was able to get teachers with outstanding credentials.[21] The non-elite schools have also hired women a bit more broadly. For example, an assistant professor at one of these schools had gone to Vassar College and then to Brooklyn Law School, where she became editor-in-chief of the law review. She noted in an interview that the major law schools wouldn't look at a candidate who had not come from an institution like their own. Of course, this was true for men as well as women.

The career pattern of women professors today seems to be coming closer to the pattern typical for men. For example, Barbara Underwood of the Yale Law School went to school at the Georgetown University

*Nineteen seventy-two seems to have been the year for Columbia to remedy its selection. Kellis Parker became the first black to hold a full-time faculty seat at the law school with his appointment as associate professor. *The Columbia Law Observer* 1, no. 6 (March 1, 1972): 1, 5.

Women Law Professors

Law Center, worked on its law review, and became law clerk first to Chief Judge David L. Bazelon of the U. S. Court of Appeals for the District of Columbia and then to Associate Justice Thurgood Marshall of the Supreme Court. In her position with Bazelon, however, she served for a longer period of time than male clerks normally do.

Equality of opportunity or recognition does not seem to have been achieved on law school campuses. Most law schools, national and local, have women on their faculties, and a good many have "their" woman full professor and perhaps one or two women assistant professors, who are not expected to stay—a situation similar to that in other divisions of the university (although, in the past, it was more common for law faculties to grant tenure to junior faculty after an appropriate length of time in comparison with other departments in the university). Unlike other divisions of the university, however, the law school cannot be secure about keeping women stars since all of them possess skills that are transferable. Both Ellen Peters of Yale (for many years the only woman full professor on the Yale Law School faculty) and Ruth Bader Ginsburg of Columbia have left to assume prestigious judgeships, and although Barbara Babcock of Stanford returned to teaching, she was for several years assistant attorney general of the United States. Another tenured Columbia faculty member, Vivian Berger, resigned her post in 1980 to work on a major rackets case and do appeals work in the office of the New York district attorney.*

Most male professors do not come to law school faculties as established stars. Although they usually have outstanding student records, they typically start teaching after spending only an interim period after law school as a clerk and perhaps a few years in practice. Thus, faculty hiring committees are quick to identify "promise" in the young male candidate. Both men and women law school professors at Boalt Hall Law School, for example, concerned with the way woman candidates were assessed, indicated that women were subjected to a different set of standards both at the entry level and for tenure decisions.†

*A third Columbia professor, Ellen Oran, was at the time deciding whether or not to take time off to practice law. The hiring committee at Columbia Law told the school's alumnae group that it was finding it difficult to replace the women who had left the school.
†A delegation from the Boalt Hall Law School (University of California at Berkeley) Women's Law School Association and a concerned male faculty member came to talk with Ruth Bader Ginsburg while she and I were fellows at the Center for Advanced Study in the Behavioral Sciences at Stanford in the spring of 1978. The professor advocated increasing the number of women and blacks on the faculty, and said he had faced great difficulty in pressing this goal. Women's applications were scrutinized to a much greater degree than men's, he claimed. For example, the written materials submitted by the women were examined scrupulously by the members of the hiring committee, while colleagues' letters of recommendation were considered sufficient for the men. Further-

It is interesting to note that the proportion of minorities on law faculties is higher among women than men, and this difference appears to be increasing. Fossum's study showed that in 1975/76 approximately 7 percent of the women on law faculties (18 out of 271) and 4 percent of the men (143 out of 3,579) were minorities.[22] The AALS data on law teachers for 1979/80 showed that out of 187 newly hired women, 15 (about 8 percent) were minorities; of the 453 newly hired men, 16 (about 3.5 percent) were minorities.*

What a woman candidate looks like, whether other faculty members think they will like having her around, what her reputation is in the legal community—these are real issues to faculty who think about a long-term collegial relationship. They make decisions not only about the candidates but about themselves; each is interested in maintaining the prestige of the department. The performance of new faculty members reflects on their taste, their ability to judge, and their own reputations. Sometimes this attitude can create a virtual standstill in hiring, and it makes decisions about hiring women especially difficult. At elite schools with more concern about "image," the selection process can be excruciating.†

D. Kelly Weisberg has identified this as the "myth of the perfect candidate."[23] There is also the search for the "perfect" candidate for tenure. A woman contemplating tenure at a prestigious place remarked:

I guess I have the feeling that everybody wants very much to give me tenure if only I would make it easy for them by writing a brilliant and world-shaking article. I guess that's better than having a situation in which they didn't want to give me tenure even if I were to write a brilliant and world-shaking article. But . . . I simply, at the point of decision, have a very good and promising track record. So I don't know what's going to happen.

The evaluation process in law school may be harder for women than in other kinds of legal work. While a woman can develop a reputation

more, he said, one senior faculty woman's recommendations were disregarded because it was believed she was "prejudiced and would recommend 'any woman.' " Others expressed the belief that this woman not only had high standards but knew it would be foolish for her to recommend women who were not qualified.

*The percentage of minority women is slightly smaller in the total of women faculty than in the newly hired group: 52 of the 707 (7.4 percent) full-time women faculty members were from minorities. Ashburn and Cohen, "The Integration of Women into Law Faculties," pp. 65–66.

†Though not always on matters of principle; a member of a hiring committee at one elite school is reported to have commented that it might have been reasonable to consider one woman candidate as a law professor but he wouldn't have wanted her as his wife.

for winning cases in practice, the more subjective determination of "brilliance" in scholarly work and teaching is prone to stereotyped evaluation.

There are different stereotypes linked to men's and women's appearances. To judge from the gossip one hears, it seems that discussion of women's appearance is common on all-male hiring committees. But does it matter? It is hard to tell. To cite an exemplary comment from a male law professor:

The type of woman they like at my school is not too aggressive personally, not too threatening in a personal way. I hate to say it but it helps if you flirt and you're pretty. . . . The number one comment here when male faculty members review the file is "What does she look like?"*

In the final analysis, identifying the sex of a person is the bottom line. Pretty or not, a woman is a woman, and preconceptions about competence are attached to possession of that status.

Some male faculty members are overly rigorous in their evaluations of women faculty even when the women's reputations are corroborated outside the institution. In a moment of candor, a male professor at Stanford told me he felt that the law school had bent over backward to give tenure to a woman professor, now an esteemed member of the law school community, Barbara Babcock. As he put his reservations:

I don't think it's really fair to say that we would have laughed at her for tenure, but it would have been a tremendous fight because she'd only written one or two things . . . they weren't bad, but certainly they were in no way significant. And it sailed through. Everybody talked about it, you know. A couple of people said how good the articles were and so knowing smiles went around and conversation was just absolutely perfect. There was no objection from the faculty. The record was made. I'm not sure we wouldn't have given tenure because tenure is a tricky thing.

About three weeks later, she was appointed assistant attorney general of the United States. It would have been a bit of a scandal if we had denied her tenure in the meantime.

---

*In my own discipline, sociology, faculties in high prestige institutions often resist appointing more than one woman full professor. Although hiring committees often complain that there are few women in the pool of eligibles or that the stars they would select are already employed by other institutions (they often have become stars by being appointed elsewhere), lists of desired candidates often include names of women who are clearly unavailable. These lists are said to be padded as a bureaucratic maneuver to indicate sensitivity to the "woman problem" and to meet affirmative action requirements. Elite institutions, eager to acquire or maintain prestige by bringing in the proven, fail in their status-conferring responsibilities as far as women are concerned. Thus they limit the pool by denying recognition to the deserving.

Another Stanford law professor talked about differences in evaluating appointments. "We did deny an appointment," he said, smiling as if to anticipate some objection:

You can make a big deal out of it . . . to Rose Bird, who is now chief justice of California. I remember my wife said to me, "So she's good enough to be chief justice of California and she's not good enough to be on your faculty." I said, "First of all, what makes you think she's good enough to be chief justice of California?"

A member of a hiring committee (also in the West) felt there were further problems in locating a pool of eligible women candidates. We had a conversation about it:

"It's hard to find qualified women. One person we had in mind went to UCLA because her husband wanted to practice there. That's another problem with dealing with women. They tend to be less mobile than men."

"Are there a lot who won't come?"

"Well, we don't make offers to some because it is known they're looking for a job in the New York area."

"What do you mean by 'known?' Do you ask them?"

"No. We usually don't even get that far. They have typically told somebody that they're looking for a job in the New York area."

"Well, do you act that way when you hire men? If they're looking for a job in New York, do you forget about them?"

"Well, no. I guess the answer is we almost never hear that men are looking for a job in the New York area. We mostly hear they are in the market. The theory is that they'll go where they get the best offer."

"Is the theory that women are motivated by other reasons?"

"The women often are. And there are other problems. There was a woman married to another law professor. And we didn't want him. So, there's no point in even asking."

Limitations on women's access to faculty positions continue to be rationalized by gatekeepers with such comments and observations. But in addition to the country's changing climate and to new legal tools such as H.E.W. pressures, law schools were also motivated to find women faculty by the efforts of their women students. Active associations of women law students existed in 70 percent of law schools in 1972/73.[24] Nowhere else in the university have women students been so directed and so effective in accomplishing their goals. Although faculty members deplore this technique, which they feel infringes on their prerogative to decide who shall be among their peers, it has been effective.

There is no question that having women on law school faculties makes a difference and that the difference is multifaceted. The informal referral system, crucial in placing students in jobs, is composed predominantly of male faculty. Male professors are more likely to recruit male students, with whom they have greater contact and of whom they expect more. In addition, larger numbers of women faculty members lighten the lone woman's burden of being seen as the representative of *all* women. Token women appear to be more burdened by assignments and student demands than their male colleagues.[25] Looked upon as the advocate for women students, the only woman member on many committees, and the women students' "friend" generally, the lone woman faculty member is expected to serve in many capacities no matter what her interests or personality.

The presence of women on a faculty also insures that women's interests are protected. When women sit on committees fewer sexist comments are made and women candidates' positive attributes are probably focused upon. At Harvard and elsewhere, some new faculty members were young women active in women's groups. These women helped mobilize some older women faculty members into participation and inspired them to become moving forces themselves. But perhaps most important, the presence of more women faculty members alleviated the students' sense that the one woman in the school was unique in some manner.

When women entered law faculties in the 1950s and 1960s, their courses were clustered in family law, trusts and estates, and legal research and writing. These areas corresponded to women's specialties in law practice or their initial experiences as librarians in law schools. Women professors rarely taught the courses with the reputation of being the most intellectually challenging and prestigious (and which correspond to the most prestigious and lucrative specialties of the profession), such as business, securities, and antitrust.[26] In the late 1960s and 1970s, women faculty also began to teach courses that had not previously existed in law schools. Although all of them had expertise in some "classic" area of law, women faculty (especially the "firsts") taught courses on women and the law and on sex discrimination.

Ruth Bader Ginsburg, for example, gave the first course in sex discrimination law at Columbia and co-authored (with Herma Hill Kay and Kenneth M. Davidson) the first casebook in the field.[27] Many other schools instituted such courses in their curricula at the urging of women students. At some schools, women lawyers were brought in on an adjunct basis to teach these courses. At others, women faculty, even if they had not had experience in the field, were asked to teach them. Male faculty members were not asked, and women students probably would have been offended if they had been. On the other hand, it was found that women and the law courses were regarded as less valuable for career building than other courses, such as contracts or tax law.

Gradually, sex discrimination law became a subject area men as well as women needed to know about as the cases involving it proliferated. Professor Elisabeth Owens of Harvard, an international taxation expert, spoke of being asked to teach the women and the law course after Ruth Bader Ginsburg left her temporary position at Harvard:

Ms. Ginsburg had taught it here and then left to go to Columbia. The Women's Law Association was very active at the time and they wanted somebody to teach it, so I taught it. It's really ridiculous. It wasn't seen as a respectable subject.

Asked whether it had become more "respectable," Owens responded, "Anything you can earn money with is respectable in the legal profession."

Whatever their standing in the prestige system, women and the law courses were very important to the women students in the law schools. The courses spelled out injustices in the laws and also made available knowledge about legal redress. Many women students and professors involved in these courses became active in clinical programs that, with other legal agencies, instituted class action suits on employment issues and other matters of special concern to women. The courses also brought together many of the women who had come to law school in the late 1960s and early 1970s because of their concern with social justice. Whereas in most other classes the still relatively small proportion of women would be scattered, in these they were usually a majority and had a greater sense of belonging, purpose, and legitimation.

Although it was important to have these courses in the early days of building awareness and development of the case law on women's rights issues, it was clear that most of the materials ought to be part of the subject matter of courses already in the regular curriculum—that issues of rape should be taught in criminal law courses and due process and equal protection toward women in constitutional law. The goal of integrating the subject matter has had limited success because of the same problems that limit integration of "women's issues" in many other disciplines. Professors have to educate themselves about the new material and often think it is less important than the material they have emphasized in the past. The second-class status of women in society is reflected in the second-class status of studies focused on them.[28] The ranking of studies is also related, as Professor Owens pointed out, to their pertinence in dealing with money and power. As men have seen practical advantages to becoming informed about employment discrimination issues, so these issues have gained in prestige. Family law (like family sociology) still suffers relative prestige deprivation.

Women's organizations feel strongly that women and the law courses should continue, and the schools show a symbolic commitment to equality by offering them.[29] It remains typical that they are taught by women, and women are often specifically hired as adjunct faculty to teach them. On the one hand, this provides employment and visibility on law faculties for women and provides women students with role models and advocates for women's interests in the schools; on the other, the adjunct route is not a good one for full-time appointment to faculties. Moreover, because the courses they are brought in to teach are not considered part of the core curriculum, women may not get much professional credit useful in climbing the academic ladder.

But women faculty members are now teaching a wide array of courses and most do not fall within sex-typed categories. This does not mean that their work does not have an impact on the place of women in the law schools or in society. For example, Ellen Oran, who teaches contracts and negotiable instruments at Columbia Law School, stated in an interview with the *Columbia Law Alumni Observer* (October 20, 1978), that one of her goals was to establish a forum for women to discuss their experiences in the law school and the profession. Women faculty of all ages have shown their interest in and willingness to help women students in the past decade. Most are interested in the problems of women, and although they may seem warmer or cooler in personal interaction, they indicate a community of spirit with other women.

Do they treat the women students especially well? Most didn't think so. But they are sensitive to the problems of women students. For example, Barbara Underwood of Yale said:

I consciously feel responsibility . . . in that I find that women tend to underestimate themselves more regularly, so that when I think a woman is good I go out of my way to tell her so and urge her to apply for a clerkship, submit her paper to a journal, that sort of thing—things that men seem much more likely to do for themselves. So I guess I try to be a patron for particularly good women students. . . .

I've been in teaching long enough so that I can begin to see some fruits. I urged a woman who was quite outstanding to do some writing. I urged her to submit it to the law journal. She did and won a prize. So I urged her to apply for clerkships, and she said, "Who, me?" But she did and got a clerkship. Now she just applied for and got a clerkship with Justice Thurgood Marshall. She called me to tell me she was particularly glad because she had always thought I was her role model and the one who encouraged her so it was very fitting and proper that she had "my" clerkship.

Women faculty have also been helpful to each other. Older and more established women have been advisors for younger ones. I have heard many accounts of the helpfulness and advice given each other by women faculty. It is only recently that their numbers have become too large for all to know each other personally or even know of each other. Barbara Underwood commented:

It used to be the case that when someone said to me, "Do you know X?"—a woman in law teaching—even if I didn't know her, I at least knew someone who did. I have been surprised lately that there seem to be some women in law teaching whom I don't know and have never heard of.

There is no doubt that as the pool of women continues to grow there will be more pressure on faculties to employ them and fewer rationalizations available as to why they should not. However, the problems women face generally in combining careers with home lives, the problem of time demands for publishing and scholarly work, and the continued stereotyping with regard to their capacities, still remain issues to be solved. Furthermore, personnel committees still maintain that the pool of promising candidates remains small (for a variety of "new" reasons including competition from the large firms for the best talent), but they are being pressured by alumnae groups and women students to keep up an assertive search for women law professors.

Thus, there are forces moving law faculties to include more women while there are those that are undermining any substantial increase. The most judicious prediction of how well women will do in becoming a larger proportion of faculties in the near future is that they will maintain their numbers but there will be no upsurge, which is the usual effect when conflicting forces create ambivalence in groups.

Benchmarks: Women in the Judiciary

IF Perry Mason and Clarence Darrow inspired generations of lawyers to dream about becoming brilliant courtroom attorneys, great judges such as Oliver Wendell Holmes, Louis Brandeis, and Felix Frankfurter inspired ambition of another sort—to ascend to the bench and dispense the even hand of justice. Although the Supreme Court represents the pinnacle of such achievement and, indeed, ranks first in occupational prestige in the United States,* judges in all courts—from lower magistrates' courts up through state and federal courts—are highly regarded.

In the mid-1960s, women lawyers who were interviewed usually dismissed the idea of aspiring to the judiciary. One lawyer, who was an assistant in a judge's library which serviced the civil and criminal courts and an active member of her political club, said when I asked her about her goals:

I have no particular goals. My mother was friends with the first woman magistrate, Jeanette Brill, but she [Brill] died a broken-hearted woman. I never thought about being a judge as a result.

Another lawyer, whose son suffered from chronic illness, pointed out that this prevented her from running for a judicial post as her political supporters had urged. That both these women are judges today indi-

*Although lawyers and judges ranked below doctors, college professors, bankers, dentists, architects, and chemists (in that descending order), Supreme Court Justice ranked first according to the famous study of the National Opinion Research Center, "Jobs and Occupations: A Popular Evaluation," *Opinion News* IX (September 1, 1947): 3–13. Reprinted in *Class, Status and Power,* ed. Reinhard Bendix and Seymour M. Lipset (New York: The Free Press of Glencoe, 1953) pp. 411–426.

cates a change in the times as much as a change in each woman's life situation.

Today a judgeship is considered a reasonable aspiration for a woman, as one lawyer said when asked about her goals in the next ten years:

Some people want to be judges. I happen to be one of them.

How do you get to be a judge? Would one get to be a judge by being very active in the bar association and things like that?

People seem to get to be judges through a variety of ways. That certainly is one of them. Some through politics. Some through their being well known through the large firms. A lot of the judges come out of having substantial practices in the large firms and becoming known through the solicitor's office. But they are looking for blacks now, and I guess they are coming around to looking for women, and I'll be ready when they do.

Another woman attorney even thought it was easier for a woman to attain a judgeship than a partnership in a large firm because "there's a large discrepancy today between what you can make as a judge and what you will make as a partner in a large firm, and better qualified women than men are willing to do that."

In the past, women's chances to become judges of city, state, or federal courts were slim; no woman, of course, had ever served on the U. S. Supreme Court. Some judges are appointed and some elected, but in the United States both procedures are tied to party politics. For top federal positions offering lifetime tenure, senators nominate candidates that the president can then appoint. At the state level, leading judges are appointed by governors. Judges are elected to city courts, and candidates actively campaign for these positions. Thus, in entering the judiciary, women face the same problems they have in achieving other political positions: political parties do not support their candidacies, funds are not available for their campaigns, and they lack network connections in government and business to use in the wheeling and dealing of politics.[1] "Patronage as usual" is the habit of many senators in selecting district court judges, the panel on judicial selection of the Tenth National Conference on Women and the Law found.[2] The system of patronage affects the chances of men as well, but women make up such a small percentage of experienced lawyers that they could offer only a small pool of eligible candidates. Furthermore, prejudice against women in the legal profession keeps all but a few of the most outstanding from being nominated for judicial positions. Even when women were nominated in the past, it usually was without the endorsement of local and national bar associations.

Benchmarks: Women in the Judiciary

A few women did manage to slip through the filters that excluded women and minorities from the judiciary. Of those, most were located in domestic relations courts or lower municipal courts. They often were outstanding women, many from prominent professional families. The choice of Dorothy Kenyon as a New York City municipal court judge by Fiorello LaGuardia is an example of the appointment of an outspoken liberal attorney—a fighter for civil rights and women's rights—by an iconoclastic mayor. "He hounded me to put me on the bench," Kenyon told me in an interview in 1966. "And I told him I couldn't bear to be a magistrate—it's just the dregs. . . . And I didn't want to be on the children's court, or the family court . . . because I said that's a social worker's court, and I'm a lawyer." She added sarcastically, "And I'm not sure really that I care much about women and children." Kenyon pointed out that LaGuardia "didn't have much in his power beyond that," but she did decide to fill a vacancy on the municipal court in 1939. Kenyon ran for election when her term expired, but, as she recalled it, "the big Tammany machine rose up and knocked me down."

Dorothy Kenyon was the daughter of a founder of a major Wall Street firm and the niece of a U.S. senator. Her brothers practiced law in the family firm, but she practiced in the name of "the cause" rather than business law. Another lawyer, Justice Wise Polier, the daughter of Rabbi Stephen Wise of the prestigious Free Synagogue of New York, became a family court lawyer in New York City and then a judge on the New York State Family Court after making her mark as an advocate on behalf of children in a program later integrated into an ACLU project. Of course, family connections were no guarantee that even a gifted woman lawyer would make it to the bench. Nannette Dembitz, a lawyer well known in New York City circles for decades because of her work for the ACLU, and a niece of Supreme Court Justine Louis Dembitz Brandeis, failed to win American Bar Association endorsement for a judgeship on the New York State Court of Appeals, although she did achieve a post on the Family Court of New York.

The first few women to serve as judges were appointed to minor judicial positions nearly a century ago. Marilla Ricker was appointed the first woman U. S. commissioner in 1884 in Washington, D.C., and Carrie Kilgore, the first women graduate of the University of Pennsylvania Law School, was appointed master in chancery in Philadelphia in 1886.[3] But the following decades saw a mere sprinkling of women judges appointed in the various states. Only a few were recognized with high court posts and their names stand out.

Florence Ellinwood Allen, whose name graces scholarships and law prizes today, was elected judge of the Court of Common Pleas of Cuyahoga County, Ohio, in 1920, then was named to the Ohio Supreme Court in 1922, where she was the only woman in the country sitting on a court of last resort. In 1934, President Franklin D. Roosevelt appointed her to the Court of Appeals for the Sixth Circuit, where she served until her retirement in 1959.[4] No other woman had served in such a high post, second only to the Supreme Court. But Florence Allen faced problems of discrimination from her fellow male judges who avoided her. Shirley Hufstedler, appointed by President Lyndon Johnson to the Ninth Circuit Court of Appeals, was considered the replacement for Judge Allen on the circuit level until she was appointed secretary of education in President Carter's administration. Hufstedler was the one woman among ninety-seven appeals court justices, and only the second woman in the court's two-hundred-year history.

Judge Burnita Shelton Matthews was the first woman to be appointed to the District Court of the District of Columbia; she was named in 1949 by President Harry Truman.[5] Judge Lorna Lockwood, the first woman to serve on the highest appellate court of Arizona, was the first woman to be elected chief judge of a state supreme court. Four women have served on the state supreme courts. The first woman, Sandra Day O'Connor, was nominated to the U.S. Supreme Court in 1981, as this book went to press.

It is hardly strange that women did not attain judgeships when the legal profession was antagonistic to their participation as lawyers, but it was no comfort to women lawyers that their representation in the judiciary was in proportion to their representation in the bar. Political scientist Beverly Blair Cook has argued that since law schools admitted only those women with substantially higher abilities than their male counterparts, the pool of women lawyers qualified to enter the judiciary would be, on the whole, more competitive than that of men.[6] Thus, despite the proportionality, women's representation on the judiciary was unfairly small.

In the mid-1960s, only a few voices decried the dearth of women judges. A feature story in the *Washington Post* reported that the few women on the bench at the time could seldom expect to be elevated to higher courts, and more often than not, their vacated seats were filled by men.[7] In articles and speeches, Doris L. Sassower, president of the Women's Bar Association in New York, called for a substantial increase in the number of women judges and suggested that the dearth of qualified candidates for appellate positions resulted from the failure of federal and state executives to make significant appointments of women to the lower courts.[8]

Attorney Emily Jane Goodman has written about particular kinds of bias women faced in judicial selection.[9] Both male and female candidates for positions on the New York State judiciary are screened by the male-dominated New York State Bar Association of the City of New York, but female candidates must also undergo a screening before the Women's Bar Association. Goodman pointed out that women up for judgeships were often treated prejudicially and were questioned about their family lives and personal habits, such as what they planned to wear on the bench. For example, while she was seeking to run for a seat on the New York State Court of Appeals, Judge Nanette Dembitz was asked by the New York State Bar Association how she planned to manage her family.

The absence of women from the judicial screening panels is often cited as an influential factor in judicial appointments. The fourteen-member American Bar Association Committee on the Federal Judiciary, which advises the White House on nominees and selection criteria, is completely composed of males.[10] Women judges also face other problems. Journalists who have interviewed them report that female judges feel isolated because they work in courts where there are no other women judges. A survey of many women judges revealed male judges' objections to their being on the bench.[11]

In spite of ongoing resistance to women in the judiciary, during his administration President Jimmy Carter vastly increased the proportion of women federal judges; the number of women appointed or elected to judicial positions at other levels also improved greatly during this period. When Carter took office, there were five women (1 percent of the total) serving on federal district courts and courts of appeals.[12] As of spring 1980, there were twenty. Having followed the career of Ruth Bader Ginsburg, and consulted with her on women in the legal profession since 1970, it seemed momentous to read her commission of appointment as a federal circuit judge in September 1980:

Jimmy Carter, President of the United States of America, To all who shall see these Presents, Greeting. Know Ye; That reposing special trust and confidence in the Wisdom, Uprightness, and Learning of Ruth Bader Ginsburg of New York, I have nominated, and, by and with the advice and consent of the Senate, do appoint her United States Circuit Judge for the District of Columbia Circuit. . . .[13]

Carter fulfilled a pledge to redress the historic under-representation of women as well as minorities on the federal bench. The thirteen circuit judge nominating panels he created to recommend candidates included seventy-seven men and fifty-four women, both lawyers and non-lawyers.[14] Although partisan politics, as usual, resulted in disagreement about the quality of some appointees,* and the establishment bar had trouble recognizing the qualifications of women and minority candidates (especially with regard to the quality of "judicial temperament"), Judith L. Lichtman, executive director of the Women's Defense Fund in Washington, declared "He's put some extraordinary women on the bench, some of our outstanding women lawyers."[15]

The representation of women at all levels of the judiciary (now, even in the U. S. Supreme Court) has improved in the past few years. Women now constitute 5.4 percent of the federal judiciary and about 5 percent of the judges in all appellate and general trial courts.[16] Not only has the proportion of women lawyers increased, creating a larger pool of eligible candidates, but women now have access to positions that feed into the judiciary, such as posts in the U.S. attorneys' offices and professorships in law schools.

The states have uneven records in appointing women to judgeships, both in numbers and placement. In 1977, for example, women judges ruled in 12.4 percent of the family courts, an over-representation based on the usual stereotypes. The states have an uneven record with regard to the appointment of women judges to the major trial courts on the general jurisdictional level.[17] A few are good, but most are poor. Politi-

---

*ABA statistics indicate that the vast majority of women, blacks, and hispanic-Americans Carter appointed to judgeships were rated "qualified" by the association's Standing Committee on the Federal Judiciary. The committee investigates prospective nominees at the request of the attorney general, evaluating them for competence, integrity, and judicial temperament. However, a considerably smaller percentage of blacks and women than white males appointed by Carter received ratings of "well qualified" or "exceptionally well qualified." Of the men nominated for the circuit court of appeals, 70 percent of the men were rated as qualified in 1976/79, compared with 40 percent of the women.

One explanation for these lower ratings is that a high percentage of female and black lawyers are relatively recent law school graduates and have not yet had the opportunity to acquire the trial experience and reputation that count so heavily in bar association ratings. But some attribute the ratings to prejudice, and affirmative action advocates contend that the bar association committee's criteria have a discriminatory impact upon women and minorities.

Speaking at a February 1981 ABA seminar on the ratings given Carter judicial nominations, Marylin Ireland, professor at the California Western School of Law in San Diego, commented: "The main problem is the ABA's profile of the good federal judge. It is a senior corporate litigator at a major firm whose practice consists mostly of defending corporations, and that is not the career track of most women attorneys." Criticisms such as this one provoked William Reece Smith, Jr., president of the ABA, to appoint a woman to head the association's Standing Committee on the Federal Judiciary. The woman, Brooksley E. Landau, is a partner in a large Washington law firm.

TABLE 13.1
Judges/Justices in State Courts, 1980

	APPELLATE[a]			TRIAL[b]			TOTAL[c]		
	Men and Women	Women	Percent Women	Men and Women	Women	Percent Women	Men and Women	Women	Percent Women
Alabama	17	1	5.9	483	10	2.1	500	11	2.2
Alaska	8	0	—	90	3	3.3	98	3	3.1
Arizona	17	1	5.9	258	12	4.7	275	13	4.7
Arkansas	13	1	7.7	317	8	2.5	330	9	2.7
California	66	7	10.6	1,175	80	6.8	1,241	87	7.0
Colorado	17	2	11.8	459	10	2.2	476	12	2.5
Connecticut	9	1	11.1	240	9	3.8	249	10	4.0
Delaware	5	0	—	108	1	.9	113	1	.9
Florida	46	4	8.7	500	26	5.2	546	30	5.5
Georgia	16	0	—	2,428d	7	.3	2,444d	7	.3
Hawaii	8	0	—	43	6	14.0	51	6	11.8
Idaho	8	0	—	99	3	3.0	107	3	2.8
Illinois	41	2	4.9	677	14	2.1	718	16	2.2
Indiana	17	1	5.9	326	9	2.8	343	10	2.9
Iowa	14	1	7.1	300	8	2.7	314	9	2.9
Kansas	14	1	7.1	567	5	.9	581	6	1.0
Kentucky	21	0	—	214	6	2.8	235	6	2.6
Louisiana	40	0	—	872	6	.7	912	6	.7
Maine	7	0	—	52	2	3.8	59	2	3.4
Maryland	20	1	5.0	250	7	2.8	270	8	3.0
Massachusetts	17	4	23.5	264	14	5.3	281	18	6.4
Michigan	25	3	12.0	514	30	5.8	539	33	6.1
Minnesota	9	1	11.1	238	6	2.5	247	7	2.8
Mississippi	9	0	—	656	5	.8	665	5	.8
Missouri	37	0	—	300	13	4.3	337	13	3.9
Montana	7	0	—	224	1	.4	231	1	.4
Nebraska	7	0	—	110	3	2.7	117	3	2.6
Nevada	5	0	—	110	1	.9	115	1	.9
New Hampshire	5	0	—	129	0	—	134	0	—
New Jersey	28	2	7.1	688	11	1.6	716	13	1.8
New Mexico	12	2	16.7	245	3	1.2	257	5	1.9
New York	40	1	2.5	3,404	76	2.2	3,444	77	2.2
North Carolina	19	2	10.5	202	3	1.5	221	5	2.3
North Dakota	5	0	—	303	1	.3	308	1	.3
Ohio	51	1	2.0	1,251	24	1.9	1,302	25	1.9
Oklahoma	18	0	—	751	13	1.7	769	13	1.7
Oregon	17	1	5.9	373	4	1.1	390	5	1.3
Pennsylvania	23	1	4.3	874	14	1.6	897	15	1.7
Rhode Island	5	1	20.0	87	2	2.3	92	3	3.3
South Carolina	10	0	—	703	1	.1	713	1	.1
South Dakota	5	0	—	141	2	1.4	146	2	1.4
Tennessee	26	1	3.8	478	5	1.0	504	6	1.2
Texas	79	1	1.3	2,505	27	1.1	2,574	28	1.1
Utah	5	0	—	236	3	1.3	241	3	1.2
Vermont	5	0	—	71	1	1.4	76	1	1.3
Virginia	7	0	—	274	3	1.1	281	3	1.1
Washington	25	1	4.0	326	11	3.4	351	12	3.4
West Virginia	5	0	—	264	1	.4	269	1	.4
Wisconsin	19	2	10.5	406	6	1.5	425	8	1.9
Wyoming	7	0	—	139	4	2.9	146	4	2.7
Total States	926	47	5.1	25,724d	520	2.0	26,650d	567	2.1

[a]Appellate includes courts of last resort and intermediate appellate courts.
[b]Trial includes general and limited jurisdiction courts.
[c]Total refers to appellate and trial courts combined.
[d]Data from limited jurisdiction courts in Georgia do not include the judges of the criminal court, police court, and municipal court (other than the municipal courts in Savannah and Columbia). These data therefore are not included in any total figures.
SOURCE: National Center for State Courts, National Court Statistics Project, State Courts Organization 1980 (U.S. Department of Justice, Bureau of Justice Statistics, Washington, D.C.; U.S. Government Printing Office, 1981) Tables 1,3,4, and 5.

cal scientist Beverly Blair Cook found that in 1977 twenty states had no women judges *at all,* and another ten had only one (see table 13.1). In the fifty state systems, she found 130 women among the 5,155 judges (2.5 percent). According to Cook's analysis, New York, California, Florida, Pennsylvania, Illinois, and Michigan had considerably more women judges than did other states.

Courts of limited and special jurisdiction varied more widely in the states. Six states had no women at this level, but a few had a large proportion. Of the total of 5,452 judges with limited jurisdiction, 317, or 5.8 percent, were women.

Representation of women in the judiciary varies from city to city, with Chicago and Los Angeles at either end of the spectrum. In 1975, women judges in Chicago held posts in divorce, probate, adoptions, and small claims court, but were absent from the largest and most significant divisions, namely civil and criminal law. In Los Angeles, on the other hand, women have been assigned to criminal and civil courts on a rotation system which provides opportunities for judges without respect to sex.

Women's increasing presence on the bench is a major breakthrough, but their legacy of poor treatment by judicial peers is only beginning to be overcome. Women judges have encountered all kinds of discriminatory treatment—from ostracism by other judges, such as Judge Florence Allen faced,[18] to degrading treatment by attorneys, such as that reported by Chief Justice Susie Sharp of the North Carolina Supreme Court (a lawyer in her trial court said, "Honey, I don't think you understand my case very well").[19] Perhaps Constance Baker Motley of the U.S. District Court for the Southern District of New York experienced one of the oddest confrontations when defendants in the sex discrimination case against Sullivan and Cromwell asked her to disqualify herself on the grounds that she was a woman. Certainly the plaintiffs would not have expected a male judge to disqualify himself on the basis of his sex. But Motley is used to discrimination. A lawyer with the NAACP Legal Defense Fund from 1945 to 1965, she had won nine out of the ten cases she argued before the U.S. Supreme Court,[20] but recalled in an interview that male colleagues doubted her ability to withstand courtroom pressure.

Judge Rose Elizabeth Bird, the first women to become chief justice of the California State Supreme Court, was appointed by Governor Edmund G. Brown, Jr., in February 1977 amidst a storm of controversy.[21] Her background included a series of "firsts": first female clerk for a judge on the Nevada Supreme Court (1965); first woman hired by the Santa Clara County's public defender's office (1966); first women to teach law at Stanford Law School (1972); and the first female cabinet officer appointed by a California governor (1975).[22] Chief Justice Bird has had a highly controversial career on the bench, and in November 1978 she narrowly won confirmation of her seat, in the face of a well-financed campaign to unseat her by conservatives and law-and-order advocates who charged she was "soft on crime."[23] But Louise Renne, president of the California Women's Bar, asserts that women lawyers in California generally agree that the opposition to Rose Elizabeth Bird has had more to do with her sex than her political views.

Recently, efforts have been made to combat disrespectful or foolish behavior toward women in the judiciary. In October 1979, more than one hundred women judges met in Los Angeles for the first meeting of the new National Association of Women Judges (NAWJ).[24] For some of the participants it was the first time they had ever even met another woman judge. At the meeting, the NAWJ planned to work to establish a network or "psychological sisterhood," to encourage communication among members and sharing of concerns and remedies for the problems they encounter. The organization today represents more than half of the roughly five hundred judges in the country who are women.[25] (The total number of judges is about fifteen thousand.)

Like women lawyers, few women judges consider themselves different from male judges. Since many have come to the judiciary via civil rights work or advocacy for children or the poor, it is not surprising that they should be particularly sensitive to these issues. However, this sensitivity does not entail a sacrifice of judicial prudence.

It has been charged, for example, that Judge Constance Baker Motley acts "like a woman" because she makes it a practice to ask the defendant, "Do you understand?" to make sure the person is not denied his or her rights because of the complexity of legal terminology. But others protest that liberal male judges do this as well. It is the profession's and the public's stereotyped expectations of behavior by women in positions of authority which determine the perceived character of the woman judge. Some feel a woman will be compassionate, while others think she will be particularly tough. When one of them does not like the woman judge's decision, he or she may attribute it to her sex rather than to her personality or ideology.

Women judges have been active in appointing women lawyers as their clerks, an important stepping stone to law school teaching and also to the judiciary. For a woman to become a clerk to a federal judge or a Supreme Court justice was almost an impossibility until the 1970s.

Justice William O. Douglas appointed the first woman clerk on the Supreme Court in 1944, and no others were appointed until 1966, when one reported to Justice Hugo Black; and in 1968 another reported to Justice Abe Fortas. After 1971, although few in number, there always were women among the clerks to Supreme Court justices.* In 1977 and 1979 there were seven women in Supreme Court clerkships, a high point; but since then the number has again decreased.[26]

Because it has been mostly liberal justices who have chosen women law clerks, the developing conservative composition of the Supreme Court may create problems for women aspiring to clerkships. Whether this means that the resistance to women entering the judiciary, somewhat assuaged during the Carter years, will again develop, we do not yet know.

Yet, President Reagan is responsible for nominating the first woman Supreme Court Justice. And although his administration is hardly oriented toward women's rights, for the first time there is a roster of competent women judges with credentials rivaling those of the best male candidates, and so the outlook appears to be hopeful.

*The first woman clerk, Lucille Lomen, was appointed by Justice Douglas in 1944. The second, Margaret Corcoran, was appointed by Justice Hugo Black in 1966, and Justice Abe Fortas appointed Martha Field Alschuler in 1968. Barbara Underwood was the fourth woman clerk, appointed by Justice Thurgood Marshall in 1971. *Docket Sheet,* Supreme Court, vol. 16, no. 3, May–June 1979, p. 8. In 1972, two women were appointed clerks to Justice William O. Douglas and one to Justice Byron R. White. The following year, one woman reported to Justice William H. Rehnquist. In 1974, four women were appointed —one each by Justices William Brennan, Harry A. Blackmun, Marshall, and Lewis F. Powell. In the two years that followed, three women were appointed (in 1975, by Justices Warren Burger, Powell, and Blackmun; in 1976, by Justices Potter Stewart, Marshall, and Blackmun). The number of women law clerks increased substantially in 1977 to seven, dropped the following year to five, and in 1979, increased again to seven. In 1977, Justices Marshall and Blackmun appointed two women each, and Justices Stewart, Powell, and White, one woman each. In 1978, Justices Stevens, Stewart, and White appointed one woman each, and Justice Marshall, two women. In 1979, Justice Marshall appointed two women, and Justices Stevens, White, Rehnquist, Blackmun, and Powell one female law clerk each. (This information was obtained from the Public Information Bureau, Office of the Supreme Court, Washington, D.C., from Barrett McGurn.)

14

Professional Associations

LAWYERS meet, plan policy, socialize, and determine the rules of their profession in the hallways and bars of large hotels during the annual meetings of the American Bar Association, in the meetings of county bar associations, in the libraries and lunchrooms of the city and state bars, and in the seminars of specialized groups of tax, admiralty, or trial attorneys. Here, lawyers otherwise separated by practice, firm, or case reinforce their ties as professionals, focusing on the larger concerns of the legal profession, such as the quality of legal education and protection of the autonomy of their guild. In the course of engaging in these collective rites, private interest is also served as lawyers become visible to each other as individuals, developing contacts and establishing reputations. The professional associations provide another arena, together with the courtroom and negotiating table, in which the lawyer may develop skills and demonstrate expertise to his or her peers.* The associations also provide library facilities, lectures on specialized legal subjects, and opportunities to make professional acquaintances and to work on special legal and legislative projects. Some associations evaluate lawyers for judgeships and determine the rules of professional behavior.

There are more than seventeen hundred bar associations in the United States. First among them in membership, with 230,000 lawyers,

---

*The associations also provide an environment in which the breadth of one's intellect is developed in the course of interchange with peers. Commenting on his own rich experiences in the scholarly community of Columbia University, Robert K. Merton has reached back in history to provide an insight into the process as offered by Bishop Sprat writing about the Royal Society of London, a professional association of scientists in the seventeenth century: "In Assemblies, the Wits of most men are sharper, their Apprehensions readier, their Thoughts fuller, than in their Closets." Said Merton, "The good Bishop was underlining the importance of face-to-face interactions . . . which were being made easier and consequently more frequent by the newly invented Society." "Merton on New York: 'The City Remains a Cultural Capital'," *Columbia University Record* 6, no. 20 (February 27, 1981), p. 8.

is the American Bar Association, the largest professional association in the country, now 100 years old. There is a National Women's Bar Association, which has a few thousand members, with local chapters throughout the country. There are also specialized bar groups such as the Patent Attorneys' Association and regional and local bar associations. They differ in size, prestige, and importance, and many lawyers belong to several of them.

Although women lawyers were members of the bar associations, in proportion to their percentage in the profession they played only a limited role in professional organizations in the past. The combination of discrimination, women's areas of practice and specialty, and their family obligations, made women relatively silent and invisible members of these network-creating institutions. Because the bar associations reflect the stratification pattern of the profession, and because each association has its own hierarchy, women typically held low-ranking positions if they held any at all.*

Bar associations at all levels of presitge limited the membership on one basis or another until anti-discrimination laws and public opinion made that difficult. In its early years the American Bar Association, which claimed the right to represent the entire profession, developed the reputation of being anti-black, anti-Catholic, and anti-Semitic. Three black lawyers were reported to have been admitted to the organization by mistake in 1912, but applicants were later directed to identify themselves by race, and no more blacks were admitted until the 1940s.[1]

Although women were eligible to become lawyers in 1886, the ABA admitted women only in 1918; the ABA officially denies it has ever restricted membership to males and professes not to know when the first woman was admitted.[2] Barbara Armstrong, a Harvard Law School graduate, reported in 1951 that women's struggle for admission to the bar associations was sharper and even more prolonged than their fight for the right to a legal education.[3] The prestigious Association of the Bar of the City of New York, founded in 1879, did not admit women until 1937, and New York's Queens County Bar Association, founded in 1876, did not admit women until 1960. The New York Lawyers' Association,

*A study of the Chicago bar showed that graduates of elite and prestigious law schools and members of large firms were over-represented in the committees and leadership of the city's bar association. Solo practitioners, lawyers in the small firms, and government lawyers were under-represented. Since women and minorities clustered in the latter kinds of practices, they tended not to be in leadership positions in the bar association. John P. Heinz et al., "Diversity, Representation, and Leadership in an Urban Bar: A First Report on a Survey of the Chicago Bar," *American Bar Foundation Research Journal* 1976, no. 3 (Summer 1976): 717–85.

however, accepted women members from the time of its incorporation in 1908.

Resistance to admitting women to membership in specialized bar associations surfaced into the 1970s as well. In 1971 attorney Sybil Hart Kooper sued the Metropolitan Trial Lawyers' Association complaining that she had been barred from membership because of her sex. She and her law partner husband had been invited to apply to the association, but while her husband was admitted, her application had been put aside. The secretary of the association noted, according to the *New York Times* report, that "we do not have any members who are women, not because we discriminate against them but because so few can qualify to our standards."[4] Mrs. Kooper had credentials as the secretary of another professional organization, the New York State Trial Lawyers' Association, and was a member of the Board of Directors of the American Academy of Trial Lawyers. The president of the Metropolitan Association, Alfred S. Julian, who had proposed her, promised to try to "break the barrier" if he could, for her and other women who had been turned down. Now, ten years later, I inquired about the current status of women in the association, and found that nothing had changed— among its ninety-two members, none were women. A spokesman reported, "Well, we used to have one lady, but she became a judge."[5] The "one lady" was Sybil Hart Kooper, now a Brooklyn Supreme Court Justice.

New York City has several citywide bar groups, among which the Association of the Bar of the City of New York is the most prestigious and powerful. The New York County Lawyers' Association draws its membership mainly from smaller firms, although high-prestige members of the bar usually belong to both it and the Association of the Bar. Separate bar associations exist in the city's other counties, and most of the boroughs have their own women's bar associations.

In cases where there were no formal barriers based on sex, in these associations women were often excluded by other means. New York's Brooklyn Bar Association claims it never excluded women,[6] but several lawyers who have lived and practiced in Brooklyn for more than twenty years maintained in interviews that they were excluded until the early 1960s. The Brooklyn Bar Association, the Bronx Bar Association, and the Bar Association of the City of New York, all barred women with the excuse that no rest room facilities were available for them, as did the firms that excluded them.*

*It is also interesting that the same excuse has been employed by Cambridge and Oxford Universities to limit enrollment of women. (David Caute, "Crisis in All Souls,"

Membership

It is generally believed that fewer women than men join professional associations because of family obligations. But statistics on bar association membership indicate this is not true for the legal profession.

In 1951, when women made up 2.5 percent of the profession, the American Bar Association had 1,028 women members, about 2 percent of the total ABA membership of 44,262. In the New York County Lawyers' Association, the 252 women members reported that year constituted 3 percent of the membership. The percentage of women members was considerably lower in the more prestigious Association of the Bar of the City of New York.[7] In 1966, the percentage of women members remained about 3 percent in the New York County Lawyers' Association (277 women out of 8,370 members). Again, this reflected their overall percentage in the profession, although slightly less than the 3.5 percent figure reported by the U.S. census. In the Association of the Bar of the City of New York women represented 3 percent of the membership in 1967, an apparent increase over the sixteen-year period.[8] In 1971, women were 4 percent of the membership, and they increased to 9 percent by 1976,[9] also slightly exceeding their proportion in the profession.

A very high proportion of the lawyers interviewed in the 1960s and 1970s were members of bar associations. Eighty percent of those interviewed in the 1960s who had practiced belonged to a bar association at some point in their careers. One third belonged to bar associations integrating men and women, another third belonged to both an integrated bar and a women's bar group, and 11 percent belonged only to the Women's Bar Association. Only 15 percent reported never having held membership in any organization. Women belonged to bar associations of every rank and type. Private practitioners belonged to local bars and the New York County Lawyers' Association, and women in large firms belonged to the Association of the Bar of the City of New York and other high-prestige associations.

Although women's membership in bar associations was increasing in the 1970s, many of the young women interviewed in that decade were more interested in the new women's legal organizations that were forming, such as those that planned the women and the law confer-

Encounter 26 [1966], p. 5.) Women were accorded full status as students of the university at Oxford in 1920 and at Cambridge in 1948. (*Education in Britain,* British Information Service, rev. ed., January 1964, p. 8.)

ences held every year since 1969. However, it seemed women's participation in the established organizations was becoming regularized, reflecting their interests and proportion of the profession. Much of this was due to their insistence on being given equal access to all activities of the bar association after decades of discriminatory treatment.

Women's Rate of Participation

If women belonged to professional organizations in proportion to their numbers in the profession, they were hardly visible and could not be found among the leaders. The mere fact of membership in a bar association is not an accurate index of participation because many men and women are only nominally members. Some are active in the social activities of the organizations but not their professional activities. Some join to use the library or other facilities and cannot be said to participate. But, of course, many more men have been more visibly active than women (even if the majority of men have not been). It is generally believed that women lawyers' lesser participation on committees reflects their inability to devote extra time to the work involved and their lesser involvement than men in the organizations.* But for generations past, the fact was that they were not invited to serve on committees even if they wished to serve.

Yet for many women a low level of activity in professional associations can be attributed to their home responsibilities. Interviews showed this to be true to some extent, but it hardly constituted a complete explanation. Lawyers of both generations reported that marriage does not significantly affect participation but that motherhood does. The demands flowing from motherhood hinder full participation, especially when children are young.

A lawyer with four young children and a full-time job reported that she did not belong to a bar association at present. She said:

*Rita Simon's findings on women Ph.D.'s show that, contrary to expectations, women participate in professional organizations no less than men, although men give greater priority to committee work. About 90 percent (both married and unmarried), as well as 90 percent of the male Ph.D.'s, belonged to at least one professional organization. However, the participation rates of unmarried women are greater than those for men (who are almost all married) and married women. Only the unmarried have the time available which comes from freedom from family demands. Rita James Simon, Shirley Merritt Clark, and Kathleen Galway, "The Woman Ph.D.: A Recent Profile," *Social Problems* 15 (Fall 1967) p. 234.

I was a member in the American Bar Association, but I like to do a thing well and I don't believe in being a member of anything in which I can't spend the time to participate. And I find it's too busy with the children and the housework, and here (the office) to have any time. And any extra time I have, I want to do with it what I want.

Many women's activity in bar associations followed a discontinuous pattern reflecting their discontinuous participation in professional life. The 1960s interviews showed that many women left law and returned to it for periods of time, or concentrated on building their careers at certain times but not others. Some women retired from active membership in the bar during the period when their children were young, and others were continuously active. One woman's account of the "life-cycle" of her law membership was typical:

I'm a member of the American Bar, the State Bar, the New York County Lawyers' Association, the New York Women's Bar. I was president of the Women's Bar for two years, and I was quite active in all of them for a long period of time. Now, I just retain one committeeship with the New York County Lawyers. I stopped being active when I had children. I found that since I was devoting most of the day—or, as a matter of fact, up until recently every day —to work, that the evenings belong to the family.

I enjoyed my association work very much. I'm sorry that I had to give it up. I hope to resume it. This is a hiatus.

Women in practice with husbands reported that although they dropped out of bar association activities when children came, their husbands continued to attend meetings. Some women, however, maintained membership in a bar association long after they had left active practice. Some women interviewed for the 1960s study reported that they retained membership in the women's bar associations, particularly in suburban areas, and attended meetings though they had not practiced in years. Some felt it was a way they could "keep their hand in" the profession, hoping to resume practice one day, or because they enjoyed the social contact with their friends. This assumption was right. By 1980 some of the women interviewed fifteen years before had returned to active practice as their children grew older and opportunities developed.

A discontinuous pattern was not only representative of the married women with children. Several single women in their fifties and sixties said they did not attend bar meetings because they were afraid to go out at night, and therefore were only nominal members.

Because the lawyers interviewed in the mid-1960s were on the average in their late forties, most had older children and this may explain why they were an active group. More than half were members of bar associations and were actively involved in them in some way. About three quarters of those who belonged to bar associations said they attended meetings and lectures on subjects in their specialties. When asked about activity on committees, a sizable percentage—close to 60 percent—said they were active but only on the few committees believed to be appropriate for women.

One bar association that did not follow the pattern of the establishment bars was the National Lawyers Guild, an organization noted for civil rights activities and the left-wing politics of its members. Many of the lawyers interviewed for this study, especially those who identified themselves as feminists involved in movement law, belong to the lawyers guild. It is striking that many women became visible in that organization in the early seventies and have remained so. They are prominent in the listing of the guild's Speakers Bureau; they are on the platform at important meetings.

But the role of women in the guild was not always so visible; it has evolved over the years, aided by the membership's ideology of equality. Women have had to fight to avoid being relegated to traditional assignments, and until the 1970s they did not always succeed. One woman interviewed in 1971 said:

I would not join the lawyers guild because the same problems women face in a traditional organization are to be found in the radical ones. There are the same male ego problems to be faced there—the women do as much work as some of the stars, but it is the Kuntzlers and Kinoys who get the credit.

Nevertheless, this bar association was far less resistant to using the talent of women members, and the women themselves were assertive in seeking active roles within the organization. Probably the larger role of women within the organization was facilitated by the ideology of equality which characterized the commitment of lawyers guild lawyers in their work, the fact that the women members had played leading roles in the political activity of the 1950s and 1960s, and the fact that the organization was small and independent of corporate interests.

Committees

Permission to enter the bar associations did not mean enthusiastic reception; up until the 1970s and into the 1980s, women were barred from full participation at decision-making levels, and they were seldom elected to prestigious committees or to executive posts. It was only in February 1981 that a woman was elected to the twenty-three–member board of governors of the American Bar Association. No woman has chaired its policy-setting House of Delegates, nor have any of the association's 104 presidents or other officers been women. Today, ten of the Association's committees, commissions, sections, and divisions are headed by women.[10]

All bar associations are prestige-ranked by the profession, and within them there is a ranking of the committees dealing with legal and administrative problems and areas of the law. While a lawyer joins a bar association by choice (admission usually is by a process of sponsorship and election), committee membership is the object of intense competition within the prestigious associations.

A young male partner in a high-prestige New York law firm described his bar association experience. He said that although lawyers are often asked to serve on committees of the New York County Lawyers' Association, membership in a committee of the Association of the Bar of the City of New York is generally conferred only after the lawyer has proved his skills and standing in a prestigious firm. He recounted that

On joining the Association of the Bar, I was interviewed at length on my special interest—Constitutional law—I was then told that it might be ten years before I would be offered membership on a committee.

He went on to comment wryly that when the offer came, nearly ten years later, it was for membership on a committee dealing with the specialty in which he made a reputation, not the committee on Constitutional law. He added that committee membership was invaluable, not only because it marked the lawyer's "arrival," but because it brought him into close personal contact with men in other firms and provided insights into the methods and personalities of the lawyers with whom he would be dealing throughout his professional career. He said that committee membership often resulted in friendship that brought younger attorneys into the networks that interlace the upper strata of the law profession in New York.

254

Women's second-class status in the bar associations was emphasized and reinforced by the inappropriateness of their committee assignments. One lawyer who did corporate work in a high-prestige firm said she had quit the New York County Lawyers' Association after a year because she found that women were assigned only to the matrimonial and house committees. A specialist in admiralty law, she asked, "What would I do there?" She said that some women joined hoping they would be assigned to a good committee if they demonstrated their worth within the organization. "But," she said, "they never can." She noted that at the time she was a member, the association had offered an insurance scheme that did not permit women to obtain the same amount of insurance as men. (However, women fought for and won a change in the plan.) By 1980, the New York County Lawyers' Association had women represented both in its important standing committees and the special committees; and although no woman sat on its executive committee, there were two women among the directors of the organization.

The prestigious Association of the Bar of the City of New York did not contain one woman's name in its listing of officers from 1870 to 1967.[11] However, women participated in the association's eighty committees to a far greater extent in 1967 than they did in 1951, when Barbara Armstrong reported that only 10 women were participating committee members.[12] In 1967, 40 women's names appeared in committee listings. Women did advance considerably during the 1970s and by 1977 there were 179 women committee members out of a total of 1,523, and there were 8 women chairing committees.

Women bar association members continue to serve in disproportionate numbers on the committees concerned with family law, children's rights, and civil rights, reflecting their concentration in these specialties. But women have expanded their presence into non-traditional fields in the 1970s, serving on committees concerned with the military, arbitration, and nuclear technology. A prominent attorney, now well placed in a prestigious position in the Association of the Bar of the City of New York, told me "They make a practice of having one woman on every committee. I'm chair of the [X] Committee."

New committees were also formed, such as the "Sex and Law" Committee of the Association of the Bar of the City of New York, reflecting the new fields women attorneys were involved in and the number of cases growing out of anti-discrimination legislation. Women were over-represented on these committees, although there were a token number of men as well. By the 1970s, all important bar associations had some kind of committee on the status of women. These committees have also been responsible for reviewing association materials to ensure that they use non-sexist language and for developing programs of concern to women at their annual meetings. However, these sessions are attended mainly by women. There is still some segregation in the bar associations although it is informal, and women are certainly more comfortable in them today than in the mid-1960s when these interviews were begun.

Women said repeatedly that as young lawyers they often faced problems of personal acceptance in the bar associations. They were forced to cope simultaneously with the problems stemming from their sex-status and from their status as initiates in the profession. To some extent this is mitigated in the large bar associations by the formation of special "young lawyers" committees. The younger women report that age peers tend to be more accepting of their presence than older men, although even the young men questioned the womens' motivation in joining the association. One lawyer recalled: "I was accepted more when I married, because I was no longer considered a predatory female." Being married is a distinct advantage for the lawyer who wishes to be integrated into the association.

Local Bar Associations

Membership in local and suburban bar associations poses special problems and provides special benefits for women. Their acceptance depends on how well known the lawyer is in the community and how much she is considered a part of it; whether or not she is married to a lawyer who is also a member; whether or not members of her family have been or are members of the bar; and, of course, her attractiveness and personality.

The bar associations found in small towns and suburbs are quite different from those in central urban areas. Persons in small towns are more apt to be assessed on the basis of composite information—how well they did in school; what awards they have achieved; how hard-working they are known to be; and whose sons or daughters they are. This is so even in the borough bar organizations of New York City. One lawyer who lived on Staten Island commented:

Professional Associations

In our Richmond Bar Association we make no distinction as to whether it's a male lawyer or a female lawyer—we are lawyers. Sex is left out completely. That's unlike other boroughs in the city, isn't it? In the other boroughs they do have separate organizations, and I think it's a bad thing. . . . The Richmond Bar is an independent group because it's insular.

We're very much of a family type on Staten Island. Everyone knows everyone else. Usually they know all of the families and, of course, my first two children were very well known to the entire legal profession over there because we lived just around the corner from the Borough Hall and the court. So the lawyers all saw the children as they were growing up.

Smallness and closeness can also intensify exclusiveness, however. The local bar associations have their "in" and "out" groups. Women are out-group members unless they can ride the coattails of an in-group member. A Brooklyn lawyer commented that

The Brooklyn bar association is a glorified frat house. There is more politics going on there than in a political club house. What opened doors for me is the fact that my husband is so well liked.

Women's Bar Associations: Segregation or Solidarity?

Bar associations through history were marked by informality because they were tightly knit groups in which men could gather to socialize in a male society. A quotation from an old British newspaper suggests that the pattern was equally true of other Western societies:

"Full-blooded hilarity" which had so often attended the meetings of the Bar, would be inconsistent with admittance to the Bar of the mass of gentlewoman-kind. And yet it is frankly conceded by most members of the Bar that to exclude them entirely would be a denial of the rights or privileges which are common to all. . . . the main body, it would appear, are favorable to a middle course, a sort of compromise, mainly, full right of admission for all purposes save dining and participation in the subsequent exercises . . . and while the ladies would share in all the strictly professional benefits, the men would preserve to themselves the sanctity of the recreation to which they had grown accustomed and regard as a special heritage.[13]

In the United States, male resistance to female membership in bar associations was summed up in the rejection of four women's applications for membership in the Baltimore Bar Association in 1928. The arguments, which still sound familiar, were as follows: "their meetings would not have the same freedom, they could not tell the same stories or have the same refreshments . . . if the women came they would try to run the whole Association . . . women's place was in the home."[14] Mindful of members' attitudes in this southern border city, the association contended that if women were admitted, Negroes must also be admitted. The rejection was so clearly communicated and rationalized that at least some of the women accepted it without further resistance.

The exclusion of women from bar associations by informal means resembles the avoidance mechanisms used by in-groups against outsiders. Hubert Blalock points out that avoidance is one of the most pervasive and subtle forms of minority discrimination, "particularly in situations implying social equality or involving potential intimacy."[15] Although individuals tend to avoid certain categories of persons because of status, psychological discomfort, or lack of common interest, he notes, "when these avoidances become patterns . . . they result in systematic and widespread segregation that is particularly difficult to analyze because of its 'voluntary' nature."[16]

To retreat to a parallel women's organization or not join a bar association at all is discrimination by self-exclusion. It is behavior that reinforces minority self-hatred by accepting society's low evaluation of one's group, in this case by acknowledging women's incapacity to follow the same professional norms as men.[17] In the past, even some women favored the exclusion of all women from the bar associations. They accepted the image of a male profession and its male-specific behavior, particularly in informal contexts. As one woman lawyer described her feelings about the Queens County Bar Association:

There was a camaraderie in the County [Law] Association, a terrific spirit. In other associations the members are very staid . . . but there everybody knows one another and they joke. They were prejudiced against admitting women but I think they were justified. It's not the same with a woman around. They aren't free to express themselves, to tell off-color stories—they should have that.

At the same time that women's bar associations have helped women to participate in professional life they have diverted them from full integration in the profession. Some women's bars were organized early in the century in response to exclusion and discrimination by the bar associations dominated by men. It was felt that only through a separately organized bar could women lawyers advance their professional position and also assist women in attaining social and political rights such as suffrage and the right to sit on juries.

Professional Associations

The Woman Lawyers' Club (which became the National Association of Woman Lawyers in 1911) was founded in 1899 because the American Bar Association did not admit women.[18] The women's bar associations in New York City also were formed in response to exclusion by the male-dominated associations in the Bronx, Queens, and Brooklyn. Similar groups were established in other parts of the country. Women's professional groups have been formed in other fields for similar reasons, even though in some, women have not been excluded from the main professional society at any time.*

In their early years, the women's bar associations made substantial contributions to the legal structure by their activity in promoting local legal aid for indigent persons accused of crimes. In part they did this by tapping members who were not in active practice but who were happy to volunteer their services on a part-time basis. In the past, the programs of the women's bar associations were oriented around the major "female" specialties: probate law, taxation, administrative law, domestic relations, and juvenile delinquency.[19] They have now been expanded to deal with the wider range of women's interests and to more self-consciously act as a support group for women.

By the time the American Bar Association ended its exclusion of women, the National Association of Women Lawyers (NAWL) was established firmly enough to continue its independent existence. Today the NAWL has a membership of 1,600 lawyers,[20] a number that has grown over the years but represents a declining and tiny percentage (2 percent today) of America's women lawyers.†

The NAWL's supporters assert that the organization's continued operation is justified by the ABA's persistent discrimination against women, especially at the higher levels of the organization. The lawyers involved in the NAWL claim that their organization assures women of representation in the ABA because a representative of the NAWL sits

*The National Council of Women Psychologists was established in 1942 after it became clear that the Emergency Committee in Psychology of the National Research Council was omitting women from its plans for the wartime use of psychologists. Mildred B. Mitchell, "Status of Women in the American Psychological Association," *American Psychologist* 6 (1951): 193. There are also separate professional associations for women in the fields of medicine (founded 1915), dentistry (founded 1921), engineering (founded 1950), geography (founded 1925), and certified public accounting (founded 1933). It is interesting to note that in professions typed as female, there are no separate men's organizations, nor have men been formally excluded from membership at any time.

†In 1968 the NAWL had 1,200 members—16 percent of the country's women lawyers —and in 1951, 1,072 or 21 percent. Women's Bureau: U.S. Department of Labor, *1962 Handbook of Women Workers*, Bulletin 285; (Washington, D.C.: U.S. Government Printing Office, 1963), p. 168; Faye A. Hankin and Duane W. Krohnke, *The American Lawyer: 1964 Statistical Report* (Chicago, Ill.: American Bar Foundation, 1965).

on the ABA board. They also back women candidates for political office and for the federal judiciary.

There are a number of reasons why New York City's women's bar associations remain active although women have been permitted to join all local bar groups for at least seven years. Because of their small numbers, women lawyers tend to know each other and be part of the same or interwoven networks. Their organizations thus serve as friendship groups and centers for mutual support in a sometimes hostile professional world. They act as clearing houses for contacts and clients, in much the same way as the other associations serve male lawyers. According to Diane Wilner, president of the New York Women's Bar Association, her organization is growing because of the lively committees that have been created to deal with issues of particular concern to women lawyers and women in society. These include the Committee on Equal Opportunity in Employment, the Legislation Committee which helps draft bills, and the Committee on Battered Women which considers policy issues and helps set up shelters for victims of battering. The associations also provide a supportive structure in which women's professional identity is not challenged but reinforced and a network of lawyers who consult each other about mutual problems.

Further, women, who cannot easily rise in the male-dominated bar organizations, can climb to positions of leadership in the women's bars. Although the prestige of the women's groups counts for less in the profession, some of the prestige attached to high office in them may be carried over into the male organizations and into the profession. It does act to make some women visible, and the recognition a woman gains from office in the women's organizations may be sufficient both to promote her practice and to give her a sense of achievement.

Younger women today who never considered joining traditional women's groups are also finding the local women's bar associations helpful in establishing their new practices. One lawyer, an announced feminist who was starting out on her own in Westchester County having formerly practiced in the city, saw membership as a partial replacement for a partnership:

As a result of not having a partnership I became involved in the Women's Bar here, which I found is a good group. Many of the women are ten, fifteen years older than I am, but many of them are feminists nonetheless. . . . For some it was under their skin all the time but now it's totally permissible to show it. I find many of them are really fine women and having a relationship with them gives me some sense of what other women in the legal profession have to face.

260

Professional Associations

The women's bar associations may serve women well, but at the same time they maintain separation. They deplete the pool of potential female leadership in the male-dominated professional associations and divert the energies of women members. They reinforce the impression of the public and of male practitioners that women lawyers are a breed apart, with unique qualities and interests. One of the older women made the point clearly, in an interview fifteen years ago:

I don't like women's bar associations because they underline law as a man's profession and I would much rather belong to a bar association where there are both men and women. I would hope there would be a natural give and take among lawyers. Why isolate yourselves and say you're *women* lawyers? You're only focusing the whole problem—the problem I would like to see erased—the consciousness of the fact that a lawyer is a man or a woman.

But that was in the 1960s, before the rise of the modern women's movement. Women banding together to take care of the concerns and interests of women had a more positive image in the 1970s. No longer were the women's bar associations only residues of a prior time without legitimate functions.

Women are now part of the establishment bar: they are more visibly active in it, and they should become more important to it as time goes on. Their presence will legitimate their growing numbers and greater power in the profession. It will help to create a visible pool of women candidates for the judiciary, particularly for federal court appointments —including the Supreme Court—and will help women attorneys in the mundane matters of establishing competence and creating referral networks. The bar associations are the places where the old boys have traditionally gathered in the legal professions. The admission of women to these professional and social circles is an important move for their true integration into the profession.

IV

OUTSIDERS WITHIN

15
Ambivalence and Collegiality

BELONGING, the sense of being an insider at ease with colleagues and clients—no matter what kind of practice a woman engages in, the prejudices and practices that have kept law a male profession for so long make such a feeling rare for the woman lawyer. The theme that runs though this book is the change from the virtual exclusion of women from the profession to their entry into even the most prestigious and elite domains of law. But it is one thing to be employed, and even paid well, and another to be a true working partner in the camaraderie of the legal community. The structure of the profession and the cultural views about the nature of men and women often prevent women from becoming fully integrated into the legal profession. While most men can make their lives as lawyers mesh with their private lives, their values, and their styles of behavior, few women escape contradictory pressures and expectations within the profession and in their lives outside it. These pressures and contradictions—the products of ambivalence on the part of male gatekeepers and other men and women who do not believe that women belong in the law—create ambivalence in the minds of women.

Being an insider is not merely a matter of happiness or personal adjustment for the woman attorney. The old assumptions and social customs about sex differences build barriers between men and women and prepare them improperly to work easily as colleagues. Furthermore, the commonality of understanding, trust, respect, good will, and tutelage that male lawyers expect to enjoy as benefits of belonging to a brotherhood do not yet extend to a good proportion of women attorneys.* These barriers make it difficult for women to learn the subtleties

*Rosabeth Moss Kanter points out in *Men and Women of the Corporation* (New York: Basic Books, 1977) that especially in work settings where there is a lack of assurance about diagnosis of issues or problems (like the professions), trust becomes of paramount importance and thus there is a particular reliance on bringing in co-workers with similar backgrounds.

of their professional roles, to develop their competence as lawyers,[1] and even to develop ambition properly.

There are barriers that are not to be remedied by legal action, but rather by informal means. They come from prejudices and stereotypes lodged in habits of thought, unstated agendas, informal verbal and non-verbal communication, and from the nature of the practice of law in this country: confrontational and combative. The barriers come from the culture of the society as a whole and the culture of the profession in particular, both of which create images of who the lawyer ought to be—and that is male; of what the work of law ought to be—and that is largely business; of how the work ought to be carried out—and that is with primary allegiance to work; and of where it ought to be carried out —and that is in a social environment associated with male preferences and habits, often club settings actually or symbolically restricted to men.

Of utmost importance to women's ultimate equality within the legal profession is the extent to which the patterns of social structure that cement the brotherhood of men at the same time reinforce the outsider role of women. Another crucial determinant is the extent to which our culture is geared to maintaining women's subordination to men, so that even when men and women rank equally on the job, norms persist that call for women's deference to men. The web of ambiguity thus created does not allow these problems to be sorted out, but keeps them alive. Ambivalence is built into certain social roles or social settings, so that anyone, no matter what his or her personal qualities, would be perplexed about the right thing to do.* Women lawyers, like women architects, senators, or school administrators, receive a large dose of ambivalence from contradictions built into their professional roles and contradictions in their different roles as women and as workers. How does this operate? A professor, for example, faces contradictions when he or she is torn between the norms for two or more roles—the need to be freely available to students and also to cloister him or herself for research and writing. Or the conflict experienced by the parent whose time and attention is required by both a sick family member and a pressing deadline at the office. Merton and Barber distinguished between sociological ambivalence and psychological ambivalence, showing that one was the product of the social structure and the other of internal psychic states (as when a person is offered two equally attrac-

---

*This follows the seminal analysis of Robert K. Merton and Elinor Barber in "Sociological Ambivalence," in *Sociological Ambivalence and Other Essays*, ed. Robert K. Merton (New York: The Free Press, 1976), pp. 3–32.

tive jobs and cannot decide which to take). But social patterns may induce psychological states of ambivalence. The norms for teenagers, for example, who are defined in U. S. society as neither children nor adults, require them to be independent, self-reliant, and responsible, and at the same time appropriately dependent on or accountable to their parents. Thus many teenagers feel psychological ambivalence, or simultaneous feelings of love and hate toward their parents. Anger is another consequence of being placed in situations that create ambivalence.

Members of certain groups, such as women or ethnic or racial minorities, may face more ambivalence than others. Groups that are secure in their domains have greater access to social ways of resolving ambivalence.[2] For some, ambivalence is a product of a particular stage in their lives. Teenagers, for example are permitted to grow out of the state that produces ambivalence at some culturally agreed age—say at eighteen or twenty—when they are legally free of parental supervision. Interns and apprentices pass from the stage of learning to the stage of autonomous decision making so they can practice independent judgments while still under supervision.[3]

The sociological ambivalence inherent in the situations of professional women can be viewed as historic, caused by the separation of work and the home, the market and the family, and at the same time as a mechanism to ensure they remain marginal in competitive and valued spheres. Gatekeepers have indeed conspired to keep women out of law firms and have told them they were taking bread from the mouths of male lawyers' families. And men or women who believe that assertive behavior necessary for the trial lawyer is unseemly if the trial lawyer is a woman may be unaware they are engendering an impossible bind, a "double-bind" in Gregory Bateson's terminology,[4] for women attorneys. The consequences of these situations are similar: they undermine women's ability to act the way other lawyers—male lawyers—do.

Impediments of the Culture

Our cultures provide us with guidelines about appropriate behavior for the sexes. Cultures are belief systems that form the basis for individuals' notions about what they should do and should expect others to do. Assumptions of what women and men "are" or "do" pervade so much of social life that most people take them for granted. Anthropologists were among the first to alert us to what is now commonplace—that each culture's views about human nature have consequences for the ways in which societies honor, educate, rule, or assign tasks. Views of man's nature and woman's nature further structure the ways in which each sex is exposed to the rules which pattern behavior.

But culture may embody values that are not consistent and may even be contradictory. Robert Lynd, four decades ago in a book called *Knowledge for What?* showed how Americans wrestled for closure in a system that presented them with such contradictory values as "everyone should be judged on individual merit" and "we should treat all people equally," or "women are the finest of God's creatures" and "women are inferior to men in reasoning power and general ability."[5] Merton and Barber showed later that contradictions in cultural values are one of the sources of sociological ambivalence.[6]

Cultural myths and stereotypes also create problems for women whose presumed qualities of intellect and emotion are regarded as unsuited to the demands of professional life or suited only to particular professional spheres. As participants in the culture, women often cannot help but believe these myths about themselves and thus contribute to the self-fulfilling prophecies in which stereotypical bias creates the situation it has predicted.[7]

The Notion of the Good Woman

A pervading myth about women is that they possess or should possess higher moral standards than men.* Although the culture is not consistent on this point (there also occur themes posing women as seducers and temptresses, as witches and devotees of the irrational when it comes to supporting larger group interests, for example), the myth of the "good woman" has been exploited in professional life to serve the

*Such norms as the "double standard" of sexual behavior exemplify this view in the modern era. Historian Nancy Cott, interpreting sexual ideology of the eighteenth and nineteenth centuries, points out the roots in the writing of the clergy; for example, "Thomas Gisborne clearly considered women moral beings responsible for society" (p. 225). "Evangelicals . . . [proposed] that the collective influence of women was an agency of moral reform." In the United States, New England ministers "renewed and generalized the idea that women . . . were more pure than men. . . ." (p. 227). "Passionlessness: An Interpretation of Victorian Sexual Ideology, 1790–1850," in *Signs,* 4, no. 2 (Winter 1978): 219–236.

interests of particular groups. Ironically, this stereotype has been used not only by male gatekeepers seeking to keep women out of the profession, but also by feminists employing it out of a sense of mission to achieve social goals.

A step back into history documents the way in which this view set the stage for law:

Nature has tempered woman as little for the judicial conflicts of the courtroom as for the physical conflicts of the battlefield. Woman is modeled for gentler and better things. . . . Our . . . profession has essentially and habitually to do with all that is selfish and extortionate, knavish and criminal, coarse and brutal, repulsive and obscene in human life. It would be revolting to all female sense of innocence and sanctity of their sex . . . and faith in woman on which hinge all the better affections and humanities of life.[8]

So wrote Chief Justice C. J. Ryan of the Wisconsin Supreme Court in 1875 when he argued against admitting Lavinia Goodell to the bar of that state. Ultimately, of course, women were admitted to the Wisconsin bar, as they were to the other bars in the country, but the views about women as unfit for courtroom strife or at least more suited to "better things" still were prevalent in the 1960s and were only seriously challenged in the 1970s.

The notion that women should do good, while a worthy sentiment, inflicts a special burden on those who become lawyers, adding to their responsibilities while lessening their privileges.[9] In the view that women are good or ought to be so, "good" usually translates as "too good"—too good for politicking and therefore governing, too good to make deals and therefore to enter business, too good to be tough-minded and therefore to make good scientists, physicians, or lawyers. As a result, women have been encouraged to perform "good works," which, typically, are low in prestige and poor in career potential. Thus the assumption is used to legitimate the restriction of women to positions without power or prestige.

The set of social controls that forces women to be moral is so internalized that women themselves assess their decisions about work and family in terms of whether or not they are the "right" or moral choices. The cultural appointment of women as upholders of virtue continues in the version of women's movement ideology that advocates changing institutions to rid them of the "male model," which is seen as placing economic profit over humanitarian concerns, and to substitute a "female model" that is morally responsible and responsive to people's emotional needs. Some feminist colleagues also argue (as do other people) that women develop a special sensitivity to issues of "right" and "wrong." Anyone attending meetings and conferences on women's issues hears the ideology expressed in an insistence that women be especially morally accountable in exercising professional and personal options. Betty Friedan, the founder of the modern women's movement, has warned about the faction within it who were most

scornful of "male ego trips," "power trips," and "elitism" when it comes to mundane matters such as fighting for a job or professional decision-making opportunities. They make women feel guilty who are advancing in their own profession . . . by claiming the main concerns need to be racism, poverty, rape.[10]

It is all a matter of degree, of course. Men are also asked to exercise morality and altruism in the church, in the army, for the welfare of their communities, and to be decent in everyday affairs. But in the ordinary work world, they are not usually pressed to devote themselves to the public good and do not have to apologize for working for their own interests. Women, on the other hand, are often pressed for such devotion or such apologies, and the attitudes that stress their morality have negative consequences for their exercise of the career options available to men. Choosing high-powered work that lacks a public service component often makes women defensive and creates anxiety.* The unfairness in the demand that women should more strictly devote themselves to good works is compounded because lawyers who engage in public

*Matina Horner has shown that many competent women college graduates demonstrated high "fear of success" scores in a study reported in 1968. ("Sex Differences in Achievement Motivation in Competitive and Non-Competitive Situations" [Ph.D. dissertation, University of Michigan, 1968]), which drew a great deal of attention as an article in *Psychology Today* ("Fail: Bright Women," *Psychology Today* 3, no. 36 [November 1969]: 36ff.). Although the motive to avoid success was proposed by Horner as a "psychological barrier," she has recast her interpretation, now emphasizing that the motive is aroused as a function of an individual's expectations regarding the negative consequences of being successful ("Toward an Understanding of Achievement-Related Conflicts in Women," *Journal of Social Issues* 28, no. 2 [1972]: 57–176.) For women, this is apt to be a realistic assessment borne out by studies which show that males tend to be less favorably disposed to female achievement than to male achievement. Cf. the following: Lynn Monahan, Deanna Kuhn, and Phillip Shaver, "Intrapsychic Versus Cultural Explanations of the 'Fear of Success' Motive," *Journal of Personality and Social Psychology* 29 no. 1, (1974): 60–64; Adeline Levine and Janice Crumrine, "Women and the Fear of Success: A Problem in Replication" *American Journal of Sociology* 80, no. 4 (January 1975): 964–74. L. Robbins and E. Robbins, "Comment On: Toward an Understanding of Achievement-Related Conflicts in Women," *Journal of Social Issues* 29, no. 1 (1973): 133–137; B. Rosen and T. H. Herdee, "The Influence of Sex-Role Stereotypes on Evaluations of Male and Female Supervisory Behavior," *Journal of Applied Psychology* 57, no. 1 (1973): 44–48.

Marlaine E. Lockheed refined Horner's findings to show that it was not just "success" that was a problem for women, but being successful in an activity described as typical for men but deviant for women. Marlaine E. Lockheed, "Female Motive to Avoid Success: A Psychological Barrier or a Response to Deviancy?" *Sex Roles* 1, no. 1 (1975): 41–50.

service law such as legal aid make a financial sacrifice and usually do not get rewarded in prestige terms either.

The law is a particularly interesting context in which to explore the way in which broadly esteemed virtues such as service (which all lawyers are supposed to subscribe to) or improving the lot of the poor become negatively valued attributes that define their practitioners as not very successful, as in the case of public interest lawyers.* When women hold jobs that are low in prestige and rewards, the stereotyped view that they are unambitious to begin with is reinforced.

Indeed, the nature of the legal system itself creates certain conflicts for people with liberal values, as Ralph Nader, Jerold S. Auerbach, Gary Bellow, and other critics of the legal system have pointed out.[11] Just as student debaters learn to research and argue any resolution they may draw, lawyers learn to build cases in support of their clients, in most cases, whatever the issues. Lawyers are trained to argue either side of a case: they never present all sides of an argument or even the smallest piece of evidence that might detract from a client's case. The adversary system of law is supposed to safeguard against inequities—as the lawyer for each side collects and presents all the evidence relevant to his or her case, the "whole truth" is supposed to emerge.

Of course, many lawyers hold to a standard of ethics which precludes defending clients they believe to be unworthy, but it is my impression that women find this especially difficult, not because they are more moral than men but because their attitudes and behavior are controlled by constant reminders that other people expect them to display such special sensitivity.†

A lawyer who wishes to move ahead may join a public service institu-

---

*Robert K. Merton has identified the general process of unanticipated consequences of social action in "The Unanticipated Consequences of Purposive Social Action," *American Sociological Review* 1 (1936): 894–904, and again in *Social Theory and Social Structure* (New York: The Free Press, 1957).

There are other negative consequences of this presumption of women's virtue. Several authors have pointed out that the behavior of women criminals has been incorrectly or incompletely interpreted. Women are not believed "capable" of committing crimes for pecuniary gain or in reaction to political or social injustice. Rather, theirs are believed to be crimes of passion, incited by men and/or love for a particular man. Marcia Millman and Rosabeth Moss Kanter, *Another Voice: Feminist Perspectives on Social Life and Social Structure* (Garden City, New York: Anchor/Doubleday, 1975). See also, Frances Fox Piven and Richard A. Cloward, "Hidden Protest: The Channeling of Female Innovation and Resistance," *Signs* 4, no. 4 (Summer 1979): 651–70.

†Jerome Carlin, in *Lawyer's Ethics: A Survey of the New York City Bar* (New York: Russell Sage Foundation, 1966), found that lawyers in the elite bar could be more ethical since the kinds of cases they dealt with did not expose them to the moral dilemmas that lawyers practicing in the lower strata faced. (Of course, one could argue that although the upper strata might not deal with strictly criminal situations, defending large corporations on issues such as pollution is itself unethical.)

tion to train for a practice that will later position him or her on "the other side." For example, the lawyer working on anti-monopoly cases for the government may, after some years, move to a private firm that defends large corporations; or attorneys who prosecute for the government may later argue cases for the defendants. And, of course, with less opprobrium, lawyers from large firms may go into government service, although this is usually at the top where prestige is high though income is low.[12]

The process is so common that a term has been coined for it: "The revolving door." But no matter how common, it violates the value-consistent behavior societies require of their citizens if they are not to be labeled as hypocrites. When this happens, attorneys may be asked to explain their ideological turnabouts, particularly when the turn is from the public good to the private pocket. But at the same time that consistent behavior is expected, certain social mechanisms provide justification for inconsistent behavior; for example, society provides a ranking system that permits deviating from one type of behavior if the change serves some higher priority. Thus, the lawyer can plead practicality, the need to support a family, or the need for more intellectual challenge. When lawyers leave practices representing indigents, it is commonly presumed that there is little one can do, that the system is generally corrupt, and that the work is too emotionally taxing.

Yet women lawyers who abandon public good for private gain are especially frowned upon, and indeed, some frown upon themselves, particularly when the move promises substantial personal gain. The common view is that women are secondary breadwinners and so economic need does not justify such a change. Furthermore, because women are placed on moral pedestals to begin with, any fall is evidence of greater capitulation than that of a man who is viewed as more worldly and thus subject to human frailties. For a man to enter the corporate law sphere is appropriately masculine, but a woman entering that world is guilty of exhibiting masculine, and therefore bad, behavior. Merton has noted the phenomenon in which in-group virtues are regarded as out-group vices. Thus the in-group hero is judged thrifty while the out-group villain engaging in the same behavior is labeled miserly; Abe Lincoln is praised while Abe Cohen or Abe Kurokawa is damned.[13] When men leave public practice for private gain, it is for the family or for the intellectual challenge; when women do, they are selling out.

Women seem to have more "explaining to do" to justify choosing to practice in large corporate firms. The opinion of one woman partner in such a firm was typical:

In law school there were a number of people who seemed committed to working for legal aid, doing something which on the surface it's much easier to justify as "doing good." Maybe a lot of people feel differently, but I happen to think there are all kinds of ways of doing good. If you really have an interest in making changes, it often can be more effective to go into the established institutions, become part of the management of them, and try to change them.

Many women who go into big firms acquire ambitions for success in terms of prestige and money which they did not have as students.* Some of these women feel uncomfortable about liking work they once viewed, or continue to view, as socially unworthy. One young associate pondered:

The type of work I am doing is incredibly stimulating, but in terms of my ultimate goals, and how I see myself as affecting society, I wonder. The clients I have are all big companies, very rich, and I know society is not bettered by killing myself for [client's name]. I would feel better in the long run if I worked for government or something like that.

Men as well as women associates, especially those who went to college in the 1960s, often described how they expressed guilt among themselves about working in the large corporate firms, but a few of the women seemed more ready to go on to other, less lucrative work, or so they said. One lawyer indicated how she saw the change:

We all took these jobs for the training, and then hoped to do something like work in the district attorney's office. That's the way everyone rationalizes the sellout. We all talk about it; even the people who have been here five years talk about the sellout. When we were poor starving students the only thing that mattered was Constitutional law, but in the meantime, you had to make money. So I'm getting the training. But I can give it up to do something else.

It is hard to pin down the "public opinion" that disapproves of women's participation in the business of the professional world, but one can speculate. Part of it comes from the ambivalent position most people find themselves in in regard to professionals generally: needing their services, having to defer to their expertise, but resenting dependence on them and their high financial rewards.[14] It is bad enough when the professional is male, but to many it seems unsupportable

*See chapter 11.

when the professional is female. Women are supposed to look up to men and not to be in a position to lord power and money over them.[15]

Women who showed ambition in the large firms puzzled their male associates. A young man from the Columbia Law School described his bewilderment:

You so rarely found a woman in law school who would say, the way men will tell you, "I'm interested in making a lot of money as soon as possible." And I don't believe there were many who thought that. . . . Now, when I run into a woman in the firm who is very committed to making money and becoming a partner, it seems weird.

The feeling that ambition in women is "weird" has its consequences. This associate thought the partners in his firm had "this picture of the guy who is in there wearing his vest for dear life, but they don't have that picture of a woman." As for himself, he didn't think Wall Street was wonderful, but he believed that "only women could afford to take the public interest and government jobs." The clear-cut motivation of a woman aiming for the top was headlined recently in the *New York Times* in a way that revealed popular ambivalence: "A Professional With No Apology for Ambition," was the way Wall Street law partner Barbara Thomas, who became commissioner of the U. S. Securities and Exchange Commission, received top billing.[16]

Women may find themselves lumped together as a class,* sometimes by men but also by women. This makes them particularly subject to the power of "public opinion," for example, to accusations of disloyalty by other members of the group. A highly successful attorney, who left a large corporate practice to teach law, spoke of her growing consciousness about women's interests and her growing discomfort with the kind of legal work she had been doing:

Women will discover in many of the practices in which they're now engaged that they are ultimately working against their own self-interests. For example, those who work for drug companies. On my own part, I had had a growing awareness of working against my own interests, in the sense of being a woman, through the years in a big Wall Street firm.

Because men are generally viewed as persons, not a class with a similar set of "male interests," a political or professional attack on an-

*Since *class* used in the sociological sense implies ties and unity among people who belong to the same category and since all women do not regard themselves as tied to each other, the term *category* might be more technically correct here. But even if categories are not classes, they are viewed by others as classes, and this fact "is real in its consequence" (to borrow William I. Thomas's phrase; see p. 362n).

other man is not seen as an act of disloyalty. Many women, particularly those identified as feminists, have, like blacks, a sense of group consciousness and feel strongly that women ought to support women's issues and causes. Charges of cooptation by the system, or "Aunt Tomism" are made against those whose behavior is viewed as against the cause.

One feminist lawyer felt that women were used by "the other side":

On women's cases they always trot out women on the other side. Always! I find it oppressive, horrible! There was a time when I had three women's cases all of which had a woman on the other side, at the state and city levels. One time I said to one of the women, "Look at what they're doing to you!" She said, "No, it's just a coincidence." One day there were four women's cases. They were all cases of women against the government. All the government attorneys were women. And she could no longer say to me it was an accident.

An equal rights advocate in San Francisco reported that when her firm was fighting in a case for special benefits for pregnant women, the attorney arguing against them was a pregnant woman. A number of woman attorneys interviewed felt that corporations, law firms, the government, and individual men (in divorce and custody cases) tried to seize a psychological advantage by having a woman attorney represent them when women's interests were at stake.

Women lawyers with a crusading spirit about defending other women against exploitation or discrimination have been especially angry about meeting female lawyers on the other side. One feminist lawyer described her chagrin at finding a woman lawyer representing a male client on a child support issue:

It's shocking to find that women will argue cases which are against the interest of other women. For example, another woman attorney is opposing me on a child support issue and trying to decrease the amount the husband will pay my client, the wife. I find it incomprehensible.

Another said:

I'm disappointed in women lawyers I meet as opposing attorneys. I've had occasion to see this in cases which were both matrimonial and criminal. I don't know whether its imposing an unreasonable double standard to say that they were worse than men, or I only felt that because I wanted them to be better.

An advocate for prisoner's rights and the rights of the poor said that women who prosecuted such cases as district attorneys and U.S. attorneys were immoral:

"I have a basic disappointment in women being on that other side. I realize there are terrible flaws in that, but it's that you want people you identify with to identify with the things you do."

"Aren't there quite a few women assistant district attorneys?"

"Yes, more and more, and I sense that I wouldn't want to be friends with any one of them. They are worse than the men."

For many ideologically inclined women lawyers, there is little danger of having to oppose feminist concerns. Their reputations as feminists are so established that no one comes to them with cases that might put them into a moral dilemma. Women lawyers who are free from economic pressure also feel freer to act on principle. A lawyer interviewed in a large corporate firm felt that as a woman she could risk her career rather than choose the wrong moral alternative because she was less financially dependent on her job and therefore less ambitious for its highest rewards. But she also felt, echoing a popular view, that as a woman she was also specially attuned to what the moral issues were:

Women can take more risks if they are married.... I feel that I've taken so many other chances in my life, if this doesn't work out, who knows? There are so many other things that I might possibly do. You're more likely to stand up for what you believe than someone who's going to be a company man.... Ethics. I think I've put across to you that I'll be goddamned, even if my career at this point is at stake, that I will not do something unethical. And I'm not sure that a man who sees partnership as the goal might not contemplate it.

The Consequences of Sex-Typing

"Sex-typing," the labeling of social roles as appropriately performed by men or by women, creates ambiguity and ambivalence for those who choose a profession traditionally associated with the other sex.*Women who have gone into law, long regarded as a male profession, are entering a non-traditional work role for women and are subject to the suspicion, hostility, and incredulity society attaches to any nonconformist.[17] As one attorney recalled:

*In Robert K. Merton's analysis of the dynamics of status sets, occupations are described as sex-typed when "a large majority of those in them are of one sex and there is an associated normative expectation that this is as it should be" (unpublished lectures at Columbia University 1960–1964).

Ambivalence and Collegiality

I grew up having to be different in the law . . . it was always "look at the lady lawyer," and at parties people would come over and say, "I hear you're a lady lawyer."

Because "male" and "lawyer" make up a familiar "status-set,"[18] when a lawyer is a woman, attention is often drawn to her sex status rather than to her professional status. Thus the status-set typing of lawyers is violated.* The circumstances under which "functionally irrelevant" statuses are activated in a social context have been explored by sociologists.[19] Everett Hughes wrote of the predicament of women physicians and engineers in the 1940s, when there were so few practicing that their presence was unique in their professions.[20] Focusing on inappropriate statuses rather than those that are appropriate may be interpreted as a deliberate or unwitting effort to remind those defined as inappropriate to the group that they are outsiders. Thus, when a hiring partner in a law firm sees a woman law student as a "sweet girl," he is not thinking about how best to assess her competence as a lawyer. A lawyer commented on the other negative results of such inappropriate behavior, in its impact on the woman's authority and her self-image:

Some judges call you "young ladies" and do not take you as seriously as they take male lawyers. There's nothing you can put your finger on and it's nothing where I can say I'm being mistreated, but the case may be coming out the wrong way because of it. It is a very unpleasant situation. What I've always found is that when the other people are not taking me seriously, it makes it that much harder for me to take myself seriously.

Not that it is always improper to focus on the "irrelevant" statuses a person holds in a social setting that calls for activation of another status. Far from it. References to shared statuses as sportsmen or members of political parties often help smooth social relations among professionals —as when doctors kid each other about their golf or tennis in the operating room to reduce tension. It is only when primary attention is paid to the sex status (or minority status, or whatever else makes the person seem inappropriate) that the person is prevented from performing as required.

---

*Status-set typing refers to the phenomenon that exists when a class of persons who share a key status (for example, lawyer) also share other matching statuses (for example, male, white, Protestant), and it is considered appropriate that this be so. Cynthia Fuchs Epstein, "Encountering the Male Establishment: Sex Status Limits on Women's Careers in the Professions," *American Journal of Sociology* 75, no. 6 (1973): 966, n.4.

"Proving Themselves"

Women lawyers, sensitive to the generally negative evaluation of their competence, have in the past reacted overwhelmingly to the need to prove their performance. This often resulted in an intensity of over-work that others regarded as over-reactive. The mid-1960s study of older women lawyers showed them, strangers in a male profession, reacting as persons subject to status uncertainty.[21] They made frequent unsolicited comments about their need to prove themselves and, in-deed, to be "better than a man." Here are some of their comments:

I think that any woman who wants to go into any profession unfortunately has to constantly compete. She's got to show she's as good as the men. She's not weighed just on the level of being a doctor or a dentist or a lawyer. And this puts a greater strain on her. You've got to either be better or want to be better.

If you're a woman, you have to make fewer mistakes. . . . A woman must put greater effort into her work . . . because if you make a fool of yourself, you're a damn fool woman instead of just a damn fool.

We have to work harder . . . we have to be on the ball, or we have to have someone pushing our cause a little harder—and that somebody is ourselves—that's the difference.

Over the past fifteen years, the need of women attorneys to prove themselves seems to have abated somewhat. Numbers have made a difference. Young women in law school today do not feel quite so visible or under constant evaluation as their predecessors did. In the late 1970s they were also confident that the strong demand for lawyers meant they did not have to be better than the men. But once these women had left school and gone to work for firms in which they were a distinct minority, proving themselves again became important. Several young men in large corporate firms complained that while men would write fifteen-page briefs, women would write forty pages in an attempt to be thorough. Women doing litigation interviewed by Frances Coles in her study of California attorneys said they worked harder than men.[22] The young radicals I interviewed in the early 1970s said they worked night and day not only because of their political commitment but to "prove themselves," and a legal secretary at a Wall Street firm told me that the other secretaries did not like to work for the women associates because the work load was heavier than that of men, since the women lawyers felt they had to do more to make partner.

Ambivalence and Collegiality

Women have had to prove themselves because their commitment is considered dubious by male colleagues. A partner in a Wall Street firm gave this report to his colleagues around a lunch table about a woman applicant from a top law school, who first had worked as a secretary to put her husband through law school:

They [women] prove time and time again that they put their husbands and children first, ahead of their careers. . . . They just don't have what it takes to become a Sullivan and Cromwell partner—they're not 100 percent committed.[23]

For some women, working extra hard has been a way of showing that they are not "like other women" in that their commitment is no less than a man's. Thus, a woman intent on achieving partnership reported:

I felt you had to be super terrific to make it as a woman in the years 1970 to 1975 [when I made partner]. The hours I worked, the deals I took on, the strains were tremendous. I thought you had to do if not double, at least time and a half of a man to make it.

Why Can't a Woman Be More Like a Man?

So asks Henry Higgins in *My Fair Lady.* If women only behaved like men, he reasoned, all would be sensible and comfortable. But when women do act like men, it doesn't work either. "When a man acts rather aggressively, other colleagues refer to him [with awe] as a real bastard. There is no such term for a woman," bemoaned a young woman associate in a large firm. In fact, the male lawyer is praised for being tough and driving a hard bargain, but the tough woman lawyer is considered difficult to work with. Younger women lawyers today are often stereotyped as "tough" and "humorless," and this perception is an example of how the process of ambivalence acts to restrict women's integration into a society of peers.

Professor George Cooper of the Columbia Law School noted ironically in a recent interview that in the past hiring partners often told female law students they probably were not tough enough to practice law. Women law students applying for jobs with the New York district attorney's office were also challenged about their ability to act tough with the criminals they would be prosecuting as well as their ability to impress the court personnel and police they would be working with. But women lawyers in the past were damned for being assertive. One commented pungently:

... you hardly ever met a man practicing law who didn't regale you with stories of the horrible experiences he had with ballsy, nasty, aggressive women lawyers and how different you were.

The indictment of women lawyers as "too tough" persists from the past. Even an older feminist woman lawyer thought that "the women coming into law are too intense and humorless" and asked "whether the law schools are producing a kind of woman who is so competitive that she can't relax a little." "Stiff" and "inflexible" were other adjectives applied to young women lawyers by men and older women in law. One described a young woman lawyer:

... entirely inflexible with respect to points any objective person ... would agree to ... inflexible on points that were just totally unreasonable. ... She just wanted to show us that she was tough. That she was equal to negotiating with the men. She was a very good attorney—there's no doubt about that. But just being inflexible, consistently, was just a shock to us all.

Being too serious made young women lawyers "creeps," a woman lawyer in her fifties told us in the company of her lawyer-husband who nodded in assent. A young male associate, who was considered so sympathetic by women co-workers that one of them recommended that he be interviewed, described women lawyers as "too harsh and unbending." He thought that many of his male colleagues felt the same. "Women lawyers are more intransigient than men. They want every *i* dotted and every *t* crossed," he said. Another male lawyer commented, "They don't operate in the spirit of compromise that male lawyers do."

It seemed odd to me that the bright young women lawyers of this generation would have such bad "press." I found some of them warm and some cool, but all within the range of acceptable and pleasant behavior. Thus it appeared that not only toughness or inflexibility, but any characteristics women were seen to have or to lack underscored for men women's incapacity to do exactly what the professional lawyer ought to do in every context. Part of the problem may lie in difficulties of interpretation arising in a perplexing time of transition, when both men and women are unclear about how to behave and what to expect of each other. A male partner in a large firm who regards himself as a supporter of women shared the harsh judgments of his peers when he explained why women were not meeting his standard of professional behavior:

My own feeling is that women in law are less human, less compassionate, and less accommodating.... And I don't know whether that's because they're afraid to be tagged with that image of womanhood that they're trying to separate themselves from or how they can sort it out. Maybe it's simply because they don't have the tools of communication because it's still a man's world.

"Maybe its simply because they don't have the tools of communication" was certainly a possible explanation for behavior of those women who were not just strong but really inappropriately inflexible. Communication is on everybody's minds these days in social science circles, in corporate settings, and in social movements. And women have special difficulties in learning and mastering the subtle processes of interaction and role virtuosity in male-dominated professions. One woman lawyer reported:

As a matter of fact, I do think there are a few things that you have to be aware of as a woman lawyer. One that I noticed, one thing that happens to me all the time, is that I always manage to get on pretty bad terms with the opposing counsel. It just always happens. We're sort of fighting. And I was at a trial last week, it was another case, but the law firm that was representing the defendant I've had in another case. Well, we went in there, I never said hello to them, I almost always say good morning, how are you, good afternoon, etc.... and after a while, I realized that I was angry with them for not being courteous to me and at the same time I saw that in a way, that they didn't know how to deal with me. And the one thing I want to do in terms of developing my ongoing work relationships is to deal with the situation of being friendly in court. It's for me to always go over and shake hands, to take the initiative, that responsibility, because if I don't they won't and if they don't I'll sit over there and sulk and think they're being mean to me.

The professional role ideally is a balance of detachment and easy rapport. It is the detachment that is most visible, however, and outsiders are often unaware of the lighter side of collegueship or its important functions on the job. As outsiders, women have stereotyped the behavior of men in male-dominated occupations as cool, detached, and emotionless. This probably does characterize some segments of male professionals, and it may even constitute an ideal for those in the elite strata of the professions. But at all levels, professional work requires fellowship and a commonality of understanding. Thus joking, informal language, and discussion of outside interests serve to smooth the course of interaction.* Those women professionals who behave self-consciously or with coolness, detachment, and strength may feel that to appear professional or to counter men's stereotypes of them, they must act in ways they consider representative of men's behavior. One woman admitted that an attorney on the other side of her current case might think of her as hard-nosed:

> He thinks I am a pain in the neck. Maybe when he first heard a woman was working on the case he figured, no problem—a pushover. That is what I think a lot of men believe initially—Oh, it is a woman, yes you can push her around a little bit. So I don't give in on issues where a man might. It's hard to know where to draw the line.

Sometimes women observe men's behavior and try to imitate it. But observation is always selective, and their translation of behavior may be too literal. Translations of literature and poetry aim to capture the sense and mood of the original, even if it means substituting different words. When women do the same things men do, they may lose the essence of the interaction. For example, if a woman were to slap a man on the back to show camaraderie, her gesture might be viewed as awkward. A big smile or even a hug might serve the purpose better. Translation is a talent, but it is also a craft that can be learned if there are models to emulate or teachers to teach it. An attorney doing constitutional law in the public interest exclaimed:

> It's real hard to be back-slapping members of the club. Men would take it the wrong way. On the other hand, men don't know how to slap women's backs. They only know how to grab-ass them.
> They're [the men] all jocks together, and we cannot go into that locker room and we cannot be like them.

Since they can't go into the locker room, or in some cases the lunch clubs that are limited to male membership, women have difficulty learning how to pick up the nuances of informal behavior.

---

*Joking to reduce tension in the operating room is one example of behavior, in this case among surgeons, of which the public has little knowledge (despite the fact that sociologists have written about it). See Renee Fox, *Experiment Perilous* (Glencoe, Illinois: The Free Press, 1959).

Rose Laub Coser has pointed out that although there are considerable advantages to using humor for tension management, women and other people in subordinate positions are constrained from using humor and are rather supposed to receive the messages humor communicates from superordinates. She points out in a study of psychiatrists in a mental hospital that, while women staff told jokes in informal settings, in formal meetings they held back. In Rose Laub Coser, "Laughter Among Colleagues," *Psychiatry* 23 (February 1960): 81–95.

The Club

Exclusion from the informal organizations, the clubs, has been a powerful way of maintaining separatism. There are the clubs with formal memberships, meeting places, and dining rooms, and there is the club without walls, the male "culture." These complement each other. Long preserves of male exclusivity, clubs such as the Downtown Athletic Club, the Racquet and Tennis, the Knickerbocker, and the university clubs in New York—and other clubs in other large cities—had as members venerable alumni, invariably men of upper-class status who gathered for sports and for conversation about business, politics, and other subjects of interest. In towns and cities across the United States, country clubs or the Elks or Kiwanis performed similar functions for those whose fellowship was based on broader background qualities.

Most women did not care or dare to join these institutions in the past; indeed, most never even imagined that they ought to enter the male preserves until they realized that important discussions about work and the politics of work were going on there. Two women attorneys spelled out the costs of remaining outside the male network:

I know there is a certain camaraderie that is missing when you can't have lunch with the people. Certain conversations where one discusses certain clients. You know, office gossip that you miss out on. There are certain things that I won't hear about and the people will say, "How come you don't know?" Not that any attempt was made to keep it secret, but I realize that it was probably something that was discussed at lunch.

It's amazing. Now, I think that the associates in my department have been really extremely nice about including me in lunch groups from time to time, but it is amazing how much you pick up in forty-five minutes of conversation at the lunch table. It may be just one conversation, say, having to do with a new tax ruling that's come out, but four heads are always better than one in examining one regulation, and so you've got the chance of four ideas whereas I, going out to lunch by myself or with, say, a friend who is not a lawyer, miss the other three ideas.

Exclusion from clubs has not been casual; men have been intent on perpetuating the image of their professions as societies of men. This was dramatized to women by symbolic devices, among them separate women's entrances to clubs usually limited to male members or restaurants limited to male patronage during lunch hours. Of course there were public restaurants in the Wall Street area or midtown for male and female law colleagues to lunch together, but men in the profession seem to favor the club setting. Two women in Wall Street practices in the sixties observe:

There is a lot of business that is done and contacts that are made at luncheons, and most of the downtown lunch clubs are for men only; there is a lot of business done at athletic clubs. I am speaking now not only of country clubs, where people live in the suburbs, but the Downtown Athletic Club, the New York A.C., and that sort of thing. Even within the firm itself, and in contacts with other firms, the men go out to lunch with each other all the time, and there is a great deal of information that is exchanged this way.

There were problems of being the only woman in a male world. The men had to make accommodations for a ladies' room. When there were dinners they always managed to pick a club that didn't admit ladies, so they had to make special arrangements. At the outings and so forth, they always had to make special accommodations.

But in the 1970s considerable pressure was exercised on important clubs to admit women. The pressure came from the firms who had made women partners and were faced with the problem of not being able to hold a partners' lunch in their favorite clubs; it came from club administrators who realized that the economic problems facing the clubs because of changing life-styles and leisure habits might be helped by admitting women. Furthermore, women in the profession were making louder and louder objections to the discriminatory practices that reinforced an outsider status on them. Some clubs changed their rules easily and others only after a struggle with older men who coveted the leather-and-oak quiet of the clubs.* Still others didn't change at all because anti-discrimination legislation does not affect private clubs.

The younger men interviewed supported women's entitlement to participation in those clubs that remained segregated. But older men seem to feel it is important to be alone with other men from time to

*The Yale Club of New York began admitting women in 1969, the same year the university became coeducational, and the Harvard Club followed in 1973, partly as a result of pressure from Radcliffe and professional school graduates (with Harvard degrees) who refused to go up the back stairs to enter the club. In 1977 the Downtown Athletic Club dropped its barriers against women with a vote of 297 to 5, showing how much the tenor of the times had changed. The club's leaders maintained that it was economic pressure that made them do it. Yet other posh athletic clubs in the city still deny women access (Nadine Brozan, "Downtown Athletic Club Admits Women," New York Times, December 20, 1977 p. 41; Carey Winfrey, "In New York, Private Clubs Change with Times," New York Times, January 5, 1978 p. 81, 85). The courts have upheld the right of private clubs to bar whomever they wish. However, the Department of Labor had issued a proposed regulation that would prohibit Federal contractors from subsidizing membership in private clubs that have discriminatory rules ("White House News on Women," The White House, Washington, D.C., vol. 2, issue 2 [February 1980], p.7).

time, unencumbered by women who "might giggle in the library or drink too much and fall off a bar stool." "Before you know it, they'll take over the swimming pool," objected one old grad from Harvard in 1972, before the Harvard Club opened membership to women (although a keen reporter noted that the Harvard Club had no swimming pool).[24] Even today, men as prominent as President Ronald Reagan and U.S. Attorney General William French Smith insist on belonging to clubs that exclude women.*

One young male associate told an interviewer he thought the women made too much of a fuss about clubs these days. He claimed that clubs were not so important since so many had been integrated and women now had many places to go with colleagues and clients. But they were important, he maintained, because they served the purpose of permitting courtly, aging men to "have a place to go where they can let down their hair and tell the dirty and bigoted jokes they wouldn't dare tell anymore in other social circles."

A woman partner in a large midtown firm had another view of the clubs:

In our firm, not many people belong to the clubs. A few belong to the Harvard Club, but that's not like the University Club where other firms take their clients. Our firm takes a table in a particular restaurant. Anyone can go to the table and sit there. Certainly I go to it just as a man would, but if the firm had its "place" at the University Club where its partners go to lunch, I couldn't go.

I had lunch there just a few months ago with a young partner from a large firm who told me the firm pays for their memberships at the University Club. Women cannot belong to the University Club. They may go as guests and then are seated in a special dining room. So we don't do that, which is important because we have a lot of young women coming up in the ranks and many of them have to take clients out to lunch, and they can't go there. Only when a firm doesn't care whether or not their women advance will they use the University Club† as a meeting place.

For the reasons outlined by this woman partner, lawyers handling sex discrimination suits are asking for settlement provisions that businesses

*Both Reagan and Smith belong to the Bohemian Club of San Francisco, which confines membership to men. Smith also belongs to the California Club of Los Angeles, which also bars women. William Robbins, "Philadelphia Battle over All-Male Clubs Reflects a Wider Dispute," *New York Times*, February 1, 1981, p.20. In hearings held prior to his appointment, Smith said that he saw no reason to withdraw his membership in these clubs.

†In 1980 the University Club of New York voted to continue excluding women. The same action was taken by the Cosmos Club of Washington, D.C., a favorite club of distinguished professional men. (Robbins, "Philadelphia Battle over All-Male Clubs," p. 20.

and law firms not hold business-associated lunches in private clubs that discriminate against women. With more and more women joining the firms, partners are realizing that this will have to be done in any event. But local differences remain across the country. Businessmen and lawyers still meet with no embarrassment in places that exclude women and will continue to do so until some legal provision bans this behavior. And women who are denied access as members will miss out on the talk, the social learning, and the contacts made in the relaxed and pleasant settings that clubs provide.*

Making Conversation

In the same way that women's behavior is partly a response to stereotypes, so men's behavior and women's responses to it are also partly based on stereotypes. For example, women who hold back in mixed groups are considered to be timid; men who act assertively in some business settings do so because they know that this will be regarded as both appropriate and good. Many women in law feel that real male behavior conforms to the ideal typical† behavior of the professional—to the abstraction, not the reality. Ideal typical behavior, according to Talcott Parsons's scheme characterizing the normative core of traits expected of the professional, encompasses the "functionally specific"; that is, attention is focused on the task of the discipline.[25] For example, a doctor displays functionally specific behavior by treating a patient's wound, but not by inquiring about his or her rose garden. Many women who discuss the professional behavior of men—students in law schools and feminists at conferences—accept these norms and compare notes about how men are emotionally detached, how they are not friendly, and how difficult it is to engage with them in small talk. Of course, men do engage in small talk, but they may not do so as comfortably with women as they do among themselves. When women exhibit behavior

*Clubs are also important because they permit the member to develop acquaintances with colleagues across firms and across age groups. Mark Granovetter has shown that mobility in professional and technical jobs depends on building an acquaintance network, because most job offers come through "weak ties." Mark Granovetter, "The Strength of Weak Ties," *American Journal of Sociology* 78, no. 6. (May 1973) pp. 1360–1380.

†I use the term ideal-*typical*, following Weber, as 1) abstract generality and 2) an exaggeration of empirical reality. Talcott Parsons, *The Structure of Social Action* (Glencoe, Illinois: The Free Press, 1949 ed.), p. 604.

that seems to be detached and cool, they are criticized by other women as well as men for engaging in "mannish behavior." Aside from the fact that some women are naturally distant and cool, as some men are naturally warm and engaging, women may seem stiff or inflexible because they fear that small talk or joking might be considered frivolous or open up the possibility of unwanted intimacy.

Three participants interviewed at a meeting of women lawyers in Washington, D.C., reported their problems with informal interactions.

If you are informal then you are cute, or you can talk about baseball; but that is the only topic you can talk about.

There are a lot of mechanisms one can use to be informal that simply don't work in heterosexual situations, although in all-male settings they work very well. Because if they don't know about baseball and don't want to do the drinking routine—because Lord knows what you might get yourself into [laughs], you know, it is a problem as to how you do loosen up.

Well, you could talk about kids, but it is a conversational problem as to what you have in common with that guy across the table: you have to work to find it.

Of course, some women lawyers act cool because they are wary of being considered unprofessional. This wariness is well-founded, considering that in the past their competence often was called into question. Like the lawyer quoted above who was wary about being regarded as a pushover, other women lawyers feel they ought to adopt a professional mien. Some do, as she admitted, over-produce a "professionalism" that can be interpreted as stiffness. I was told that women lawyers who happen to be serious, who do not smile and ingratiate themselves, find some men are upset by it, although the same men would probably interpret smiles as an indication that women are too soft for the rough and tumble of the law, as did the lawyer at Sullivan and Cromwell whose negative evaluation of a young woman's potential for partnership was that "she smiled too much." Then there was the observation by a woman to *National Law Journal*'s reporter David Margolick that a senior partner described a woman associate as having "the build and ferocity of a song sparrow" and that Sullivan and Cromwell "wants no sparrows—only eagles."[26] Ambiguity is created in these situations because there is often no "right" way to act. Damned if she is friendly, and damned if she is serious, it is difficult for the woman professional to win. By behaving "like a man," women are blamed for interpersonal incompetence. But it is often difficult for women to gain access to an experienced person who can teach them the informal dimensions of professional roles. Many men have been loath to take on women as professional protégeés, through prejudice, or because they think women are poor investments, or because they fear accusations of sexual intimacy.[27] And because until recently so few women held high-ranking positions in law and other male-dominated professions, there were only a tiny number who could be mentors. Socialization to the professional role can be accomplished through emulation of role models. Without women models to guide them, women have depended on male role models—useful for the formal performance of roles but full of traps for those learning to navigate the uncharted waters of informal behavior in the male-dominated community.

Role Alternation

The ability and desire of women lawyers to move easily from one role to another in the professional setting is another measure of their acceptance as members of the legal community. The lawyer must interact well with clients, fellow lawyers, and judges,[28] altering demeanor and tone as befits the demands of each role. More complicated is the movement from one kind of role behavior to another with the same persons, as when members of the same club compete with one another in debates or sports but activate the colleagial dimension of their relationship later in the clubhouse. Lawyers also expect to alternate roles as opponents in the courtroom with roles as colleagues when a court goes into recess or at the close of a day. Many male lawyers claim that women lawyers are inept at or lack this role virtuosity.

One male lawyer in a large firm described women's difficulties in the legal sphere in this way:

> This lawyer that I met in California said he hated women lawyers because they have no capacity for controlled belligerence. I thought about it when he said it and he's right. The women attorneys who I think are unpleasant to be with and to be in opposition to involve themselves so completely with the case that they're furious with you in court and they're angry with you outside.
> You couldn't say, come have a cup of coffee with me, because the sparks are just flying. I don't think this is true of most male lawyers I have seen and heard

and watched. They can fight like the devil in court, and then it's "Yeah, come on out and have a drink."

Many women lawyers, especially those in public interest and feminist practices, seem to resist such role alternation, objecting that it conflicts with their values or is uncomfortable for them. One lawyer prominent in movement law summed up her feelings about this:

There really is this whole professional subculture; lawyers tend to have more in common with other lawyers than they do with their clients. And women— and, of course it's hard to know how much of it is that they're women and how much of it has to do with the fact that they have a particular political stance —don't do that . . . get friendly after a day in court. . . . Partially, we're not led that way. Partially, if that starts happening, it starts happening on a sexual level.

This lawyer felt, as did colleagues in her firm, that women tend to take the advocate-client relationship more to heart than the colleague relationship, while male lawyers move between the two with more flexibility.

Perhaps the "we-ness" of the legal community is more salient to the man who is truly part of it and thus, after a bitter day in court, can shake hands with his opponent and have a drink with him. Most of the women lawyers interviewed who commented on it felt their outsider's behavior was due to a combination of factors, their early socialization as women and the denial of interaction with professional role partners. New recruits can be made welcome in groups or made to feel excluded, depending on whether members inform them of the group's formal and informal rules. Established members provide information not only about how to act, but how not to act; they can bend the rules to accommodate someone who is different or make the rules even more stringent, as they like.

Some women interviewed were perceptive (and some were not) about the ways in which men could make them comfortable or uncomfortable in working contexts. But many found some male behavior hard to interpret. Among lawyers, when men joke with women colleagues, or make jokes about other women, it is hard to know whether the men are being good-natured or hostilely laughing at them to punish them for entering the male domain. A prominent woman attorney recalled the ambivalence-producing behavior of a client:

I remember in the late fifties coming out of an elevator with a client who met a friend of his and said, "Look what's my lawyer!"

Joking is often used to mask aggressive hostility,[29] and when women do not respond to a joke because they perceive hostility, they are informed that they cannot take a joke, and they are also being told they do not belong.

Doing Interpersonal Work

Sociological studies suggest that men usually work less than women at keeping social interaction alive.[30] In a group discussion among Washington, D.C., women in various kinds of practices, the question of "Who carries the conversation?" came up spontaneously in reaction to a comment that men felt women lawyers were stiff. One woman responded:

> I think what happens in a lot of social settings is that the woman assumes the burden of making a conversation and feels guilty if the conversation isn't going very well, when in fact a lot of stiffness emanates from the other side, although the men fault the women if the conversation does not go well.
>
> Yes, it can be men's awkwardness in not making conversation as much as anything else, more than the fact that we are stiff.

In the past few years I have spoken with friends and colleagues on many campuses about men's reluctance to talk as openly and freely as women do. Both men and women seem to agree with this observation. Of course, these observations tend to be stereotypical and do not take into account individual variance, yet "evidence" from at least one study shows that females disclose more of themselves to others than do males.[31] Conversation need not always involve self-revelation, but openness is one mechanism for creating social warmth. If we assume that men do less interpersonal work in interaction with women it may be because they are unwilling and unmotivated, or simply ignorant. Ambivalence about women joining "the male club" probably increases reluctance to work to make an interaction run smoothly. Or perhaps, as Nancy Henley has proposed, professional men particularly those of upper ranks, are used to maintaining a "cool" or "controlled aura."[32] After all, it would be foolish to propose that men generally lack interpersonal skills. Men "work" at interaction in many settings: effective salesmen do it; Henry Kissinger, in negotiating with heads of states, certainly did it; men sitting around the cracker barrel in the old country store did it; and so do men chatting in the bar or at the gas station today —all engage in interactional activities for their own sakes.

So it is not so much that men and women differ in their ability or disposition to engage in interpersonal interactions. Men may be as facile in initiating communication in cross-sexual interaction as are women, but may simply choose not to do so in certain situations, especially when they can depend on being the object of someone else's efforts. Men do work at interaction when they are friendly with other men who are their equals, when they can expect profit from interaction from other men, when they are in subordinate positions vis-à-vis other men, or when they wish to court or charm a woman. Because women are generally in subordinate positions vis-à-vis men in this society, men learn to expect women to carry on most of the interactional work,[33] and this expectation then puts pressure on women to do it.[34]

Maintaining Social Distance

Issues of hierarchy also come into play. To make women feel at ease would be to concede their equality in situations where men might prefer to maintain rank discrepancy.

If making women feel at ease decreases the social distance between the sexes, it must be assumed that there are men who wish to keep up the barriers. Howard Newby's work on deference builds on the work of a long line of sociologists and social philosophers who have pointed out that because face-to-face interaction tends to reduce social distance between persons of different strata, social mechanisms are brought into play to maintain status differences and preserve traditional authority.[35] Historically, he points out, etiquette in various societies has required the subordinate to curtsey, bow, or touch his or her forelock.[36] The significance of these rituals was noted by Park in his essay on social distance:

Aristocratic society maintains itself on an insistence on social distinctions and differences. The obeisances, condescensions and ceremonial taboos which characterize a highly stratified society exist for the express purpose of enforcing the reserves and social distances upon which the social and political hierarchy rests.[37]

291

Goffman also points out that "any society can be profitably studied as a system of deferential stand-off arrangements".[38] Conversational ploys such as remaining silent and determining the subjects of conversation are some of the distancing arrangements characteristically used by "traditional authority" to protect them from the "potentially polluting close-knit interaction."[39] Another is the insistence that women assume a different demeanor than men. "I have had judges tell me to smile because women are supposed to smile when they greet men," reported a veteran attorney who tended to be serious but certainly was not gloomy.

Class Differences

Class differences between male and female lawyers can aggravate problems in the social side of professional life. The stratification of the bar is such that lawyers of higher socio-economic status tend to practice in higher state courts and in federal courts while lawyers of lower socio-economic rank tend to be found in lower courts, such as criminal courts and family courts.[40] But this does not seem to necessarily hold true for women lawyers who, studies show, tend to come from higher socio-economic strata than their male colleagues, with the discrepancy broadest at the lower levels of the legal hierarchy.

People from the same social classes tend to speak the "same language," and misunderstandings are apt to occur when they come from different social strata. In big city lower courts, such as criminal court, male attorneys who engage in informal banter with women attorneys exhibit behavior they say is meant to be light, but may be considered rude, often calling the women "honey" or "sweetheart" on first meeting. But even when women attorneys do understand how these remarks are intended, they detest the condescension and misplaced familiarity in them. Women in large firms and those who work on appeals cases say that upper-court and upper-class lawyers are courtly and gracious even if they don't go out of their way to put women at ease. The men practicing in lower courts are said to fall back on their customary ways of treating women as subordinates because that is the position of most women they come into contact with: wives, waitresses, saleswomen, women whom they see as subservient. The disparity in backgrounds between the men and women in the lower courts may have been narrower in the past when many women attorneys came from the same kind of immigrant families or the same communities as their male colleagues (although they did not report being treated any better).[41] Today there is apt to be more social distance between male and female lawyers in the lower courts. Women who work in legal aid or legal services in the lower courts tend more than men to come from higher status backgrounds, and thus are different in class as well as sex status from the men with whom they work.[42] Howard Erlanger found that women lawyers working in legal services came from families with higher social statuses than the men who worked there and had attended nationally recognized law schools.

Inevitably, these differences create potential conflict. When a woman is considered inferior because of the classic cultural bias yet holds a higher rank than a man, animosities may readily surface. One woman lawyer reported it was common practice for male colleagues to address her by her first name while introducing themselves by last name: "Hello, Joan," one would say, "This is Mr. Smith." She found it offensive, but when she countered with the same treatment he bristled and felt she was a humorless prig. This anecdote is told in many versions by women in professional life who find that they are addressed by their first names in work settings where the men are addressed by formal titles and last names.

Sexuality in the Professional Context

Men make sexual advances to women in the workplace ranging from verbal flirtation to proposals for assignations, sometimes obviously in the hope of making sexual connections and sometimes as a mask for inhibitions or ineptitude in communicating naturally and seriously with woman of equal occupational status. Of course, women make sexual advances too, but since norms permit men to initiate sexual contact, it is more of a problem for women than men. To some extent, sexual advances may be the result of women's messages that they are available. "I don't get as many propositions as I used to," one woman lawyer told me recently; she had complained five years earlier that men were constantly "on the make" with her. "I don't get them because I think I don't send out so many vibrations now," she explained. Some women inadvertently seek sexual attention at work but may not be comfortable when they get it. On the other hand, men may misread women's friendliness as an invitation to sexualize their relationship. The women interviewed related examples of sexual advances, physical touching, and innuendoes, but few complained about outright harassment. A few told of judges who almost chased them around the courtroom before the 1970s, when such behavior became more obviously objectionable.[43] Several lawyers talked of clients pressuring them to have a social relationship, but sex "in the air" was far more usual in the lives of these women than the kinds of harassment reported in the press recently.

I suspect that some women lawyers tended to take sexual overtures for granted in the past. Some dismissed them as a "normal problem" not worth mentioning, and of course, most found adequate ways to deal with the problem. Some women lawyers welcomed the sexual component in relationships with the men they worked with. Their work could be accomplished even with an undercurrent of sexuality, provided it did not become a basis for exploitation. Sexuality is often activated to charm and seduce for reasons other than the physical act. "Men use sexuality to convince juries," Ruth Bader Ginsburg told me, "it's not something only women use."

Further, some persons can engage in flirtation or sexual encounters without letting either interfere with their work or the tasks at hand. As with other social roles, people can and do compartmentalize their emotional feelings when necessary.* Dedication to the task at hand, respect for the other, and a sense of what passes for "good taste" usually keep sexual attraction and expression from creating problems at work. On the other hand, sexual feelings may be difficult to keep under control. In the past, male lawyers objected to employing women lawyers saying they feared introducing the possibility of sexual relationships; they did not, however, report the same fears when hiring secretaries. Most of the women lawyers interviewed were wary of permitting sexuality to enter their relationships with men on the work scene and attempted to limit exchanges with men to the task at hand. For example, a young lawyer in trial practice, whom I interviewed in the 1960s, expressed her concern at avoiding "advances" and noted the problems and consequences that could come from these encounters in a comment that is timely today as well. In the ethnic professional network she worked in, she said:

---

*A graduate student on my staff admitted to lusting after an attractive male lawyer she interviewed, but she managed to keep the interaction on target and free of sexual messages.

Ambivalence and Collegiality

I'm very careful. I entertain no one. If one or two of the lawyers who are very close personal friends come in and if my brother is around, we might offer them a drink. But other than that, it could be a very awkward situation and I'd be opening the door to all kinds of things that could be misconstrued. It might not happen all the time, but why put myself in such a position? I just never get involved. I must work with these men.

It's not bad. You'll encounter four thousand gentlemen and one idiot—but if the idiot wants to make a pass you have to think of a way to say "no" and also let him save face because you have to work with him again!

And he's apt to know a lot of people—you can't sit down and explain to the assistants or to the detective what a character the guy is if word gets around . . . it's embarrassing. Who needs all this nonsense . . . and the small gossips?

So you just slide away and make believe it's a big joke. And you stay as far away from that kind as you can—that's what I do.[44]

There are certain kinds of sexual communications women cannot avoid. One woman lawyer spoke of such dilemmas and her attempts to deal with them:

A lawyer that I like a lot that I work with from time to time would say things like, when are you going to come down here and sit on my lap and talk about this case? And I would just say to him over and over, God damn it, you really wind up pissing me off about this and you can't talk to me like that because I get angry because I start feeling that you aren't just interested in debating the case with me, but that what you want to do is just carry on this kind of flirtation and it's fun for you to have me down there. I don't think the guy ever had any romantic interests in me. It was just his way of handling it. Probably because of problems he had.

So I just kept hammering away at it and finally got the point across. With the clients it's a little harder because they think they're being nice. It's real hard for them to get the idea that it's not working that way. And one of the things I've done with them, probably a good way to handle it, is to get angry, is to just pound my fist on the table and get real bossy and assertive and yell at them.

What she and others like her may not have known was that their male colleagues interpret such behavior as a character problem and unpleasantness, and may have denied them opportunities as a result. But I would be surprised if many women who resist sexualizing of informal interactions are met with threats of dismissal or demotion. Sex can be a vehicle for smoothing communication, for ego gratification, and for power, for both men and women—some of whom are aware of what they are doing and some of whom are not. To be rebuffed is unpleasant, whatever the intention of the overture, and those rebuffed may respond with psychological defenses that can cause further unpleasantness for their objects.

Frustrated by their dealings with women colleagues of higher social class who often do not appreciate their friendly gestures or who even see them as vulgar, some male attorneys display anger. An extreme reaction was the startling display of a criminal lawyer who, infuriated at the voice and manner of a woman attorney who reprimanded him for his behavior toward her, pulled out a gun and waved it at her in frustration.

Women's position is undermined by jokes in which they are characterized as sex objects, a kind of joking usually done among men while not in the presence of women. Among themselves, women in professional life also make sexual jokes and bawdy remarks about men. Some of their joking is motivated by the same intentions as those that motivate men who tell jokes about women—to undermine their dignity and to degrade them.* Much of the humor of this sort is defensive—an attempt to turn the tables on men, to cut them down to size. But joking directed by the powerful against subordinates, as women in these situations typically are, make some women who are already vulnerable even more nervous and self-conscious. It is not the currents of sexuality in themselves, I believe, that cause problems on the job for women. The problems arise when men and women cannot put sexuality aside in an inappropriate context and when sex is used as a proxy for power.

Ideological Separateness

In addition to the divisions created by differences of sex and class, differences of ideology and politics can create a gulf between colleagues. For example, the women who tend to work for legal service agencies are usually liberal or, as in the post-1960s period, radical. In a study of legal services attorneys, Erlanger found that women lawyers were more likely than men to have engaged in prior reform activity and to come from families with liberal orientations.[45] The men who worked with them were more likely to practice a politics and a profession of expediency. Men practicing in the lower courts who are hard-bitten, cynical, and sometimes reactionary clash with women who are more

*People who are the objects of degradation are first "made strange," and "placed outside" writes Harold Garfinkle—a process that fits the situation of women. See Harold Garfinkle, "Conditions of Successful Degradation Ceremonies," *American Journal of Sociology* 61 (March 1956): 420–24.

idealistic. Thus ideological distance compounds the problems of communication. An example comes from an account of a plea-bargaining negotiation. A male legal services attorney from a lower income ethnic background argued with a woman lawyer from an upper middle-class family with radical leanings: "Joan, what do you want from this case? They're just a bunch of *schvartzes.*" Angered, she shot back, "That may be a part of your vocabulary, but it's not part of mine." He retorted, "What do you expect from a woman?"

Reducing Ambiguity

The ambivalence that interferes with women's competence and limits their inclusion in interpersonal situations can be reduced when one set of values is stressed over another, "proper" statuses are activated in appropriate contexts, or a perception of reality supplants myth.

The reduction of ambivalence is an obligation built into certain roles. Leaders of social movements address themselves to it, as do organized special interest groups. And judges in courts should attend to it. But ultimately it is when individuals interpret their own roles that the shifting of the balance from ambivalence to clarity takes place.

Women lawyers have often managed to gain a foothold in professional circles with the help of strategic persons. For example, women with power help smooth the way for their younger associates. A partner in a major firm pointed out how, as "a member of the inner councils," she was able to intercede on behalf of other women and to orient male partners to their roles in instructing up-and-coming women:

I think there is still skepticism about women's skills and talents. Many women come into the corporate department completely oblivious of the business world; not that all the men come knowing of this world, but at least someone has talked to them about business. . . . One male partner commented on this at a partner's lunch, as a joke. He said, "What she [a certain woman] needs is a dose of the *Wall Street Journal* every morning. So I said, "Tell her!" He thought that was an odd idea. He told the story around. About how I said, "You can take her to lunch or just talk to her. Tell her to get the *Wall Street Journal,*" and everybody chuckled.

In the course of these interviews, women often told of individual men who helped them deal with incidents that undermined their right to "belong." Men in Wall Street firms have, for example, insisted that clients recognize the competence of their women associates. In one case, typical of many, a young woman lawyer discovered that her client had asked she be replaced by a male lawyer. The senior partner in charge of the case refused the request and delivered a polite lecture stating that "This firm only hires the best, and if Ms. [B] was hired by this firm, she is among the best. If we have confidence in her, so should you." The client submitted because the woman attorney was legitimated by a person of higher status.

A number of young women lawyers and their young male associates report that younger men are more aware of women's problems in the profession and they are helping them build proper and natural modes of interpersonal behavior. One young male lawyer told, with sympathy, the story of a woman in her sixth year in a large firm who had not mastered the problem and who was told she would not be made a partner:

There was a woman here who other people thought was abrasive. I could see why some people might call her that, but I didn't feel she was. The rumor was that she would not make partner because she was loud and pushy. She was definitely wrapped up in her work and had a pretty good opinion of herself but that's true of a lot of the men! I would say this woman would fit the mold of the hard-nosed litigator, and she was in the litigation department.

Young men and some older men have helped the women in their firms face inappropriate behavior from opposing attorneys or clients. A young woman lawyer told of clients who cursed or used "four letter words" but then excused themselves profusely, thereby making the point that they were unaccustomed to having a woman in on the proceedings. A young male colleague, also present, put an end to the embarrassing behavior:

Every time they said something like "shit" or "damn" they would say, "Excuse me, [L]." This went on and on. At some point they discussed sending their papers to their opponent to start a suit. One of the clients said, "All right, send the fucking subpoenas . . ." and turning to me, said, "Excuse me, [L]," My colleague had had enough and retorted to the client, "She has heard of subpoenas before."

Another woman associate told of an incident in which, after a long work session, the opposing attorney stretched and put his arm around the back of her chair. He then took her hand and put his other hand on her knee, saying, "Are you having a good time?" This provoked the wrath of her male colleague, who took the attorney aside and told him, "[K] is a dedicated and fine lawyer. You don't have to play these ridiculous kinds of games with her."

Of course it helps to have a friendly man around to rely on, but women have to find ways of dealing with the problems themselves. Merely turning away objectionable comments does not work, since when they do so, women are thought of as "stiff." When they use humor, women must be cautious lest they be misinterpreted or provoke hostile reactions. Furthermore, although some women use humor to get themselves out of sticky situations, others still think it is improper in the work setting. Rose Laub Coser found in a study of hospital staff that women who were quite witty at other times hardly ever used humor in formal staff meetings, though men did.[46] She pointed out that women are supposed to "receive" humor, not initiate it, and they were in subordinate positions vis-à-vis the men at the meetings. An attorney in the Manhattan district attorney's office, told law students at a meeting at the Columbia Law School in 1980 that "It's necessary to be aggressive with the officers in order to get your cases moved along, but if you're aggressive in a way that is perceived as too masculine, then you're censured." She suggested that a hearty, good-natured approach is an appropriate manner for women attorneys. She said that when she was made the object of flirtatious behavior, she tolerated it. She said that "coming on like a cheerleader" helped her because it implied a positive attitude and was feminine without being sexual; it didn't make people angry or uncomfortable, or imply an interest in sexual exchanges. "Yes," another lawyer agreed, "with the men I'm kind of 'hail fellow, jolly well met' . . . I'm very open. I take jokes and I can be one of the boys very comfortably. . . . I'm also safe. I've been married twelve years." The "cheerleader persona," especially when displayed by a woman known to be married, helped men in resolving their ambivalences about women in the professional sphere.

A Wall Street attorney pointed out that in each new situation with men she didn't know, many of whom ignored her, she made it her business to "work the room." She went around the room, introduced herself to each of the men, let them know she was an attorney, and made it clear that she expected to participate in their informal conversation.

The new women of law have interests they share with their male colleagues, and this tends to give them more shared statuses as well. They are far more interested in sports, both as spectators and participants (tennis, squash, and running), than women were before. Some of this comes from the recent emphasis on healthy living, but for some it is motivated more by the desire to be true colleagues, as this lawyer from a large firm explained:

Here basketball is important. In firms where they do softball and volleyball—a lot of women do that. It is a big thing, and they know it, and that's why they do it. Now, who knows, I might be inclined to learn basketball. If I thought it were important, I might strongly consider it.

Another lawyer reported on sports at her firm:

We used to have a good team. We have lost some of our stars. There is a softball league and a basketball league, and we participate. The women are welcome too. I do not. Mostly because I am not very good. If I were good I would definitely participate, and some of the women do.

Women in the large corporate firms also share class and educational statuses with their male colleagues; it is likely now that they too have been undergraduates at Yale or Princeton and have gone on to the top law schools. As an experienced woman lawyer assessed the future:

There are always those people who are not let in. However, in a way it will be easier for women . . . because they have been in the country club and come from families in the right corporate circles. The woman who's grown up in Greenwich, Connecticut, played tennis at the right tennis club . . . when she goes to Harvard Law School and then to a Wall Street firm will be very comfortable.
And she will know exactly what to say to the client, and she will know what he's talking about; she will understand the social frame of reference.

Another thought that women from upper middle-class backgrounds would not suffer from the same kind of exclusion as other women or members of minority groups, at least in the large firms.

Sometimes you see these sort of social announcements in which so-and-so is marrying so-and-so. Her grandfather came over on the Mayflower and was the chief of something or other. His grandfather also came over on the Mayflower, and his father is a big time lawyer. And the bridegroom is going to work for Sullivan and Cromwell, and the bride for Winthrop Stimson, and she is retaining her maiden name. Right? If you're in the *real* club, then being in the locker room is not the major problem.

Women lawyers can also now talk about office gossip more comfortably with the men. When men were on the mobility track but women stood on the sidelines, they weren't privy to the information that makes professional gossip possible. Now they can discuss opportunities in the profession, who is going where, what their chances are for success, and what alternatives there are for those who fail.

In firms with a number of women, rather than a token few, men are less apt to attribute one woman's performance to women in general.[47] A woman partner claimed that because there are "just more of them," women lawyers now entering the profession come in "with a wider range of personalities and a wider range of intellect." Her perception was that

Men can see that women are like everyone else: some are good, some are bad, some are mediocre, and each meets different needs in the system. . . . Less and less do you see the idea of one woman sort of blowing it for the others. . . . I don't hear people saying, "That's a woman for you" anymore.

Another woman in private practice reported how informal interaction had changed in the professional associations:

I used to go to bar association meetings in which I was the only woman there . . . and there was a lot of "Isn't it nice to have a woman" and "Watch your language," but now the bar association committee I am on is close to 50-50 . . . and nobody has the nerve to pull all that cutsie stuff they used to . . . pointing to the woman, poking a colleague in the ribs and saying "Oh, look what's a lawyer."

The integration problem has not disappeared, but it has been assuaged as men deal with more and more women in the workplace. People's awareness of the factors that impede easy communication is also an aid to change when they wish it. (Henley and Jourard and Rubin, for example, show that women tend to be physically touched more than men—both by men and by other women—and that men monopolize verbal space so that women have trouble getting a group's attention.)[48] Few of the problems of communication can be solved unilaterally because they are usually interactional in nature and change with particular situations and people. But not all hinge on interpersonal negotiation. Even micro-choices, such as whether to smile or frown at a particular remark and whether to make a quip or be silent, are determined by people's relative power and position in the social structure. While interpersonal power must still be negotiated, women are developing the ability to communicate in an easier and more relaxed manner in the formal exchanges of professional life.

Experience, power, and expertise all help women to move less self-consciously in the professional community. Women partners and woman judges—gatekeepers themselves—are now involved in setting the informal norms and the tone for the legal community, as men in power have always done. But it remains to be seen whether women in the legal profession will be fully accepted as partners their colleagues can be confident of and whose behavior they can "read."

As Judith Blake and Kingsley Davis, two noted interpreters of behavior, noted some time ago:

Behavior in a given situation tends . . . to be strongly affected by individual interests, to be unpredictable from a knowledge of the norms alone. Far from being fully determinant, the norms themselves seem to be the product of constant interaction involving the interplay of interests, changing conditions, power, dominance, force, fraud, ignorance and knowledge.[49]

V

MANAGING AND COPING

The Self: Confidence
and Presentation

S ELF-CONFIDENCE, assertiveness, and self-assurance are traits women vitally need in order to crack the male professional establishment. There has been a lot of talk among women in the past decade about the problems of being assertive, of feeling self-assured, and of conveying that self-assurance to those they work with. Articles on the subject and advertisements for workshops teaching women these skills appear frequently in magazines and newspapers geared to the working woman.

Studies have found that women have lower self-images than men. This may originate in girlhood; research indicates that teenage girls are less secure about themselves than are teenage boys.* This holds for the post-graduate years as well. A 1976 study found that self-esteem of male law students was higher than that of female students.[1] Different socialization makes for different modes of personality, but personality can change when the conditions of life are altered.

Confidence may also be situation specific. Aside from the fact that some people grow up more confident than others, some may feel competent in some situations but not in others—in short, their self-images may be linked to specific situations. Unfamiliar or uncomfortable situations—for example, when one is isolated from one's peer group or left alone in the uncharted social territory of another social class—may make even the most self-confident person tremble.

*In a study of teenage youths from twelve to fourteen years of age conducted in 1968, 43 percent of the girls reported highly unstable self-images in contrast to 30 percent of the boys, and 32 percent of the girls scored very low on self-esteem compared to 26 percent of the boys. Roberta Simmons and Florence Rosenberg, "Sex, Sex Role, and Self-Image," *Journal of Youth and Adolescence* 4, no. 3 (1975).

16

Many women who enter male-dominated work settings, such as law, have either overcome the pre-existing problems of a socialization that tends to make them less confident than men or have been able to avoid situations that cause problems of confidence. But after the point of entry they must still overcome the problems of working in a setting that can easily alter their confidence.

There were considerable differences in the way women lawyers interviewed felt about themselves—ranging from high self-esteem to only modest self-esteem—but most of them tended toward high self-esteem. Unconventional upbringings and idiosyncratic experiences may have developed in them personalities with the strong "survivor" components typical of women in untraditional fields. Self-confident women are not likely to recognize the punishing or uncomfortable social dimensions of their work worlds or, if they do, they find ways of avoiding them—slaloming in and out and around the barriers like skiers. Single-minded persons, those geared toward particular goals, and persons who are known to be thick-skinned or, as David Reisman called them a generation ago,[2] "inner-directed" may survive and then thrive in groups where they are outsiders.

It is not that these individuals work better than those who may leave the profession or the firm, but that they are able to withstand, avoid, or even be oblivious to resistance from the establishment. A professor at an Ivy League law school talked about how her self-confidence served her in getting through law school and a law career:

When I was young I was pretty impervious to what happened to me . . . I'm just extremely inner-directed. It takes a real hit in the head [to see that something is wrong]. It can be a good attribute but it can also make me look pretty foolish. But I think it probably did save me a lot of unhappiness that I might otherwise have felt.

Furthermore, where women have become established, many have found that male colleagues bend to the inevitable and respond positively where before they would have been rejecting.

Interviewing in the 1960s, I was impressed by the poise and self-confidence of the women lawyers. They did not agonize over decisions or spend a lot of time ruminating. Unlike friends and colleagues in academic life, they weren't always wondering whether they were making the right decisions or ought to be where they were. Although these were only impressions, they were supported in later years when I met again with many of these women at conferences and conducted follow-up interviews.

Other evidence also indicated that most women attorneys were not given to self-analysis. Although I had feared that the interview questions might be too superficial in the matter of personality assessment, some women commented when we were done that they felt the interviews had been soul-searching experiences, something on the order of a psychoanalytic interview. There was further evidence of this attitude in the interviews done in the 1970s, although many of the younger women, along with their educated and urbane sisters in other spheres of life, were immersed in a culture that emphasized exploring their feelings and self-images, even those who claimed they were still "searching" or who wrestled with value conflicts exuded confidence at being competent lawyers.

How did these women get to be so confident? In the 1960s, the successful women lawyers were those who had surmounted cultural discrimination or discriminatory actions in their profession. But the women of the 1970s gained confidence from the knowledge that their participation and ambition in the field of law were bearing fruit. Nothing succeeds like success, and when women were finally permitted to go into court as lawyers, make appearances, argue cases, and meet with clients, they found they were not only capable of doing these things, but liked doing them.[3] It is more and more evident that the reward structure of work may be an even stronger determinant of behavior than occupational socialization.[4] As one lawyer put it:

I started out thinking very hard about whether I would be able to do it [be a good lawyer], but the more I did the more capable I felt and the more I developed a clear sense of career goals. I had a fantasy of becoming a judge in the past, but now that it seems it is something of a possibility, I am definitely thinking that is what I probably will be in ten or fifteen years.

Some lawyers drew a sense of confidence from juggling many different roles successfully. As one lawyer commented:

I feel successful this year because I did what I set out to do—to have a baby and still practice law and make some income out of it. As a bonus to that I succeeded in getting some very interesting cases. A year ago I was not feeling very confident.

Working for firms whose established reputations elicited an almost automatic trust from their clients also gave women confidence. One lawyer imagined her troubles without the firm's backing:

If I were on my own, they would be browbeating me into the ground. I demand certain documents before I will let my client sign an agreement. The other side just accedes because they realize our clients are paying for us to be meticulous. Then the clients are awed. It is wonderful when not only the client trusts you, but the opposing counsel treats you so seriously. It makes you regard yourself differently.

Another attorney said that success in high-powered cases made her feel like a "real" lawyer:

A client came to me who had problems with a trade agency in Washington. He had already retained lawyers who were in two big seats-of-power Washington firms—older men. But he took me on because he thought I could look at the case from outside the establishment. I just went in there and turned the whole thing upside down. At first the firm lawyers resented me because I was cutting into their fees and the work I was doing was infinitely better than theirs. But they ended up really respecting me. After it was over I sat in the bar at the Mayflower with this seat-of-power lawyer who told me, "It has been such a pleasure to work with you and you are so terrific." On the plane coming back to New York I felt: This is the . . . big time and I really can do it, and it is wonderful and look, I can deal with "real" clients.

Women lawyers found that their male colleagues also had problems with self-assurance. (We found this, too, in interviews with male lawyers.) Women had felt that men would not be apt to reveal their insecurities, so learning about them was reassuring:

I found that everyone else, including the men you were working with, weren't confident in themselves in the beginning in anything they were doing. I got scared when I went to court in the beginning; but they got scared when they went to court, too. And that was really great!

Knowing about the problems they would encounter in a male world and even facing them personally did not seem to undermine many women's self-confidence at whatever level of law career they had chosen. This is not surprising because for all of them, older and younger women, just being a lawyer was to have high status and, for a woman, to be outstanding. "A high-level job may be an important route to self-esteem" found psychologists Grace Baruch and Rosalind Burnett in the course of studying women between the ages of thirty-five and fifty-five, in a project started in 1978 and still going on at the time this book was being written.[5] Baruch and Burnett pointed out that their research dovetailed with another study of stress and psychiatric disorder by Dr. Frederic W. Ilfeld, Jr., a professor of psychiatry at the University of California at Davis Medical School. Ilfeld's study, described by Baruch and Burnett, showed that women exhibit twice the number of symptoms as did men, unless the men were poor, black, widowed, or single. The only group of women who had as few problems as white, well-off men were employed women whose occupational status was very high.

Stress was not a common complaint of the women interviewed, although pressure was. The women lawyers, like busy male executives often found their roles demanding and exciting. What stress they did face was created by the unsympathetic attitudes of their husbands or parents. But when husbands supported them and parents indicated how proud they were, most lawyers did not complain of stress. These findings really amount to common sense; people feel good about themselves when they are praised.[6]

My findings are somewhat paradoxical. On the one hand, much of a person's confidence depends on the treatment he or she gets; respect and honor breed confidence, diffidence and discrimination undermine it. Yet even women lawyers who traveled an uneven road generally felt good about themselves. It seems to me this is because women lawyers have high self-confidence that comes from attaining goals unusual for a woman in this society. But the paradox resolves as we see that although self-confidence carries the women a long way, when the "significant" people in their lives—colleagues, friends, and spouses whom they count on for affirmation and approval—do not reinforce their self-confidence by formal or informal rewards, the woman ultimately will be blocked. She may have a positive self-image, but without reinforcement she may put limits on her career ambitions. Women who have gone far, in elite law firms, to the judiciary, or in appellate defender work, are those whom the system has affirmed, as lawyers, as women, and as persons.

The Role of Appearance: Does Attractiveness Count?

A new associate in a Wall Street firm described herself in an interview a few years ago as not having had a particularly distinguished record in law school. She candidly admitted that she thought she had gotten the job by being attractive. Few women had graduated in her year, and the firms were looking for women to hire. She thought she stood out because she put a picture on her resume—the law school had recommended it so that interviewers would remember the applicant.

Unlike other women, who reported they felt offended when interviewers from the firms commented on their attractiveness (and indeed, it could become cause for a discrimination complaint against the firm), this attorney said:

I didn't feel ashamed at all when an interviewer said: "we would love to have you in the firm; it would be so nice to have you around." I didn't consider that a male chauvinist attitude or anything like that. It was going to be money in the bank.

Male partners often commented on the changes in their firms' hiring of women—that it was nice to have attractive women in the office. This was seldom stated suggestively. They meant just what they said: it was pleasant to have attractive women around, in the same way that it was pleasant to work in a place with good services, pleasant decor, and gracious people.

That was also the opinion of a male Columbia Law School Professor:

When the firms became more heterogeneous it was easier for the women. Take the Jews, when they started to enter the firms no one talked to them. At least women you talked to. I mean, nobody minded having women around. That's what has made it easy for women, nobody minds having them around. . . . They really minded having the Jews around, they minded having the blacks . . . but having the women around . . . that's at worst a decoration for the office, and at best it may produce good lawyers, so that's good for the women.

Though men like to have attractive women around, there may be problems at the polar extremes, as one woman partner observed about a former colleague in her firm:

She's extremely attractive, to the point where it can almost be a detriment. I think for women who are with men a lot, particularly in business, it helps to be attractive. I think a really homely woman has just an uphill battle. It helps to be attractive. But when you're really, really, attractive, I don't know. I don't have that problem, but I see there's another level of problems.

Everyone does not respond to the same physical attributes in others (and, of course, there are non-physical components to attraction), but there do seem to be shaped cultural standards on physical beauty.[7] Because the physical attractiveness of women is believed to be more important than that of men in courtship situations (a woman's physical attractiveness is considered more important to men than a man's physical appearance is to women),[8] it seemed interesting to consider the ways in which physical attributes helped or hindered women in professional life. Women often brought up the issue of physical attractiveness themselves, and several male lawyers alluded to women's "looks" as an important dimension in their ability to get jobs.

The lawyer husband of an extremely good looking woman lawyer once asked rather cynically if I had encountered any unattractive women lawyers in the course of my studies. The lawyers I had interviewed had ranged in age from the early twenties to late sixties and in looks from quite plain to beautiful. I had not thought to correlate beauty with job, but as I thought about it, I recalled the successful were markedly more attractive. The husband's question had underscored his view that male attorneys only hired attractive women lawyers with the hope, fantasy, or intention of pressing a sexual relationship or because they felt a good-looking woman in the firm would be pleasing to colleagues and clients. Competence, he inferred, ranked second to beauty. The wife agreed. So did a number of other women attorneys to whom I spoke, and some brought up the issue quite independently.

But there may be more to the process, if indeed it is biased toward physical attractiveness. Social psychologists Ellen Berscheid and Elaine Walster have reported that a number of studies have found that attractive people of both sexes were expected to be more likely to possess every personality characteristic considered socially desirable. Physically attractive people were perceived to be more sensitive, kind, interesting, strong, poised, modest, sociable, and outgoing than persons of lesser physical attractiveness. They were expected to have more prestigious occupations, to be more competent. They were liked more.[9]

The Berscheid-Walster studies indicate that attractiveness probably counts most at the point of initial contacts and thus may in fact be important for job offers. As people get to know each other better, other attributes, particularly their competence and interpersonal style, matter more, although the halo effect of physical attractiveness may persist. Women and men in law firms admit that there is discussion about the attractiveness of their associates in informal conversations, if not in formal assessments. We do not know whether in the end, when promotion or bonuses are considered, attractiveness counts or unattractiveness is penalized. We do know that appearance is important to success in a world where attracting clients, inspiring confidence, and persuading others of one's competence rests on presentation of self.

Of course, other elements that contribute to appearance also count. Women (like men) who are tall, or who are large-boned and thus perceived as having "substance," may have an advantage over the short and slim.[10]

It is certainly true that every woman who attained partnership in the last ten years to whom we spoke ranged in appearance from quiet, understated good looks to very lovely. Apart from that, their manner was typically pleasant and smooth. These were not women who were unduly aggressive, nor were they shy. They were attractive people who probably would be considered as good to have around at work. They would be an asset to the firm and could be depended on to operate always in good taste.

The Self Is Presented: "Clothes Make the Woman"

It is curious that the way lawyers dress—their clothes, the way they wear their hair, and the way they use make-up—is an important component in the presentation of a professional self. Not only men, colleagues, and clients, but women too make judgments about their competence and attitude by the way they dress. Some women attorneys are keenly aware of this, though others dismiss the problem by not attending to it or consider it an inappropriate concern.

Do clothes matter? It is clear that they do. Style in the boardroom may not be considered something very important, but that is usually because the men there have internalized the dress codes sufficiently to appear in bankers greys or pinstripes. They learn to shop at Brooks Brothers or J. Press, or perhaps the well-traveled are outfitted on Saville Row. Style in the courtroom may vary from district to district, from upper to lower courts, but the observer tends to recognize a pattern. The basic suit for men is of conservative cut, unless the wearer wants to convey the message that he is a cut above or different from his peers. In business, men dress to impress or to fade into the woodwork, depending on their position and their aspirations.

There is a lot of attention devoted to what women aspiring to executive types of careers should wear. Books on the subject have appeared and reached the best-seller lists, and magazines for the working woman have had features on the subject. In fact, *New York Magazine* recently published an article on what women lawyers should wear,[11] warning that an easy casualness in smart clothes was the look lawyers ought to strive for, rather than the very formal three-piece suit suggested by articles written by others on how to look successful.[12] Not only popular magazines but professional publications have turned their attention to attire. The April 15, 1978, issue of *Professional Responsibility Update,* a newsletter that reports on news items of interest to attorneys, such as on conflict of interest cases and fee settlements, used two and one half of their ten pages on "Practice Pointers; Attire for Women Attorneys," which led off with the question, "What should a woman attorney wear to court and to the office?"[13] Replies to a questionnaire sent to a sample of women attorneys between the ages of twenty-six and forty, practicing various kinds of law in cities across the United States, informed the reader that pants suits were fine or forbidden (depending on the jurisdiction of the court in which they practiced); that hats were fine or foolish; that long hair was in or out.

The women lawyers interviewed in the 1960s exhibited a range of styles, and few matched the stereotyped views of women in male professions as mannish. Most of the successful women looked chic and were dressed modishly in the style of the day. There was also room for women who affected a casual style. Income seemed to have as much to do with the fashion-consciousness of the woman attorney as it did for her stay-at-home counterparts. Women attorneys in small practices in suburban areas, who practiced out of their homes, often wore simple and not particularly fashionable clothes. Women in midtown or downtown seemed to care more whether the skirt length was the requisite number of inches above or below the knee.

The same was true for women interviewed in the early 1970s and in 1977. Young women lawyers who were in the big firms talked of rushing to Ferragamo during their lunch hour to buy those fashionable and expensive Italian shoes, and matching silk shirts to dark wool suits to achieve just the right efficient yet feminine look. The most chic were sensitive to the fact that the men in the office talked about what the women wore, and they saw this as part of the "different way" in which women and men were regarded in the firms. "No one comments on what the men wear," said one respondent, "but they are always talking about a woman who doesn't dress well."

Dress is also often a key to a woman attorney's politics. Women lawyers who were political activists or dedicated feminists used to dress in the late 1960s and early 1970s to indicate where they stood politically. Women wore jeans to the office and pants to the courtroom as they proclaimed they were not going to be pushed around by men or conventional codes. Yet these lawyers often found themselves in a perplexing situation because they feared antagonizing the court and thereby allowing their clients to suffer as a result of making a point about their own politics through their attire. They also found that if they looked unprofessional in the eyes of their clients it made matters more difficult.

Those whose clients were poor found that they needed to dress more conventionally to set their clients at ease. One woman attorney, radical in her politics and in her commitment to feminism, worked out a style that seemed to work for most situations she faced. She wore blue jeans to the office, but they were perfectly tailored and pressed. With them she wore a tailored shirt, a fashionable blazer, and a scarf at the throat, which tied all the colors together. Simple earrings completed the outfit —fashionable yet professional, and at the same time acceptable to a radical clientele who couldn't object to her appearance because she was wearing jeans.

Last year I was asked to testify as an expert witness in a Title VII sex discrimination suit in Dallas. The women students of Southern Methodist University, together with the Equal Employment Opportunities Commission, had sued a Dallas law firm for sex discrimination in employment of women. In the courtroom, five out of the seven women attorneys appeared in navy-blue suits with flowing bow-tie blouses. They laughed together during the recess at their common understanding about what was proper and attractive to wear to court to show they were women of substance. Norms have been established.

As women lawyers are becoming "normal," along with women elsewhere in the male world of occupations, industry is responding with the provision of clothing which is becoming the functional equivalent of the Brooks Brothers suit—that is, attractive and proper for women. The Texas lawyers in their suits and bow-tie shirts are an indication of that. Furthermore, smart briefcases and handbags, English raincoats, and so on are becoming dependable for those who find these appurtenances appropriate and comfortable. Fashion, at least in this sphere, is following function.

So Many Hours in the Day

17

THE CONFLICT between the obligations of work and the absolute limitations of time is a problem all professionals face.* Accordingly, the true professional's involvement with work is measured in the long hours he or she puts in. It is generally believed that women cannot be as truly professional as men because they do not, or cannot, put in the same hours of work.

How does the woman professional's work load compare with that of the male professional? The popular view turns out to be roughly correct according to census figures that report the comparative working hours reported by men and women in law and in the scientific and technological professions (see table 17.1). But the picture is more complicated than it appears.

According to the 1970 figures, women lawyers worked a mean 38.7 hours a week while male lawyers worked a mean 45.8 hours a week. Of the women lawyers interviewed in 1965, about two thirds held full-time positions and a little more than one quarter worked part-time. It was difficult to learn the precise hours worked by them because the women themselves did not always have a clear idea of what they were. Almost half of the women interviewed were engaged in private practice and could and did manipulate their schedules to meet personal needs, working at night and on weekends, taking off for a day or a few hours for home-associated activities. Private practice, by its nature, intersperses lulls in activity with periods of intense work. These fluctuate with the court calendar (some courts close during the summer), the type of cases a practitioner handles, and whether or not she has many clients at the moment. Comparably, 21.3 percent of the women lawyers sampled for

*This dilemma is only one example of the observation William J. Goode made some time ago, that the total role-structure of individuals is more commonly over-demanding than under-demanding. However, he poses as a corollary the notion that strain is likely to be associated with mechanisms for reducing it. William J. Goode, "A Theory of Role Strain," *American Sociological Review* 25 (August 1960): 483–96.

TABLE 17.1

Weekly Hours Worked by Employed Lawyers, Engineers, Scientists, and Technicians, b
Sex, 1970

	Mean Number of Hours		Number of Hours Worked (in percent)					
	Men	Women	Men			Women		
			1–34	35–40	41 plus	1–34	35–40	41 plus
Lawyers and Judges	45.8	38.7	8.9	32.6	58.4	21.3	47.9	30.7
Engineers	42.2	40.4	7.0	62.2	30.8	10.2	69.3	20.4
Scientists	41.6	37.7	8.5	64.5	27.0	19.1	66.2	14.6
Biologists	42.1	37.2	9.1	58.2	32.7	20.8	64.1	15.0
Chemists	40.9	38.4	8.4	68.4	23.2	14.8	71.4	13.8
Mathematicians	39.8	38.2	8.0	75.3	16.5	11.7	78.0	10.2
Physicists and Astronomers	41.7	40.0	8.9	63.0	28.1	24.9	57.0	18.2
Professors/Instructors	40.8	31.5	24.8	28.8	46.4	43.1	34.9	22.0
Biology	42.2	34.4	23.0	27.2	49.8	42.4	37.1	20.5
Chemistry	40.6	35.8	22.3	38.0	39.6	37.8	38.0	24.2
Engineering	35.9	28.0	34.7	36.1	29.2	55.8	33.1	11.1
Mathematics	39.7	32.9	28.8	30.1	41.1	37.2	46.7	16.1
Physicians, Dentists, and related Practitioners	50.8	40.6	9.8	20.4	84.6	26.8	35.1	38.1
Physicians, Medical and Osteopathic	56.0	44.5	6.5	12.4	81.1	19.1	33.0	47.9

SOURCE: U.S. Bureau of the Census, 1970 Census of the Population, Occupational Characteristics, Table 45: p. 747.

the U.S. Census of 1970 worked part-time and 78.6 percent of the national sample worked full-time (thirty-five hours or more).

How does this compare with the time spent by male lawyers in practice of their profession? Few men can afford to work part-time. Only 8.9 percent of the male lawyers work part-time, contrasted with 21.3 percent of the women. Furthermore, the women were more likely to work a standard work week (47.9 percent, compared to 32.6 percent of men) and were less apt to work overtime (58.4 percent of the men work more than forty hours a week, compared to 30.7 percent of women, which is, of course, not an insubstantial amount).*

When one compares the lawyers with women in other male-dominated professions, however, their commitment to professional ac-

*Although the data on hours of work don't reflect it, David Riesman has suggested that many men also work part-time but call it full-time. They also count long business lunches and activities they do outside the workplace as work hours. David Reisman, "Two Generations" in *The Woman In America,* ed. Robert Jay Lifton (Boston: The Beacon Press, 1964) p. 96.

tivity as measured by the percent working forty-one or more hours, is very high—30.7 percent as compared with 20.4 percent for women engineers and less than 20 percent for scientists. Compared with women in a group of so-called feminine occupations, a far greater percentage of women lawyers work overtime (see table 17.2).

What importance does the number of hours they work have for women lawyers' professional standing? Some types of practices and specialties make great time demands, others do not.* If one is in private practice, working less may affect income. Salaried lawyers, too, are affected by the belief that the amount of time spent at work is an index of personal commitment to the firm and an important consideration in granting partnership and bonuses.

Number of years of practice also affects the time/reward equation. A

TABLE 17.2

Weekly Hours Worked by Those Employed in Selected "Feminine" Occupations, by Sex, 1970

Occupation	Mean Number of Hours		Number of Hours Worked (in percent)					
	Men	Women	Men			Women		
			1–34	35–40	41 plus	1–34	35–40	41 plus
Librarians	32.3	32.8	36.6	47.4	16.0	33.5	59.0	7.5
Professional Nurses	41.9	35.5	11.9	58.9	28.3	29.6	56.5	13.8
Social Welfare Workers	40.8	37.3	9.0	66.5	24.5	16.7	72.7	10.6
Teachers, elementary	39.9	35.5	19.5	52.0	28.5	30.1	55.9	14.0
Teachers, secondary	41.2	36.2	15.9	48.9	35.2	27.1	56.0	16.8
Secretaries	40.0	36.1	15.2	55.2	29.6	20.7	71.2	8.1

SOURCE: U.S. Bureau of the Census, 1970 Census of the Population, Occupational Characteristics, p. 767.

*Perhaps the unusually high overtime percentages reported for academic personnel reflect different definitions of what is properly included in a work-week estimate. Academic men may be more apt to include reading of scholarly journals and books in their work week—activities often done at home. Other professionals may not include reading connected with their field as being "work." However, it is also likely that fewer professionals in other fields feel the need to read as much, or do in fact read as much, as college teachers. The lawyers we interviewed reported there were some periods when they had to read a great deal (when a body of laws changed because of new legislation, for example); at other times they felt far less pressure to do "outside" work of this kind.

317

lawyer can coast on previously-acquired expertise and work less after many years. One may make a considerable income and attract good clients while working only part-time, but this probably can only occur after one's reputation is established.

Women's sex-status has a number of consequences for the amount of time they spend at law work. One has to do with the specialities into which women have been channeled and another with the expectations of their advancement and success—or lack of it. In addition, there are the direct limitations imposed by the fact that they must also devote time to family demands. To some extent, all are linked.

First, because it was not believed that women could or would work extra hours, they were channeled into specialities in which work can be budgeted into a normal thirty-five-to-forty–hour week. Thus, women have been directed into fields such as trusts and estates in which crises (and attendant time demands) are minimal. As a result, there were fewer direct pressures on them to work beyond the normal work week. The women lawyers in both the 1960s and the 1970s groups who worked in "male" specialities—corporate work or litigation—were expected to put in long hours and did meet the taxing time demands of their jobs.

Secondly, in the past women were not expected to work as hard as men because it was believed that they would not receive the same rewards for hard work. It was commonly accepted in the 1960s that a woman would only rarely become a partner. In exploring the relationship of women's aspirations, motivations, and careers, it is clear that a self-fulfilling prophecy was at work. Some women did not expect to succeed and therefore did not put in the extra time at work necessary to guarantee success.

Because the family is a "greedy" institution[1] in which women's roles as housewives carry expectations that they will devote much of their time and emotional energy to it, women lawyers lack the legitimation as professionals within the family necessary to play "the game" involved in time scheduling. Women may work overtime because of an office crisis; they may not, however, wish to work overtime as a way of life, and they get little support from family and friends for choosing such a work style. Some women recast their sights when presented with the pervasiveness of time demands entailed by the fulfillment of certain ambitions. When asked if she had considered working for a judgeship, one lawyer said:

I had to make the decision a long time ago whether I would go into politics. Perhaps I wasn't made for it—but more than that, I would have had to be out a great deal. You can't do well politically unless you're out every night of the week in organization work, charitable work, political work. I wouldn't be able to see the children.

I decided that wasn't possible. . . . I even cut my connections with my bar association at that time.

It was the time problem. . . . I didn't want to be away from the children. To go for a judgeship would have meant being busy daytime and night-time.

But law creates time demands that are quite beyond gamesmanship in the service of ambition. There are workaday demands. Firms work to meet court deadlines; when cases are heard, the lawyer may have to spend all day in court and work at preparing briefs and strategy at night to be ready for the next day. Deadlines on presentations to government agencies and corporate clients may mean that preparations must be made intensively in a short period of time. A married but childless woman lawyer in the sample stressed:

While it's not quite as imperative as it is with a doctor, it nevertheless has something of the same imperative quality. When you need to get something done by a deadline, it has to be done. And if you can't give that sort of service, then you're not doing a true professional job. So that women with children do tend to find themselves in less demanding types of jobs.

The necessity for intensive work is therefore both an expectation and a need. It is especially part of the tradition of the Wall Street firm, but one can observe the pattern throughout the legal profession.

Small firms also require a great deal of overtime work because they have fewer lawyers to share the burden when cases come up requiring intensive preparation. Independent practitioners and those in small partnerships must carry the burden by themselves. Women lawyers in these work situations, like their male counterparts, put in the time required. Two respondents commented:

My time is mostly devoted to work. As a matter of fact, I used to work seven days a week—and many times almost around the clock.

I have been looking to go into partnership because I am extremely tired. I work too hard. Many times you could call me up here at ten-o'clock at night I would still be here, and it's too much. You have all the problems. You have everything on your head. You have to meet with your prospective clients and meet with your clients and yet be in court and yet be in your office. You saw: my phone does not stop ringing. This is it. If I am not in the office, I will get up to twenty-five calls to return when I return. It's impossible. Frankly, I live completely in this office and that's not good for anyone.

For the married woman, and especially for the one with children, the time pressures of work are a major factor in her decision of whether to work at all and, if so, what kind of work to do. Men too may choose work on the basis of the time demands involved, but this is largely a matter of the man's needs and temperament. The norms support his preference to work long and hard, even if his wife opposes it. On the other hand, if the woman wants to enter a demanding specialty, the norms are against her decision. The sanctions of a disapproving husband, family or public opinion may set a prohibitive cost for her. She may make special arrangements to help meet family obligations, but they are hers to make and cannot be delegated to anyone else. One respondent said:

My favorite case was [M]. But it took over a year. I was only home five weekends in the whole year, and my husband and I only had dinner home eleven times that whole year. But we were young and he had to work nights too. . . . It was before either of my children were born. It was possible then; . . . it would be impossible today.

Women responded differently to the time pressures of their legal careers. Many made choices to work part-time, to choose a specialty which did not usually require overtime work, or to quit professional work altogether. Others, however, often with young children and households to maintain, made arrangements that permitted them to meet professional time demands. Their specialties varied. Some women bent legal careers to meet home obligations, and some managed to juggle home demands and career demands (most fell in the latter category). However, a number of strategies were used by all to manage home and career whether they worked part-time, full-time, or overtime.

Although occupational tasks are culturally defined as ascendant over other tasks in American society, for the woman family tasks are considered to have priority over the demands of an occupation. The woman lawyer thus can draw on either set of priorities to legitimate her behavior. In times of occupational crisis, many of the women interviewed found that significant others in their network—mothers, husbands, and housekeepers—could assist them at home. (One lawyer who left an active career reported that while she worked as a lawyer her husband had been very "understanding," but that now he objected to the work she occasionally did for a voluntary organization.) Some lawyers said that they were easily able to arrange court postponements and changes in appointments with other lawyers because of family pressures. (Lawyers who were recipients of such "favors" stressed that they reciprocated; male lawyers, too, occasionally got sick or had home-associated responsibilities.) All of the women who spoke about this issue, however, said they felt self-conscious about asking for special treatment for fear of reinforcing the male view that women were not dependable and imposed burdens on others in the firm. Many prided themselves on never having requested a postponement.

On the other hand, lawyers who had limited their careers to meet home responsibilities felt easier about their decision because the priority of home over occupation saved them from feelings of guilt. Family and friends, as well as colleagues, usually gave support to the woman who returned home.

Making Time

Most of the women lawyers interviewed were innovative in scheduling the demands created by their multiple roles. They stretched their days and "made time." A Wall Street partner suggested a 7:30 A.M. interview. "Now I know something about your time management," I told her, thinking I would gain time I needed as well by starting early. One woman found a hairdresser who opened her shop before 8:00 A.M. and went there before going to the office. Another woman in suburban practice carried a shopping cart in the trunk of her car and marketed if she had time free between seeing a client and going to court.

The women were innovative in their capacity to expand the day in other ways. Many emphasized their ability to do with little sleep. The totality of role demands was often met by enlarging the work day, at the expense of evening and night.

One woman reported that in the interest of marital harmony she awakened at 6:00 A.M. and worked before the rest of her family arose. "Nobody minds if you get up early," she said, "but they mind if you stay up late."

Another who maintained an active social life so as not to "deprive" her husband responded to our query about how she managed her time, in this manner:

"How about—this is a double-edged question—you said that you tried to keep up, while you were at law school, an active social life so that your husband didn't feel deprived: Did you maintain this? Were you able to maintain this?"

"Yes, I did. We have many, many friends."

"Did you entertain a lot?"

"Oh yes. I love to entertain. People knew that we had an open house and if they dropped in at midnight it would not matter!"

"How could you do all your work and studying and so forth?"

"Well, I lived on Dexedrine for years."

The attorneys were catalogue and telephone shoppers, used caterers and restaurants for entertaining, and relied on a host of professional services and on modern technology. "The freezer has changed my life" said one attorney. "I cook and thaw. I put twenty bricks of food in the freezer and take out a brick each morning and pop it in the oven at night."

Most lawyers interviewed did not chat or gossip with neighbors or spend time on unnecessary activities. Most like to work and worked at whatever they did, whether it was helping children with homework or preparing a birthday party. Each activity was invariably defined as a task, made subject to a decision as to the best way of accomplishing it, and finally, was tackled and finished.

Lawyers in the early study were about equally divided between those who had an active social life and those who felt they had limited their social activities or had an inactive social life. It is interesting to note that only four of the married women interviewed in the 1960s said their husbands helped in the household, although some did help with children or special tasks.[2] The younger married women we interviewed in the 1970s tended more to share house tasks with their husbands.

Friendship

When I first thought about how women could achieve professional success and also have a family, I supposed they would probably be imbedded in "supportive networks." Male professionals make efficient use of their time because their friends tend to also be their colleagues. About half of the women lawyers I interviewed in 1965 said that from half to "almost all" of their friends were lawyers, and one quarter reported that their lawyer friends were women. About 40 percent said their close women friends were in some professional field, and only a small number said that just a few of their women friends worked. The latter tended to be women living in high-income suburban communities who had retired from law or were working only part-time. However, even these women said that non-working friends were looking for work (most had grown families), and several asked me if I could suggest job opportunities. Thus they usually had friends who would not deflect them from commitment to work time.

Married women in highly demanding legal work in the 1970s often had few outside friends, however, especially those who were married. For them, their husbands were their friends and they had no time for other relationships. Women lawyers involved in "cause" work often had friendships with clients or other people in the client network. In this way, women followed the pattern of male professionals who did not segregate private lives from professional lives.

Pressure and Stress

It has become common to think about women's work and family obligations in terms of role conflict and role stress. Certainly the number of hours in the day are finite and people can only do so many things at a time. Certainly the many demands of work do create problems for women who also wish to devote time and attention to their families, and who experience, as a result, a great deal of pressure.

But I found in my investigation of the lives of women lawyers that when faced with numerous demands, many did not feel a sense of strain or negative stress. Rather, these women found their lives exciting and dramatic. They developed greater energy when the demands proliferated, rather than feeling drained, and often did not define their situation as problematic. A dramatic, if not common example of this emerged in conversation with one candid woman lawyer, who had had an active independent practice and two small children, two and five years old. "You didn't ask any questions about my love life outside my marriage" she offered. "It never occurred to me," I said, "where would you find the time?" "One can always make time for what one wants to do," she said, and of course that seemed perfectly reasonable as I reviewed the cases of a number of women in professional life I have known who have worked hard, raised children, entertained graciously, and also managed to find the hours for flirtations or even full-scale love affairs. That is not to say that those who do so breezed through life without tension or strain or time pressures. But personality differences, ability to allocate time, and even the differences in place—where they had their offices and where they lived—made manipulating their lives manageable. Others might not have dreamed of attempting to juggle such a load even if they wished to.

Private Time

Combining career and family for women is usually discussed in terms of meeting professional and domestic obligations. Few social scientists have investigated the need for private time. Private time is often unavailable to men as well as women, but men at the top can engineer private time by shutting the doors to their studies at home, or even using a walk on the golf course as thinking time.

Many persons manage very well in a life filled with interaction with people. But others want and need time alone to read and think or to engage in some pleasurable activity like gardening or painting. Private time was a scarce commodity for women lawyers. For most, it wasn't mentioned, because they had no place for it in their lives. But it came up several times as women reviewed their lives in interviews. One commented:

For me, I'm there already. I've made it. My children are in good shape, emotionally, doing well in school and with their friends. I've paid a price for that. For the past eleven years, I haven't had a minute for myself.

A few managed time alone because they traveled so often that the plane trip offered time away from home and work obligations, unless they had to work on the plane. Some had private time when husbands worked weekends or evenings and they did not.

At a panel discussion at a small private law school in New Hampshire, a young lawyer talked about being pressured by both work and home demands so that she had no time to read or be by herself. Her husband, willing to share cooking and child care, expected them to work together at them. But she rebelled at togetherness and insisted after a time that she have an hour after work alone, keeping husband and children away. This was clearly a polar extreme and "outrageous" in the eyes of the young women law students she told her story to. There was murmuring about her selfishness; few of the students appeared to realize how common it is for men to come home from work and withdraw into the evening newspaper until dinner is announced.

The lack of leisure time was a major problem cited in a national survey of working women. Fifty-five percent of women in professional work responded that they experienced this problem in a survey by the National Commission on Working women in 1977.[3]

The Effect of Personality Variables and Physical Make-up

Most of the women lawyers interviewed characterized themselves as high-energy people who could "keep going." They were physically healthy and highly organized, both at home and at work. They were not procrastinators or "agonizers" over decisions. The women who lack these psychological and physical strengths probably make up most of the cohort who drop out of law, or any profession. They had the capacity not only to utilize their time efficiently but were able to seek relief from personal trouble in work rather than letting it deflect them from work. To fulfill professional norms, it appears that women need an overabundance of health, energy, and determination. There is an explanation for the vigor of the people who do.

The "stress" model is common in our culture, yet recently psychologists have noted that stress may exhilarate as well as incapacitate a person.[4] This view projects a model of human energy that is renewable and can be stimulated, unlike the usual assumption, based on economic models, that it is a reservoir "of stuff that can leak out or drain away altogether," as Steven Marks has observed.[5] "To be sure," he continues, "many people with multiple roles tend to run out of time and energy. Yet it is intriguing that a minority . . . are not the apparent victims of 'strain' or 'overload.'" Sam Sieber also points out that some statuses carry rights and privileges along with duties which help a person perform well.[6] Marks identified conditions under which people were energized or enervated.

Individuals seem to have both time and energy for things to which they are highly committed. Commitment may be rooted in personal idiosyncracy, Marks points out, but the culture may encourage certain kinds of commitment. Successful performances in the areas of commitment society rewards—such as wealth, power, and prestige—create high-energy states in people.[7] Many male lawyers have found that high-demand work done in a professional community results in such high-energy states. This is also true for women, where they have been rewarded (rather than punished) for their acquisition of high-ranking statuses. It may be one reason, aside from their individual competence at juggling the role demands in their lives, that they are identified as "superwomen" in the popular culture. Reward and recognition help support the positive attitudes toward a life in which they are operating on all cylinders all of the time.

They many even achieve personality enrichment and acquire greater personal capacity to cope well. Most of the successful women lawyers with families did not report having problems because they were geared toward solutions, as chapter 18 of "Private Lives" explains. Role diversification, Sieber points out, may be essential to mental health, enhancing one's self conception.[8] Rose Laub Coser also has pointed out that expanding one's role-set—the relationships one acquires by assuming a new status—gives one greater freedom and autonomy and, far from crushing a person with an overload of role demands, provides new options and a new breadth of vision.[9]

VI

PRIVATE LIVES

18

Husbands, Wives, and Lovers

ONLY WOMEN who have unhappy marriages are active lawyers," quipped one divorced woman attorney in the mid-1960s. It was a common view of "career women," who were judged suspect as females to have traded the joys and responsibilities of the home and motherhood for the harsh world of the professions. Betty Friedan, in *The Feminine Mystique* (published in 1963), showed how fiction in women's magazines invariably pictured career women as embittered, trouble-making, nasty persons who could be redeemed only by finding the right man and settling down to domesticity.[1]

In the decade and a half since publication of Friedan's book and the launching of the women's movement she helped create, women have fought hard for both careers and families, yet the problem of integrating private life and public life, of having good relationships with men and happy marriages, are still very much with us. Some things have changed, and some things have not. Although it is far more legitimate for a married woman to have a career, she still is usually expected to bear responsibility for the home and to work out ways of making her career palatable to her husband.

These problems are reflected in the marital profile of women lawyers. Like women in other professions, women lawyers are less likely to marry and to remain married than their male counterparts (see tables 18.1 and 18.2).* Three quarters of the women interviewed in the 1960s were married or had been married, and 15 percent of these women had been divorced. The proportion of that group who were married or had

*Women lawyers seem to marry and remain single in about the same proportions as women in medicine and engineering, as table 18.1 shows. They marry in greater percentages than women in academic life. (About one half of Rita James Simon's sample of women Ph.D.'s were married and one half single.) However, women lawyers, like other women professionals, are less likely to marry, or to remain married, than are their male counterparts in any of the male-dominated professions listed in table 18.1. Rita James Simon, "The Woman Ph.D.: A Recent Profile," *Social Problems* 15 (1967):222.

329

been was somewhat higher than the percentage reported by the 1960 census, but this might have been due to the large number of single women lawyers working for the government in Washington, D.C., only of few of whom were interviewed. James White's study of women lawyers who graduated between 1956 and 1965 listed 69 percent as married, a greater percentage than shown by the census, but less than the 83.2 percent reported for male law graduates of the same period.[2] The profile has been fairly constant. The 1960 census showed that about a third of all women lawyers were single, a figure that remained almost the same in the 1970 census. According to the census, 45.6 percent were married in 1960, a figure that decreased to 41.1 percent in 1970, with the divorce and separation rate (the census category also includes the widowed) up from 22.1 percent to 27.5 percent (see table 18.2).[3]

Was this an indication of bad news for married women lawyers? The divorce rate was on the rise throughout the country,* and the slight decrease in marriage rates probably reflected a general trend toward marrying later; furthermore, many young couples were living together without being legally married. But many more young adults were simply living alone. A *New York Times* article on a 1977 census report on people living alone quoted a twenty-nine-year-old legislative aide to a city councilwoman: "I would consider marriage if the right person came along, but I would not give up my career for it."[4] The report

TABLE 18.1

Marital Status of Lawyers, Scientists, Engineers, and Physicians, by Sex, 1960 (in percent)

		Single	Married, Spouse Present	Separated, Divorced, Widowed	Widowed	Divorced	Base Number
Lawyers and Judges	Men	8.2	86.8	4.9	1.9	1.8	202,341
	Women	32.2	45.6	22.1	9.2	9.6	7,343
Natural Scientists	Men	13.0	84.0	3.0	0.5	1.1	135,649
	Women	45.0	43.0	12.0	3.6	0.08	14,616
Physicians and Surgeons	Men	7.0	88.0	5.0	2.1	1.1	214,830
	Women	31.0	51.0	18.0	7.0	7.0	15,477
Teachers, College and University	Men	18.4	77.8	2.2	0.7	0.9	133,452
	Women	50.0	36.1	3.0	5.8	5.3	36,715
Engineers	Men	9.0	88.0	2.0	0.65	1.1	862,002
	Women	37.0	45.0	18.0	6.0	8.2	7,714

SOURCE: U.S. Bureau of the Census, Census of Population, 1960. Subject Reports, Occupational Characteristics, Final Report PC (2)-7A, 1963, Table 12, p. 174.

*The divorce rate more than doubled between 1963 and 1975, from 2.3 per 1,000 population to 4.8 per 1,000 population (*New York Times,* February 9, 1980).

showed that the number of adults under thirty-five who live alone had more than doubled since 1970 and now totals one million, reflecting, among other factors, "growing career ambitions of women," and "the easing of salary and credit discrimination against women." It suggested that women were no longer compelled to marry early simply because they could not support themselves on the minimal salaries of most women's jobs.

Certainly among the young single women lawyers interviewed for this book, marriage was something to be weighed along with career opportunities. Sometimes they decided in favor of one and sometimes the other. In the past, a woman was expected to follow "her man." Among those interviewed was a young assistant district attorney involved in a relationship with a lawyer who accepted a teaching post in another city. She decided against leaving New York, where professional opportunities to grow were much better, although she knew the chances were high that they would not remain a couple.

Women lawyers married to men lawyers usually do not have to face the relocation problems of corporate wives because law firms do not typically transfer personnel. The initial choice of job is the important one, except for those who decide to go into politics or take political appointments for limited periods. Most women lawyers interviewed in New York planned to stay in the New York area (many had gone to school in New York), and the same was true for their spouses.

TABLE 18.2

Marital Status of Lawyers, Scientists, Engineers, and Physicians, by Sex, 1970 (in percent)

		Single	Married, Spouse Present	Separated, Divorced, Widowed	Widowed	Divorced	Base Number
Lawyers and Judges	Men	7.9	86.5	5.6	1.6	2.2	264,378
	Women	31.4	41.1	27.5	13.0	9.8	13,317
Life and Physical Scientists	Men	11.6	84.0	4.5	0.7	1.8	181,206
	Women	35.2	50.9	13.9	4.4	5.8	27,651
Physicians, Dentists, and related practitioners*	Men	6.3	88.4	5.4	1.7	1.8	496,415
	Women	23.9	55.8	20.4	9.2	6.8	45,038
Teachers, College and University	Men	18.1	77.0	5.1	0.6	2.1	354,671
	Women	36.2	50.0	13.8	4.8	6.2	141,741
Engineers	Men	8.5	87.3	5.0	0.6	1.8	1,236,160
	Women	26.2	39.2	34.6	17.2	10.9	20,775

SOURCE: U.S. Bureau of the Census, Census of Population, 1970. Subject Reports, Occupational Characteristics, Final Report PC(2)-7A, 1973, Table 31, p. 539ff.
*Categories are not directly comparable to the 1960 data.

This was not the case for women living in other parts of the country or away from metropolitan centers. For example, women students at the University of Virginia Law School, or even at Stanford in California, knew they would face relocation to other cities and areas where the selection and number of law firms was greater, particularly in the East. Many of these women did not yet have major attachments to men and did not have to make job decisions with a husband or lover's career possibilities in mind. Those who were couples usually had decided to seek employment in places where both had good opportunities to work.

Although most of the women lawyers interviewed in the years between the 1965 study and the late 1970s were married, it had become much more common for young women to be living with men to whom they were not married or whom they intended to marry. Young people in that college generation more characteristically postponed marriage, and a growing number of women lawyers wondered if they would ever marry. None thought of themselves as in the classic mold of the "career woman." In the last decade, however, some began to complain of an apparent shortage of eligible men. Black women in particular found few professional men who qualified as potential mates.[5]

Yet women's movement ideology allowed, indeed encouraged, young women to consider delaying marriage or not marrying at all. Some young women were opposed to marriage, although they were having sexual relationships with men, and a number felt sure they did not want children. But none felt they had to make a choice between marriage and a career in the way women in their mothers' generation often did. Many wondered how to combine the two, but it never occurred to them that they could not work it out if they chose to.

Age at Marriage

If professional women were marrying later, they may have been following a pattern typical in earlier periods. The only difference was that now they had more company. Of the married lawyers interviewed in the 1960s, about half had married after the age of twenty-six, and nine married after thirty. A study of eminent women lawyers listed in *Who's Who* found that more than two-thirds had married after they completed professional training, and more than 80 percent after college.[6]

This pattern of late marriage is characteristic also of women in other male-dominated fields, such as medicine, science, engineering, and dentistry.[7] Their median marriage age is considerably later than that of women college graduates who do not proceed to advanced degrees, who normally marry at the age of twenty-three, or those with some postgraduate education but no profession, who marry at 24.9 years.* What does it mean? In the 1960s, some late-marrying lawyers were so relieved to be married that they were happy to assume more traditional family roles and let their professional aspirations recede. Today, women are eager to establish themselves in careers. But they often want marriage too and do not feel one choice should preclude the other.

Being married or in the marriage market is different today than it was ten or twenty years ago. Change is going on in the private lives of professional women. It is not just that women's consciousness has been raised; they expect more of themselves and seem more sure of themselves and of their ability to cope as independent human beings. This is not to say that many of the old interpersonal problems between men and women have been resolved. But the norms governing courtship and marriage are changing. For example, more women may initiate relationships with men than before, and more assert themselves within marriages. It is hard to know how widespread the pattern is. Those who monitor the activities of their professional friends see them as more liberated now than they were ten years ago. But we may not be aware of the fact that some responses are a product of maturing or experience.

One possible indicator would be age-ranking in relationships. Although some women claim they are willing to try relationships with younger men in a new gesture towards liberation, it is uncertain that more are actually doing so than before, or that more younger men appreciate older women than in the time of Colette. The problem with trying to assess changing norms is that generalizations are often built upon a single person's collection of revelations or experiments. Techniques for counting cannot yet measure qualities like attraction in any dependable way. And since magazine writers and their friends are among the people who are most likely to experiment, articles about the "new life" are also likely to be misleading signs for our times.

It is difficult to separate those changes in relationships between people which result from institutional change from those which do not. Women professionals tend to agree that on all fronts institutional directives have changed, and change by legislative and judicial fiat is power-

*The average age of all women marrying for the first time is 21.3. Men average 23.8. *New York Times*, February 9, 1980.

ful. But they also concur that they haven't noticed that much change in interpersonal relationships between men and women.

It is probably true that change cannot be substantive before both institutional and interpersonal levels of interaction between groups and people have changed. Fundamental alteration of social norms cannot be produced by a mechanism such as consciousness raising. Although that Rousseauian dream is held by some, even the simplest acquaintance with anthropological analysis makes it clear that no society exists without fairly explicit rules governing interaction and that to alter those rules requires articulated change, not merely widespread deviation from old rules. Thus, even if every man took to the kitchen sink and the shopping cart tomorrow, nothing much would happen to help the equality of women without enunciated change in the norms governing private life and public rights.

Being Single

Single women lawyers who want to meet and date men with hopes of creating long-term relationships face the problems common to women in all other careers. In the 1960s they were nervous about becoming lawyers because they had been warned by parents and friends that men would be put off by women who had chosen male occupations, fearing them to be too "brainy" or "imposing." In the 1970s women felt differently. Not that they weren't still being warned, or didn't have problems in their social encounters with men, but they were appalled to find men reacting negatively to their professional status. In the 1960s, when asked "What do you do?" at parties, women often held back from announcing "I'm a lawyer," and instead would murmer something like "I do legal work." Women lawyers in the 1970s reported they would not hedge when asked, but at social gatherings with strangers they were on guard. Many women who hoped to meet men were ambivalent about the "announcement," worrying that it might drive men away. While a male lawyer could anticipate being perceived as a man of accomplishment, and perhaps a "good catch," a woman could not anticipate a positive reaction. One extremely attractive lawyer exclaimed: "When I go to a party and people ask me what I do and I tell them I'm a lawyer, they seem shocked; as if I said I was a topless dancer. They can't receive that information." A black lawyer noted two other effects her profession had on the men she met: "Men react differently when I tell them I'm a lawyer. Some are impressed. Some want to compete."

Joseph Berger, Morris Zeldich, Susan Rosenholtz, and other sociologists of "status expectation theory" have pointed out that people's views about those of other statuses—what they are like and what they will do in social interaction—precedes any knowledge of the individual person.[8] But it is not only the "other" who brings expectations to the social situation. Women lawyers expect others to have stereotyped ideas about them, and they may alter their behavior or become nervous in anticipation of the response. George Herbert Mead pointed to this phenomenon long ago as he explored processes of socialization, and Erving Goffman illustrated how all stigmatized people feel "on" in social settings.[9]

A woman's response to these problems depends on how frequently she encounters them—whether she is shy or extroverted, offended at the sexist evaluation that it is odd for a woman to be a lawyer, or amused at others' incapacity to understand the changing make-up of the legal profession. And, as men stereotype women and women lawyers, women also stereotype men about their reactions to women in high-level careers. Depending on their own experience and personality these "others" may express surprise warmly and positively, or coolly and negatively. Among the negative responses mentioned with regularity were men's dislike of the "assertive" behavior of women lawyers and of the time limitations on their social lives, and the concern of some men that they were too imposing.

Many men certainly are drawn to the prospect of a friendship or relationship with a woman lawyer. An eligible bachelor physician encountered at a cocktail party thought lawyers were the best women to date. "If you want to spend an evening with an interesting, attractive woman, these days," he offered, "chances are she will be a lawyer. No one really bright anymore will become a school teacher or go into social work." One woman lawyer, whose opinion clearly was in the minority, was disturbed by the positive attitude some men have toward women professionals:

A problem I run into with the men I am involved with is that women's lib has ruined things. They have all been brainwashed now, and they think women can't just sit home and be taken care of; they have to go out and support themselves. This is unfair. I find it particularly among male lawyers. They really like the fact that their wives are professional; they like it very much, and so anybody who hears that I am thinking of perhaps not pursuing my career to stay at home looks at me cross-eyed. "How can you give

335

up all of this training?" they ask, because it means so much to them. It doesn't mean as much to me.

Responding to the charges of their assertive nature, women lawyers declared that they felt under strain when they were in combative situations all day, that it was hard to turn off their dominant "persona" at night, and if they didn't men felt repelled by their behavior. One black woman felt that black men were especially sensitive to "strong" women, but I heard similar complaints from white women too.

I've had a man tell me that it was easier to deal with a woman who was a Ph.D. or one who was a doctor (as long as she was a pediatrician) than it was to deal with a female lawyer. People resent lawyers. It's a castrating type of profession. You learn professionally how to take people apart. It's difficult to turn that off and you can be obnoxious.

One woman lawyer observed that some men liked to date women lawyers because they were exotic but had second thoughts about them when considering marriage.

They think it's kind of kinky. For some men, dating a lawyer, I think, gives them a certain status; because the woman they are with is accomplished in some way. But they move on to a more traditional woman when they want to get serious.

Time pressures caused problems for many women (like men in comparable situations), who often could not be available for dating or casual socializing because of work demands. Women on Wall Street, in litigation, and in other types of practices where they were required to work overtime with some regularity found that men soon tired of their unavailability. Some lawyers thought that men they saw socially might break off with them because they couldn't be attentive enough. One said sarcastically of the new woman of her former lover: "She's the Jewish mother type. You know, the kind men like. Women who bake them banana cake and have sex in the shower." The most extreme example of the time problem is probably the woman lawyer with aspirations in politics. One noted that the man who wanted to date her "has to be worked in around police precinct dinners, the Veterans of Foreign Wars, 16 Democratic clubs, block associations, the Girl Scouts, feminist causes, and her law practice."[10]

Although time pressure is a problem, it also creates an environment for romance. There is an excitement to working around the clock in the battle to win a case. One woman noted that:

336

Husbands, Wives, and Lovers

A lot of times romantic tangles develop because as young associates your whole life gets wound up in that firm. I worked twelve hours a day every day of the week for six weeks at a time. For many people the firm was their whole life. They worked so hard that the only people they ever saw were associates or the partners.

Dating within the firm, however, has its pitfalls. There are apparently few problems when dating within rank, and inevitably some proportion of young unmarried men and women associates have relationships with each other. Yet women claim that it is important to be discreet within the firm. This is because one's work relationships might be affected if others were aware a romance was going on. In one firm, discretion was so effective that one single woman couldn't identify any dating among the single women and men associates. She responded to my question on relationships between the men and women:

Not that I'm aware of. It doesn't seem like it whenever there is a firm function, of which there are not very many; it doesn't ever seem that any of them regularly come together or leave together or anything like that. They don't seem to come to the lunch room together.

Another lawyer in the same firm, interviewed at about the same time, painted a very different picture:

"Well, I have dated a lot of people in the office. On a very discreet basis, ah, I don't think office romances should be publicized at all. I certainly don't think it can do the men's careers any good. I just don't like to be gossiped about.

Several associates said they imposed one major dating rule on themselves: "Do not date partners." They assumed that should a relationship with a partner come to an end, it could be embarrassing and perhaps threatening to job tenure. For women involved in affairs with married partners, the pitfalls are even worse. One associate said:

I've had girl friends who have gone out with partners, and some have had to leave because the relationship ended. There is always a problem if there is any bitterness at all when you break up. Dating within the firm can affect your career potential.

The prospects for relationships are also affected by the pool of eligible men.[11] Male professionals date women of all occupational ranks, but women lawyers usually want to date only professional men.

I usually date doctors and lawyers. I meet them in social situations at parties, through friends. I might possibly consider a relationship with a man who was not professional, but it would have to be an extraordinary relationship.

Although the pool grows smaller for all women as they achieve higher and higher statuses (either educationally or professionally), it is especially small for black women because both the number and the proportion of black men in the professions are small, as table 18.3 shows.[12] One black lawyer expressed her concern:

Marriage is not a real thing with me. But you have this real deep fear of being alone. I think all of us do. I find it very disturbing that I don't have a permanent relationship and I don't see any of my friends developing one. I don't think it disturbs my career plans because I am going to have to work anyway. I've never really been without a man for a long period of time, but it is difficult maintaining relationships. I am particular in what I want and it's difficult to find. There are not a lot of men out there.

A substantial minority of the women attorneys interviewed lived rather celibate lives or had affairs with men which did not lead to marriage. Many threw themselves into work until it became their whole lives. Men who throw themselves into work manage to have families, but fewer women seem to find men willing to tolerate what women tolerate from "workoholic" men. Yet more and more women now expect men to recognize their work commitment, and more and more men seem to be doing so.

The women interviewed agreed that men's "egos" seemed threatened by women whose occupational status makes them achievers. But one never hears men commenting that their occupational status threatens a woman's "ego." These responses are the product of norms that maintain status inequity by making women who strive feel uncomfortable in their personal lives, punishing them by denying them personal satisfactions. It is clear that the punishments are fewer now than they were a decade ago. But the fact that more women than men are single or divorced today indicates that the norms are still operative.

Yet being single is not a punishment for all women lawyers, as one made clear:

While there are problems of living alone, the price [of marriage] in my mind to date, its too high. I've had jobs where I have been able to split on a moment's notice. I've been to Mississippi [for the civil rights movement]; When I joined a political campaign I was able to pick up and do that. When I decided to roam the Canadian Rockies this summer, I was able to do that.

338

TABLE 18.3

Professionals by Race and Sex

	Male					Female				
	Total Number*	White		Black		Total Number*	White		Black	
		Number	Percent	Number	Percent		Number	Percent	Number	Percent
Architects	54,948	52,212	95	1,159	2	2,113	1,812	85	195	9
Engineers	1,236,160	1,199,811	97	14,198	1	20,755	19,697	95	757	4
Lawyers and Judges	264,378	259,711	98	3,309	1	13,317	12,723	96	394	3
Judges	12,281	11,962	97	279	2	662	615	93	18	3
Lawyers	252,097	247,749	98	3,030	1	12,655	12,108	96	376	3
Life and Physical Scientists	181,206	170,704	94	5,104	3	27,651	24,818	90	1,433	5
Physicians, Dentists, and related practitioners	496,415	472,745	95	10,084	2	45,038	39,645	88	1,765	4
Social Scientists	90,601	87,417	96	2,145	2	20,673	19,135	93	1,328	6
Law Professors	2,808	2,726	97	62	2	197	197	100	0	—

SOURCE: U.S. Bureau of Census, Census of Population, 1970. Subject Reports, PC (2)-7, vol. 2, 1973. Table 2: Race and Spanish Origin of the Experienced Civilian Labor Force by Detailed Occupation and Sex; 1970, p. 12–13.

*Not included in the total number are people of American Indian, Japanese, Chinese, Philippine, and Spanish origin.

You may think it is depressing to decide to live one's life without love, but I don't think it's so depressing. I notice single women are doing quite well today. It's much easier for them to get along. We have built our contingents—we have other people. I think women have found support; they do things together now in a way they certainly didn't do when I was growing up.

Lawyers Marry Lawyers

A majority of women lawyers are or have been married, many of them to other lawyers. Close to half of the lawyers interviewed in the 1960s were married to other lawyers, and other studies support this finding.[13] Marriage notices of women with professional titles have proliferated in the past few years. A typical headline in the society section of the Sunday *New York Times,* on September 21, 1980, read: "Cary Kittle, Lawyer, Is Married." Ms. Kittle's wedding announcement was typical also in its profile of young lawyers working for large firms:

At Grace Episcopal Church in Cismont, Va., yesterday, Cary Marian Kittle and David Finley Williams, who received J.D. degrees last year from the University of Virginia, were married by the Rev. John Frizzell.
. . . The bride, an alumna of the Chapin School received a B.A. degree from Virginia. . . . Formerly an associate with the New York Law firm of Shearman and Sterling, she will join Steptoe & Johnson in Washington, next month.
. . . Mr. Williams, an associate with Covington & Burling, Washington lawyers . . . graduated from Yale.

Another *New York Times* article, February 1, 1971, announced the engagement of Pennsylvania Law School graduate Sara McCarty and Peter Solmssen:

Miss McCarty is a law clerk for Judge Boyce F. Martin for the Sixth Circuit of Appeals in Cincinnati, and her fiance is a law clerk for United States District Judge Clarence C. Newcomer for the Eastern District of Pennsylvania. . . . They plan to join the Philadelphia law firm of Ballard, Spahr, Andrews & Ingersoll next year.

And on January 1, 1971, the *Times* announced the engagement of Debra Sue Belaga to Stephen Stublarec, she a Stanford Law School graduate and he a graduate of Boalt Hall School of Law at the University of California-Berkeley, "Miss Belaga is an associate with the law firm of

Petty, Andrews, Tufts & Jackson and her fiance is an associate with the law firm of Pillsbury, Madison and Sutro."

The favored marriage pattern for women attorneys, like other women in male-dominated professions, seems to be what sociologists call homogamy—"like marrying like"—in this case, along professional lines. A similar pattern is seen among women in other male-dominated professions, such as medicine, science, and engineering.*

Of the thirty married women partners in large New York law firms in 1978, two thirds, or nineteen, were married to male attorneys. Twenty-two lawyer-couples teach in American law schools on the same faculties.[14] This pattern is probably a reflection of the marital "opportunity structure," in that most couples met in law school or at work. Because women have been in the minority in law school, they have been visible to a wide pool of eligible spouses. Women with professional aspirations or identities are also likely to seek men whose occupations will be of the same or higher social rank; this limits their choices to only a few occupations. So it makes "sense" that women lawyers would tend to marry male lawyers.

Whether there is an additional affinity of a common "mind-set"—an orientation to lawyer's logic—we do not know; but couples who do similar work share interests and understand each other's work worlds. Some women have expressed negative feelings about being in the same occupation as their husbands, because of competitiveness. But whatever the negative components, the total weighting of factors clearly favors a shared occupational status for women lawyers and their husbands.

Marriage and Occupational Rank

Managing marriage and a professional career is assumed to be hard on women and on both their marriages and careers. Gatekeepers like to

---

*Nancy A. Roeske reports that approximately 80 percent of female psychiatrists marry and that about 50 percent of them marry physicians. "Women in Psychiatry: A Review," *American Journal of Psychiatry* 133 (April 1976):365–72. And a survey of 815 registrants at the 1971 annual meeting of the American Society for Microbiology indicated that only 44 percent of the women, as contrasted with 90 percent of the men, were married. But of the women who were married, 96 percent had married professional spouses, half in their own fields, while only 42 percent of the men had chosen professional wives, one third in their own fields. Herma Hill Kay, "Legal and Social Impediments to Dual Career Marriages," *UCD Law Review* 12, no. 2 (Summer 1979):207–25.

attribute women's failure to attain top-ranking positions to their marital obligations. Yet one of our strongest findings was that women who achieve unusual success are more likely to be married than single. This was true both for the women lawyers interviewed in the mid-1960s and those interviewed in the mid- and late-1970s. In the earlier group, a higher proportion of women who were partners in firms or who made more than $16,000 (a substantial salary in the 1960s) were married. Furthermore, a greater proportion of women lawyers listed in *Who's Who* were married than the general population of women lawyers sampled by the census (83 percent compared to 45.6 percent), and only 16.5 percent had never been married.[15] Indeed, most women partners in Wall Street firms are married; thirty of the forty-one were married and only two were divorcées.

What might explain this? Does there need to be a man behind every successful woman? Does marriage make one healthier? Most successful men are married. He who travels farthest does not travel alone. Wives are supposed to assist their husband's careers. We expect the obligations of marriage to hold women back. But women, like men, need and want companionship and support. Wives who have both can do well in their careers.* Married women may have or develop social skills in teamwork (perhaps being married gives women the kind of training men get in team sports). But there are also very successful single women who have considerable interpersonal powers. One explanation for the fact that successful lawyers tend to be married may have more to do with the simple attribute of being married. Married adults are simply considered more normal than single adults. They pose fewer problems socially to co-workers and clients.

In a study of academic men and women, Debra Kaufman found that married women were more integrated into male networks than their unmarried female colleagues.[16] She hypothesized that unmarried women carried the double stigma of being female and unmarried. Since most professional men are married, they are more comfortable with women colleagues who carry the protective status of being married. She quoted an elder statesman in an academic department on his relationship with a particular woman colleague:

---

*Jonathan Cole, in *Fair Science: Women in the Scientific Community* (New York: Free Press, 1979), found that married women, including those with children (but small families), tend to publish more than unmarried women (p. 6). We might also note that all women Nobel Laureates were married. Cole offers as explanation Durkheim's interpretation that it is the stability and routinization of work patterns which leads to higher productivity. This may be partly true, but it would be limiting to focus only on the behavior of the woman and not on the interpretation of her behavior by gatekeepers.

We have worked on several projects together. I find her thorough, hard-working, and very insightful . . . I get along with her husband well, too, and that makes for a good relationship as couples. That's helpful too.

Furthermore, married women make contacts through their husbands or male friends. One lawyer interviewed had established an outstanding reputation as the head of an important government securities commission. She explained she had been invited to join the large firm, which had recently named her a partner, because she knew the senior partners well socially through her husband, who also was a partner in a large firm. Thus, through her husband she had the kind of visibility which her work in government might not have given her. In fact, when the firm felt the time was right to hire a woman partner, they pursued her because of their familiarity and ease with her as well as her competence.

Relations Among Spouses

Dilemmas characterize women lawyers' professional lives, and dilemmas also mark their private lives. Although much has changed over the past decade, the problems associated with marriage and relations between men and women continue to be perplexing. Special problems affect the dual-career couple. These encompass all the usual problems of married couples as well as the dilemmas posed by greater equality and the problem of how that equality should be played out in everyday life. Commitment to greater equality also competes with deeply held views about the naturalness of inequality.

New ideas compete with traditional ideas, and individuals may alternate between them or hold both at the same time. Views on the equality or inequality of the sexes are ingrained in our personalities and in our manifestations of dependency, autonomy, love, nurturance, and self-actualization.[17] Societies create sex divisions not only of labor but of emotions, classically assigning certain kinds of emotional work to men or to women.[18] These emotional assignments are in dispute today.

Views on relations between the sexes are also part of the foundation of the entire stratification system, with the bourgeois, patriarchically oriented family serving as the model in contemporary society, although the family actually exists in many different forms. Many ideological and pseudo-scientific views support that model. They assert that the bourgeois family is a natural extension of biological and psychological differences between the sexes;[19] that it is the most economically satisfactory arrangement;[20] or that it is most functional for the maintenance of society.[21] Regardless of merit or accuracy, these views support the ideal of the family and they are held resolutely by people or groups whose practical or emotional interests they serve.

Many "sides" are in competition today, yet there is still a dominant cultural ethos. This ethos generally assigns women to dependent roles and men to powerful roles in private as well as public lives in every society.*

The dominant cultural ethos, which specifies that women should defer to men, that they should be weaker than men and look up to them, that family considerations should come before work decisions, and that women should want to put their careers second, creates emotional and practical problems for many (but not all) women and men in highly demanding occupations who live together.

It seems clear that these norms are supported and maintained by processes that make it difficult for women to enter and participate in highly demanding occupations. Many are processes of simple exclusion. The ambivalence women face in their private lives is another contributing process. In fact, the problems and conflicts played out in the private sphere are deterrents that support the dominant system by constraining women who have established their competence as thinkers and workers and whom the laws have moved closer to formal equity. When women feel uncomfortable at home because they have achieved equity in the marketplace, they may feel inhibited about taking advantage of that equity. But it appears that women now are resolving some of this ambivalence. Many men are convinced that the new division of work and of emotional support is just, and they find it attractive. For some men, however, work demands and the complex task of translating ideals into reality make it difficult to carry out their ideological commitments.

---

*Some anthropologists have argued that in many societies women have power, but that it is a different power than that held by men because it is in the private realm. See Michelle Zimbalist Rosaldo and Louise Lamphere, eds., *Women, Culture, and Society* (Stanford: Stanford University Press, 1974). I do not agree: without command of resources in the public world, women have little power other than what men hand over by default. See Rosaldo's rethinking of her paper on this question in "The Use and Abuse of Anthropology: Reflections on Feminism and Cross Cultural Understanding," *Signs* 5, no. 3 (1980): 389–417.

Rank Discrepancy Between Husband and Wife

The new equality specifies that it is no longer necessary for men to outrank their wives in occupation; the two may be equal. For those willing to try it, equality works—even at the top. Most of the women partners on Wall Street are married to men who are partners in other (or the same) Wall Street firms, or who have high prestige jobs in the corporate and financial world. These marriages (at least as reported), seemed to be successful and less marked by strife than those in which the lawyers' tenures were still in question. Equity at the top is exemplified by a number of outstanding women lawyers, who hold high rank as law school professors, judges, or even cabinet members, whose husbands also hold top positions that sometimes are less visible to outsiders but prestigious nonetheless. Examples of those high prestige couples are: Ruth Ginsburg, now a federal judge, whose husband, Martin Ginsburg, was a partner in the firm of Weil, Gotshal and Manges before joining the Georgetown Law School Faculty; Herma Hill Kay, whose husband, Caroll Brodsky, is a psychiatrist and professor at the University of California at Berkeley; Shirley Hufstedler, former Secretary of Education, whose husband, Seth M. Hufstedler is a prominent lawyer in California; and Carla Hills, former HEW Secretary, who is married to Roderick Hills, former chairman of the Securities and Exchange Commission.

However, the road to the top, with its uncertainty about the ultimate fate of husbands and wives, is more maritally precarious for many, especially when it looks as if the wife might be more successful than the husband. Yet most young male Wall Street associates interviewed denied that their wives' success would be a problem to them or that they would feel badly if their wives made more money. The women, however, were more cautious about how their husbands might react.

Women lawyers in other kinds of practices, however, recounted problems created by success. Certainly there is a residue of the rule (and it may be much more than a residue) that wives should not outrank husbands; even equality of rank is a radical concession in a culture where women are still reared to look up to a man, literally and figuratively. There is, of course, generational change. In discussions with young people I have been reminded that the short, bespectacled Woody Allen is considered a romantic hero by many teenage girls.[22]

345

Yet the most frequently encountered romantic standard is still strong, tall, and determined, even if shy.

Women affected by the women's movement may not wish to be followers to husband-leaders any more, but many are alert to enduring cultural characterizations of manhood and show concern about undercutting them. A man may sweep along a wife as he goes up the occupational ladder, but there is no comparable image of a wife sweeping along her husband. Most people feel sorry for such a husband and consider him eclipsed, although were he the more successful one, they would congratulate the wife on his achievement and assume she was enhanced by it. A woman, by cultural definition, improves in image with her husband's success; a man may well be diminished by his wife's success.

The "wrong" rank disparity between a married pair may or may not affect their private interactions. Public opinion may be audible or silent on the issue, and some people just don't care. But there are few norms governing marital relationships in which the wife's job is more taxing and prestigious than the husband's.* Not knowing how to act can cause discomfort and marital disharmony. In any event, both men and women are wary of such disparity, and the number of wives who rise above their husbands in the occupational sphere was and is small.[23] Only in black middle-class communities, where women have become professionals (usually teachers) and their husbands often could not achieve comparable status because of discrimination, has such a pattern emerged. But this rank disparity has been regarded by both whites and blacks as a social pathology. Sociologists like Daniel Patrick Moynihan[24] and black leaders such as Stokeley Carmichael have intimated that it would be best for black women to stand behind black men.

But there has been a pattern in the cases of women whose accomplishments brought them recognition and rank exceeding that of their husbands, and that is a pattern of problems and marital stress.† In the past, when a woman was on a route to advancement beyond her husband, it was thought to be sensible for her to scale down her ambitions; she often chose to end the conflict by retreating altogether from the

*Norms, both roles and laws, exist in the case of a royal couple, for example, one in which the wife is reigning queen and the husband is her consort. Literature and anecdote, however, indicate that personal tensions have flared between these couples, mitigated by the deference of the queen/wife to the consort/husband in private.

†In many cases, of course, women will enter the occupational or public world either on the coattails of their husbands or make their mark once their husbands have died. Among cases where the women eclipsed their husbands while they were still married are Golda Meir, former prime minister of Israel, and Frances Perkins, former secretary of labor.

threatening career situation. If she had a stronger commitment to her husband than her work—and it was culturally defined as proper that she should—she might consider it a small sacrifice, assuming that her resentment at her husband's attitude had not eaten into their emotional relationship. In the earlier interviews, women commonly confessed to making job choices with a careful eye to their husband's reactions. This pattern is still found today.

A young woman associate on Wall Street whose husband was an associate in another firm addressed the problem openly in an interview a few years ago. Asked what would happen if she became a partner and he did not, she responded:

That would be a catastrophe! I would feel rotten about it. We joke about it all of the time. I always say, "just wait!" And he says, "terrific, you make partner and I will retire." But, you know, I think it would just kill him.

In fact, with that thought in mind, I have often told him that if he wanted to leave and move, even to another city, I would be willing to do that. Just because it is more important for me to have a happy husband.

And a law school professor told of a woman who refused an opportunity to keep peace with her husband:

I know of a lawyer who is married to a man twenty-five years her senior. They both work in the same law office. She had a chance to get out and go to a big classy firm. He went into a depression, serious depression. They had it out. She convinced him that if she wanted to go she should. But she didn't take the job. She said it was for lots of reasons . . . the offer wasn't what it should be; the prospects for becoming a partner were not wonderful. But I don't think if the prospects had been any better she would have taken that job. Her husband did not talk to her for two months. They lived in a house with the children and he would not talk to her. He told her the reason. He did not want his wife in a job with more prestige and attention and more money. She was going to be traveling around the country and meeting people and was going to outshine him. He said it was a double standard and he knew it, but he was not going to change. He said she knew it when she married him and he was not going to discuss it any more. And he didn't.

Those of the women interviewed in the 1960s who did outrank their husbands often attempted to conceal it or to imply that the husband's rank could not be compared.

Five wives in that study probably did outrank their husbands.* All

*There are methodological problems in rank assessment. For the purpose of this analysis, we judged high rank in law by income level and by acquisition of a high-prestige status: judge, partner, professor, and listing in *Who's Who*. Where the wife had one or a combination of these statuses, and the husband did not, we considered their marriage

were married to lawyers, and they obviously were uneasy about possessing the higher occupational rank. Yet only one defined it as a problem; she confessed to stress because of it and said that as a result she and her husband had divorced. The others seemed to feel pressure to explain away the reasons for the inequality. All described their husbands deferentially, and excessively praised their achievements. In two other cases the wives asserted that their husbands could not possibly make as much money as they did. One of the husbands was in social service work and one was an artist.

But the later interviews showed that even emancipated women wanted their husbands to be more outstanding than themselves. For most it was not an issue because most husbands did outrank their wives, but there were a few surprises. For example, a few adamant feminist lawyers confessed they were attracted to hard-driving men who were more likely to be tuned to their own work than to a wife's needs. The men who shared feminist ideology in being family oriented rather than career focused were perceived as unambitious and lacking in charisma.

One feminist lawyer who married a man during law school reported that she had become distressed as the years went by and she discovered he was lacking in ambition. Although he shared her political beliefs and chose to practice poverty law, she found it difficult to live at a reduced standard of living while bringing up two children in New York City. She felt he was not living up to his potential to be a highly visible movement lawyer; that is, he was neither making money nor making news. His interests were certainly not a match for her own fiery commitments or headline-making projects. Yet the problem for her, and for so many women like her, was that those men who place a high priority on family life and sharing responsibility are not apt to have the dramatic aspirations that high-powered women are conditioned to admire.

But changing norms often evoked a combination of behaviors, both contemporary and traditional. For example, an aggressive trial attorney spoke of her relationship, which was at the same time egalitarian and traditional:

---

a case of "wrong" rank disparity. Since we found early in the study that it was difficult to obtain information about the husband's income, we did not ask for it later in the study, so that one essential piece of data is missing. We were able to gauge the husband's success, however, when we knew the wife's income and could compare it with the home and location of the home (when the interview was conducted in the respondent's home) and the respondent's description of the family's style of life and characterization of family income (well-to-do, comfortable, and so forth).

Husbands, Wives, and Lovers

I can go home and be perfectly comfortable being female and in many ways
dependent on the judgment of my man in questions relating to our personal
lives, and be the boss in my office.

Yet when I asked her what would happen if his business required a
geographical move that would mean relocation for her, she said

> For both of us, the most important consideration is not business at all. It is
> that we make each other happy. We would sit down and analyze mutually,
> taking the whole package of his business and my business, how we came out
> better, and we'd answer it in the only terms that were relevant, purely eco-
> nomic terms. There would be no power struggle.
> But I know how the analysis would operate. It would never be, unquestiona-
> bly his interests come first. The question would be, "How would we both be
> happiest working?" And that's how I think it's supposed to be.

Problems for Some

Marriages are affected both by public opinion and by private needs.
There are stereotypes of what the proper relationship should be,
and what each partner should contribute to it. Cultural mandates
define what the wife or husband should give not only in terms of
the work of the household, but emotionally and sexually. Of course,
in all eras individuals have conformed to or deviated from these cultural rules according to their personalities and the interplay between themselves and their partners. Individuals may be happy following traditional rules or feel stifled by them; some couples go
their own ways oblivious to what society says they are supposed to
do. Similarly, some people define small stresses as large problems,
and others see no problems in major upheavals. It was often jolting
to hear women lawyers in highly demanding Wall Street practices,
who were married and mothers of small children, say they had no
problems at home. They didn't say they had surmounted the problems; they simply hadn't experienced any in the first place. What is
one to write about women who liked their lives, their demanding
work, their husbands and children, and handled the stress in a matter of fact way? How do women handle stress? If sociologists are
problem seekers, these women were problem solvers or problem av-

oiders, or persons who deny that problems exist as I pointed out in chapter 16.*

Time and Attention

Measuring success for a career woman usually includes an evaluation of her marriage and her ability to handle the responsibilities of the household. Yet studies of professional men almost never evaluate their marriages or their handling of household affairs—even with regard to performing such male sex-role tasks as keeping the cars in good order and attending to home repairs—nor do they assess the quality of men's relationships with their wives. But part of women's work is marriage, and women are supposed to work at marriage more than men do. This cultural mandate is embedded in the folkways of many societies. Today, however, young women are more resistant to assuming the major (and therefore unequal) burden of working to keep a marriage viable.

Even a woman equal to her husband in occupational rank is supposed to defer to his desire for attention, and taking care of the home is an expression of such deference and caring. Women's work in making the home attractive and in preparing meals often has been viewed not only as her responsibility but as an expression of love and commitment. Women and men share this symbolic meaning, and women have taken pride in it just as men have viewed home improvements as gestures of family commitment beyond their practical urgency. There are problems today because the meanings attached to these acts and their assignments to the sexes have changed.† When women shared the view that ministering to men was an expression of responsibility and of love, there was relative harmony; the emotional and practical trade-offs were viewed as just. But when women feel resentment at being assigned tasks they do not want or have time to do, and

*Rose Laub Coser documents the use of evasiveness as a way of dealing with structurally induced ambivalence in "Evasiveness as a Response to Structural Ambivalence," *Social Science and Medicine* 1 (1967):203–18.

†Arlie Hochschild notes that "as the symbolic world changes with structural change, so does the bearing of feeling rules." "The Sociology of Feeling and Emotion: Selected Possibilities," in *Another Voice*, ed. Marcia Millman and Rosabeth Moss Kanter (New York: Doubleday Anchor, 1975), pp. 280–307.

when they feel they are giving more than they are getting, hostility emerges.

Families, like other social groups, rest on assumptions of fairness and equity unless they are governed by tyrants. But men, used to the old exchanges, may find it difficult to reconcile their emotional and practical expectations when women press for a new standard of equity. Because women love these men and wish to be loved by them, they are often in a bind. Of course, these expectations are always under negotiation (no one is free from being reminded that his or her behavior is "unfair"), but when the traditional cultural contracts change, negotiations become all the more difficult.

A modern expression of the dilemma was related by a law professor, speaking of a former student:

We had the president of the law review here. I guess when she was on the staff she fell in love with the fellow who was head of the law review the preceding year. She said she had some problems, and when I asked her she said, "Well, my husband . . . (she was a senior here and he then was already out in practice) expects me to get home and cook supper for him. You appreciate—and he *does* appreciate—that the law review is a very exacting boss. You have to work long hours most evenings, of course you have to go to school as well as put out this review. He just doesn't realize I have to work the way he had to work, and he wants me to come home and cook supper."

In some families, however, the new definitions have been accepted by both spouses, and the old rules on commitment no longer mean the same thing. When a staff interviewer spoke with several women partners of large firms in their homes, she came away with impressions typified by this observation:

I must say that I did not get the sense of a "home" of the kind that I am accustomed to. . . . Part of my feeling about this is shaped by my visit to [A]'s. She and her husband had this fantastic apartment off Fifth Avenue. They'd been there almost a year, and the only place to sit was a ratty sofa in a small room. She was planning to get a decorator and when it's done it will probably be a knockout, but just the fact that they were able to live like that for so long (I couldn't, and God knows my standards are not high) tells you something about their home life.

Housekeeping is only one side of marriage, and there are other types of focus and attention that men want from women; problems, often unspoken, can arise due to the emotional and sexual withdrawal that may come when women are under work pressure.

Problem-Making and Problem-Assuaging Husbands

There are husbands who help and husbands who hinder the careers of women attorneys. Further, inconsistency of behavior in the same man often complicates the situation. But the problems women attorneys face with their husbands need not be overpowering and may occur and fade over time as the individuals and their careers and domestic situations change—as children grow older, or wife and husband advance in their careers, or the couple's income increases.

Even now that women are freer to take risks and become achievers, there is a residue of uneasiness about their new roles, and many cannot depend on the same support that male achievers can. The men in their lives—personal and professional—may wish to give them support and encourage their ambitions, and some do, but many seem unable to do so. Many men are not used to supporting roles; some think such roles diminish them; some probably don't know how to perform these roles well.

Women have come to be experts at a variety of behavior toward men, ranging from tactful retreat to active help. Women, too, have a need for supportive behavior, and the inexperience of men in providing the right kind at the right time can cause resentment and strain on both sides.[25]

One lawyer in private practice explained how the demands of her work affected her relationship with a man who was "warned" that she was going into a bad period:

When there is a period of tremendous stress, and I say to the person I am involved with—"I am under this terrific pressure and having problems. I can't think about anything else. I'm worked out, wiped out, exhausted and just feeling that everything that is happening in the office takes priority over my mind and life"—this creates a tension in the relationship. Men can't understand that.

Men feel rejected. And they don't know what you're talking about when you say "I've had a hard day at the office." I don't know whether it would be more acceptable if you said "those kids drove me crazy all day today." But if you say, "The judge screamed at me and I heard emotional problems all day," that won't do. I think that has a lot of potential in raising problems in personal relationships because it requires a different acceptance on their part that the women they are involved with have an independent and professional life but that life may infringe on their lives in very direct ways.

I remember that happening just before my book was published. I said to my friend, "I know I will be impossible in the next few weeks. I won't be impossible to other people but I will to you because you're close. So whether you want to be around or not is your option . . . if you stay and things are tense, remember, I've told you so." He stayed, but he had an enormous amount of trouble being supportive. He didn't complain about the work, but he would raise other things at the time which just created extra hassles for me that I didn't want to deal with then. And I said, "It would have been better if you stayed away for three weeks, I told you that." That sort of thing doesn't go over well. It takes a very exceptional man.

But another lawyer, who wasn't willing to take the risk of alienating the man in her life, also felt the giving was reciprocal:

The sphere of loving is so very different, is such a different priority, that there is no way in the world that I would ever tell him, "Go away, I have a headache because I worked hard today." It doesn't enter into it. What he gives me is the comfort and strength to shut out the rest of the world. The sphere that I have with him excludes everything else, except to the extent that we want it to enter. So rather than letting those considerations ever enter our world, they enter only when we let them, and when there's some way that he can help me or I can help him, or we give to each other to push those things back out.

The issue of the wife's work commitment was a difficult one in some marriages. Some husbands wanted to support their wives' commitment to work, but found it difficult in practice. One young law professor spoke of the course of her marriage—a marriage that ended in divorce:

When I met my husband, practically the first conversation we had was "what do you think of working women and smart women and ambitious women?" And he reacted positively to all those things. I had strong ideas then about independence and nobody holding doors for me and that kind of thing. But much later when it all fell apart, and it seemed to be falling apart because of the amount of time and energy I was putting into work, I would say things to him like "this should come as no surprise; there was never any deception about who I was from the beginning."

He answered, "I thought that what you were was what I wanted; but I found I couldn't live with it."

One young lawyer expressed her ambivalence about wanting to do well at home and to be a good lawyer too:

When I was starting on my job I wanted to put my all into it. Everyone else who had come in stayed late and came in on weekends, but I felt pressured and pulled to go home and make dinner at night. I felt guilty about not wanting to go home. After a while I hated being married.

She sorted out that problem by talking about it to her husband, who understood the dilemma because he had started working two years before. Yet for the future, she speculated she would have to take a less pressured job because she knew she would feel guilty about not leaving work to be with her children. Yet she loved her work.

Women lawyers worry about long-term commitment. Yet thousands of women have stood by husbands who were doctors, lawyers, or executives because they had no other option and the men's high intensity commitment provided the women with social status and high income. Men in careers are not motivated to make the same investment in their working wives because they do not benefit from the same rewards or because those rewards they do get do not compensate for the cost to their own careers. However, for the first time, some young wives work in large firms and bring home incomes large enough ($30,000 to $60,000 a year in 1980) to raise the family's standard of living to a level substantially higher than if the men were sole breadwinners or the wives secretaries. Husbands of these highly-paid women tend to be very supportive of their wives.

But economic interest was only one reason husbands supported their wives. Many believe in them, are proud of their accomplishments, and have encouraged their climb up the ladder. Many lawyers remarked that it looked as if behind every successful woman lawyer was a supportive man. Most married partners we spoke to had very positive things to say about their husbands. "He was always there, just 100 percent supportive," was the comment of one lawyer, but it typified the response of most of the others.

There were clear differences between women who met their husbands in law schools and those who had been second deciders and were already married when they reached law school. Women in the latter group reported more disruptions, both from the pressure of work and from their husband's ambivalent feelings toward their careers. Husbands of these second deciders were often faced with the "new woman" their wives had become. These couples had to renegotiate their relationships as well as redivide home responsibilities. They did so either through open discussion or by subtle changes in behavior. Young women lawyers who married after entering law school, usually to lawyers, seemed to expect that each partner would have heavy work demands and would try to arrange time together within the limits of those demands. They tended to have lower expectations with regard to housekeeping and entertaining and adapted to a life-style that gave work high priority for both husband and wife.

Husbands, Wives, and Lovers

Many young couples were concerned about the additional problems that would come with children, if they chose to have them, and some seemed ready to delay having children or not to have them at all because they did not wish to change their life-styles. For some young couples, the question of child-bearing was often posed as an economic rather than an emotional choice. For couples with family money with which to buy a large apartment or house and to afford full-time help, having one or two children was deemed possible.

Women lawyers in their forties and fifties expressed concern about how their homes looked and devoted some attention to the housekeeping. This attention to aesthetics was often reflected in their beautifully decorated offices. Their attitudes contrasted with those of the younger women with equally or more successful practices who lived in the bare apartments my staff and I observed. Three or four of these young women looked forward to a time when they would "get to" fix the apartments that they had moved into years before—many of them large and lavish and situated in elegant sections of New York—with the help of a decorator. Using decorators, that is, "delegating" responsibility for the home,* was more acceptable to young women than those interviewed in the 1960s, who made it a point of pride—a test of femininity —to display their homemaking skill by furnishing their homes themselves.

Class position makes a real difference in dual career families. Our young associates were often at a cutoff point. Most didn't have capital —they were very well-to-do but on salaried income.[26] High salaries meant they could insulate themselves from many of the problems faced by most working women. They could hire housekeepers and decorators, and other help they needed. But their standard of living depended on their continuing to work (and at a high salary), giving both husband and wife a strong incentive to keep the wife professionally successful.

The husbands of women who had tended the home and had been gracious hostesses in the past often experienced confusion and hostility when their wives began schooling for their new careers. A woman in her late thirties told us that her husband, a wealthy lawyer, became angry at her going to law school and refused to pay tuition. She found it embarrassing and difficult to secure loans because her family income was so high, but managed to get through school. The marriage didn't last; the couple could not regain the balance of the relationship in the face of the new dimensions of a dual career. Another lawyer reported

*This is one of the mechanisms Goode points to in alleviating role strain, in "A Theory of Role Strain," *American Sociological Review* 25 no. 4, (August, 1960): 483–96.

355

that her husband gave her full support when she was a student and tolerated her burning the midnight oil to study for exams, but he was unable to adjust to her working late into the night as a paid professional. She explained that her student status made him feel nurturant and helpful, but as she began building a career and moving toward equality in income, he began to feel competitive.

A particularly difficult equation existed among the couples in which the wife was hard-driving and the husband was not. One lawyer we interviewed had separated from her husband when he, like many young men, decided that the hard-driving life did not satisfy him, and he wanted to de-escalate. He was a few years older than his wife and had succeeded in becoming a partner in the firm in which they both worked. But she felt she had to work extremely hard, and although she also became a partner some years after he did, she still felt driven to prove herself at work. In cases like this it was not atypical for the wife to revise her sights downward. This woman was re-thinking her career path, hoping to reunite with her husband.

Men and women make choices in their work lives in response to their private lives. There is agreement in the popular culture that women think about their private lives more in making work choices than do men. Many women in academic and work circles confirm this. Cultural stereotypes assume further that women's career choices are made primarily with regard to their personal relations with families or with men.* Some women share this view, but it should be questioned as too simplistic, even when women themselves say it is what motivates them.[27] Some women lawyers' accounts of what they do "for love" may really be masking their desires for acknowledgment and fame or their retreat from work pressures they are not up to.

A leading movement lawyer explained that she took on a consumer advocacy case that required considerable entrepreneurial skill—forming an organization, raising money, and then preparing the legal work —because it was a cause important to the man she loved. But she made an outstanding and visible reputation on it. Another lawyer in private practice said that all of her major career decisions in one way or another had been made around men. She left a large firm when she broke up

---

*Marcia Millman's analysis of cultural stereotypes of the motives of women criminals dramatizes this view. She points out that women are believed to commit deviant acts in the service of love. This has the consequence of undercutting the acknowledgment of their ability to take risks or engage in antisocial behavior as an expression of social protest. Thus, the *evaluation* of women's actions are low, compared with men's, because they are said to be guided by emotions while men are not. "She Did it All for Love: A Feminist View of the Sociology of Deviance," in Marcia Millman and Rosabeth Moss Kanter, *Another Voice*, pp. 251–79.

with a man who worked there; but when she got back together with him she was glad to be away from the pressures of the firm because "if you have two people with careers, there has to be one person with flexibility in terms of schedule, and I was that person." But others claimed she left the firm when she was sure she was not going to be recommended for partnership. Yet many women who downscale their ambitions "do it for love," while men are likely to try to aim their career sights higher for love. Maybe men know that flying high will make them more desirable, while women accept the popular notion that flying high will make them less desirable.

Where women have had no choice but to aim high—for example, when they are in large law firms where the work demands are assigned and constant—love does not seem as important in their calculations. Women in other kinds of practice reconcile their ambivalence, the stress of practice, the difficulty in making money, and other problems by retreating from highly demanding careers. In the words of one lawyer:

"It is a more secondary thing for many women."

"Secondary to what?"

"Well, to families and very traditional patterns. Obligations regarding children; of income; law becomes a part-time, secondary thing."

But remember that men, too, give their families high priority when it comes to making career choices. One radical attorney talked about how the men she knew had retreated from movement law:

A lot of the men that I know who think the way I think about legal issues also happen to be the beneficiaries of trust funds. There are quite a few like that. Others have dropped out. . . . The men have families. They feel they have to bring home the bacon.

Some of the problems women lawyers face in their relationships with men close to them might arise in the lives of any working couple. Some are due to the additional voltage of a highly demanding, prestigious occupation, women's position within that profession, and men's uncertainty about what all of this means about the women in their lives. Should law become "normal" as a career for women, some of these problems will fade; but others may continue. Some women may decide they cannot face them, but my impression is that most women lawyers are learning to cope and live with them. Problems are, after all, part of the human condition.

19

Children

FOR MANY women lawyers the big management problem is not their love lives or their husbands, but how to care for their children. Fear of the demands of child care has made many young couples put off having children or decide not to have children at all.

A Wall Street lawyer commented: "I see more and more young women deciding it can't be done—that the only way to resolve the conflict is to avoid it, not to have children." One of those she was referring to remarked: "I think of my daily routine, and I wonder how I would cope. I wanted a kitten and my husband said, 'but we're never home, the kitten would starve.' Well, he may be neurotic, but that's how he feels about a child, and I agree." And the same analogy was offered by a lawyer whose work load on behalf of civil rights issues was clearly as taxing as that of any corporate "sweatshop": "I don't know how people do it. I can't even have a pet, much less a child. I mean, I couldn't take care of a cat or a dog."

To some extent, the views heard in interviews with women and men in the legal profession are views being expressed by other young people of their generation. The 1980 Virginia Slims Poll of attitudes of women (the only major assessment of attitudes toward issues relating to women in the United States) shows that four out of five women say that children are not an essential ingredient of a full and happy marriage.[1]

A young associate in a large firm talked about the discussions about children she had with the other women in her peer group:

We're all questioning. We are wondering whether if deciding to put off having a child for five years may result in deciding not to have one at all. Then they wonder how about twenty-five years later. Are they going to feel lonely? Or is it that you are born in a society where you were supposed to have children? We wonder whether we'll be good mothers. Debating the relative merits of having a child who is at least on a daily basis left in the care of a nurse or nanny. Wondering how much input you'd have to a child. Whether it's just an ego trip to have a baby, and what are the motives for having a child.

A typical response of women in their twenties and early thirties was to defer the decision: "We plan on having children, but we don't want to have them now. Probably within the next few years, partially because I'm not getting any younger. The years go by."

When *Fortune* magazine sampled twenty-five-year-old men and women "on the make" in 1980, it found that few had children and some doubted they would ever have them because they couldn't spare the time.[2] Said one of the *Fortune* sample: "I'm too selfish to give what's necessary to raise them properly. Eventually I'd resent their taking me away from other interests, just as I'd be upset that I wasn't devoting enough attention to them."

But there were some respondents who wanted children, the standard two of them, although they said they were postponing them until their thirties, when they seemed to think there would be less disruption to careers, marriages, budgets, and free time. They are a generation "consumed by work"—couples like an engineer and her husband who together earned $60,000 a year, thought they might have children by 1985, when they would be making $120,000, because "we want to make sure my career is well established, that we have all the material things we want . . . and good child care."[3]

Lawyer couples on Wall Street, with joint starting incomes of $60,000 to $70,000, said they were delaying children at least until each reached partnership. As one young women partner put it, a woman whose husband became a partner and anticipated making $100,000 or more could consider cutting back if she wanted to and could take a salaried job with the government or a smaller firm and still not have to worry about paying for good household help.

The discussions about whether to have children or not represent a marked change in the attitudes of young women. Only a few years ago it was not proper for women not to want children, and certainly it wasn't proper to admit the fact. Other people pitied the few who did not want children, or felt they were selfish.[4] Although in the past career women tended to have fewer children than the average[5] and black professional women tended to have an even lower number,[6] a majority of married professional women did have children.

Of the married women lawyers interviewed in the 1960s, 85 percent had children and 60 percent had two or more,[7] although they tended to have their children later than most women. One third had their first child after the age of thirty, when most American women had borne their last child. This was partly due to the fact that these women married later than most of the women of their generation.

The problems of managing, of feeling guilt at not being home when children were small, or of finding adequate help—the problems that plague women today—did affect the careers of some of these women, but there was wide variation. Although 85 percent of the mothers in the group were working when they were interviewed for the first study, as many were working at part-time as at full-time jobs. There was no relationship between number of children, the age at which the women had their children, and their work status.*

White's Michigan study showed a strong association between maternity and retirement from law in the late 1960s, yet close to half of those with children worked full-time, as in our group.[8] The other half was divided between part-time workers and those who were not working at all. The White study did not show how many women had retired from active careers and resumed them later. But our interviews with older women lawyers indicated that many resumed active careers once their children were older. In the 1960s we found that although maternity was probably a primary reason for leaving practice or for curtailing their working schedules, it did not necessarily lead to retirement. Many women who desired to continue practice found ways to do so, or had family assistance that provided income or help with the children so that the pressure of child care did not overcome their motivation or capacity to work.

Motherhood did not negatively affect the success of women who achieved partnership in the New York Wall Street firms. Of the thirty married women partners, twenty-seven had children and one who did not was planning to have a child.

Women in other male-dominated professions also have had a pattern of work involvement and curtailment almost unrelated to number of children. A study by the Society of Women Engineers of its members showed that they were equally likely to be married or single, but that if married, they had an average of three children.[9] Married women physicians studied by Rosenlund and Oski similarly had an average of three children. The women physicians working forty or more hours per

---

*Women who dropped out of legal work entirely because of motherhood or other reasons would not have been listed in the principal lawyers' directory and were not covered by the study.

week had as many children as those women working twenty hours or less.[10]

As the flow of women into law increased in the early 1970s, there were an increasing number of role models to show that the career and personal costs of having children might not be too high. Despite the often negative response to the idea, there were lively discussions of the options open to women, and some confessed to having second thoughts. One young attorney who had said "no" told how she changed her mind:

A few years ago, when I was pregnant, both my husband and I got very depressed. Having a baby would interfere with our careers. I had an abortion. Today, I go into court and see a lot of women in their thirties, with pregnant bellies out to here. And I think they look just beautiful, handling their cases with confidence and professional skill, and also confident to have the baby too. And I think, I can also do my thing in court and have children too.

The role models themselves feel the pressure of the continuing debate, and it makes some women even more dedicated to managing well. Most older women tend to be encouraging; after all, they did it. But the judgment is far from unanimous. One lawyer was more ambivalent than most when she was invited back to her law school to address women students:

I'm invited back to my law school as a role model, to tell the young women, who are nearly half the class now, how to integrate commitment to one's profession with family life. I see more and more young women deciding it can't be done —that the only way to resolve the conflict is to avoid it, not to have children. The doors are wide open to women now, even in the top Wall Street law firms. But to be honest, the way things are right now, I'm not so sure they can have it all. Women are afraid to take the risk of what might happen if they didn't take an opening, and didn't buck for partner. There are so many people coming out of law school behind you. The only hope is that as many young men come to these sessions on combining work and family as the women.

Those young women who have rejected the idea of having children seemed to worry the most about managing. Those who planned to have children seemed to downplay the problems, and those contemplating motherhood and a demanding career seemed to feel that the demands of motherhood were the least problematic. One lawyer said: "I'm not sure we have decided. I think I would have full-time help. I now see it as someone who comes in the morning and leaves in the evening. I certainly see myself as continuing to work full-time, and I don't see it as a problem."

Perhaps one answer to the question of why successful women lawyers are usually married and often have children is that good lawyers are problem solvers. Many of those interviewed attacked the problems of managing home and work in the same direct, matter-of-fact way they managed their offices. The theme of defining problems and solutions as real and therefore making them real in their consequences was certainly characteristic of the coping and non-coping women and deciders and non-deciders.*

Some women "decided" in a systematic way. They were invariably the younger ones, women who had mapped a direct career line—a "status sequence"—from college to law school to employment and were capable of making rational calculations about the ways to achieve career and personal goals. And, of course, some women had children before they even considered entering law, and by the time they had focused on a career the children were there and needed to be fitted in.

But for those who were thinking about the "best" time to have families there were various schools of thought.

Fitting Them In

The young women and men interviewed on the question each felt there was a best time to have children to minimize career problems. But the times recommended ranged from early—before or even during law school—to late, after making partner in a firm.

In the early decider school were women who felt that if they had children early, patterns for child care would be in place by the time they started up a career ladder. They also thought that way they would be spared the years of agonizing over when the proper time would be.

A young Wall Street associate pondered that problem and said: "There are a couple of women whom I regard as having the perfect situation: those who married in law school and had their children before

*The observation rests, of course, on the quote by W.I. Thomas: "If men define situations as real, they are real in their consequences." Widely cited, the Theorem appears once in the corpus of W.I. Thomas's writings, as I have been informed by Robert K. Merton, in the book he wrote with Dorothy Swaine Thomas, entitled *The Child in America* (New York: Knopf, 1928) p. 572. I am grateful to Professor Merton who not only provided me with this information and was responsible for framing it as a theorem, but who also points out that although the theorem appeared in a jointly authored book by the Thomases, Dorothy Thomas insisted to him that both idea and formulation were entirely W.I.'s.

they started to work. Because if you're ambitious, and if you like the job, you know time slips away."

The later deciders felt it was important to root themselves in a career so that when they had a child they would have surmounted the problem of time pressures at their worst, in the years when proving themselves made such a difference. Women felt that time pressures ought to ease with promotion to partner, when they could delegate much of the tedious work to associates.

Coping and Managing with Children

Women lawyers interviewed in the 1960s who had children invariably had worked out the problems of child timing and child care. Most had fairly continuous work lives, taking out only a few weeks or a few months while their children were infants.

One Wall Street lawyer, mother of three children, reported:

"I had my first in 1961 and the others on summer vacations."

"Did you take off the summer?"

"No, I took five weeks instead of four weeks when the babies were born."

Similar solutions were found by mothers practicing law before the 1970s, although a greater proportion of the younger lawyers were willing to enter into highly demanding practices, and did so with somewhat less guilt than the older generation. Coping strategies were both of a psychological and behavioral kind.

Some women commented (though few complained) that children altered their careers:

My child developed a chronic illness just about the time opportunities seemed to be opening up. I continued full-time practice (I was divorced) but I was also up all night so I could not take anything which involved going out of town, or anything which was too intense, either because of its difficulty or because of its involvement. Because if I was up with my child all night, when I came to the office it was all I could do to stop from falling on my face. So I had to handle it in a pragmatic way to avoid very difficult things.

I try not to go out each night, but I should go more places and be seen more places. In New York City, if you disappear for a couple of years nobody remembers your name.

Many successful lawyers did not experience problems either with their careers or private lives. Children did not create problems for them, they reported. At the very least, they found those problems normal, solvable, and not a substantial burden.

One case in point was a partner on Wall Street who was married and had three children. I asked her how she managed. Her answer was, "No problem with managing." "No problem for anybody?" I asked. "My children never had any problems," she said. But it turned out that her view of what constituted a problem was different than other people's:

"I don't mean: NO problems. My eldest child had a learning problem and we had to do a lot for him, but he got over it and he did beautifully. But there's never been a problem of the kind where you might feel you've got to be home. My husband did work many hours with my child to help him overcome his problem. I didn't. I was mostly working."

"Was that all right with him?"

"Yes, he loved it."

Part of coping can certainly be accomplished by defining a problem as solvable, or as a condition of life one must attend to without becoming troubled.

Lawyers who managed well usually found their solutions outside the workplace rather than within it. Law firms, like other employers, generally are not anxious to accommodate women who want part-time employment. Furthermore, women who are helped by their firms often find their colleagues are resentful.

Time Management

Many women lawyers chose specialties that have time schedules that fit their personal needs and wishes. For some this meant the choice of a specialty with stable hours and no overtime requirements. For others, specialties with flexible and irregular hours were preferable. Most married women interviewed in the 1960s did not work in specialties that demanded that the practitioner work a full week and overtime and weekends with regularity, unless their husbands' work made similar demands. These lawyers met the very taxing time demands of corporate specialties early in their career but found that the pace was difficult to keep up as time went on. They pointed out, however, that many male lawyers also find that they are unwilling to continue to work at specialties that demand "total commitment"—the rewards simply may not compensate for the costs. A number of women in the older group who started legal careers as corporate specialists changed to trust and estate work later because of its more stable time demands.

A lawyer who was in joint practice in a litigation firm adjusted to time demands by selecting those aspects of her firm's work which would give her freedom, such as preparing briefs and doing research. When asked if she wanted to do trial work in the courtroom, she replied:

No, I didn't. I did a little bit in the beginning. Well, quite a bit in the beginning. But once I got married and the children came, I didn't want to have to be at a set place at any set time. If I were in the midst of a trial, and a child gets sick, I would naturally rather be home. So I haven't got myself involved in litigation —active litigation—since the arrival of my children. But I am here [at the office] every day, and I go to trial for pre-trials. . . . I go out a great deal to the courts but do not begin a trial which would oblige me to be there every day.

A number of women who felt strong commitment to their roles as wives and mothers told how they managed to work as well by severely limiting their hours. These were "family first" lawyers:

None of my children was going to be raised by a baby sitter. So I make it my business to be home by three, unless something extraordinary happens. And I try not to go out at night. Of course, if I have to, like with small claims where you just can't fit it all into the day, I work at night. But I try to make it a practice that I work just as a teacher would: between nine and three.

Another lawyer said:

I never stopped—I never stopped entirely, but a day here, two days there. With the assistance of grandma and the assistance of a good maternal babysitter, I never really stopped. As a matter of fact, with the second child labor pains began in the office and I went from the office to the hospital.

On lawyer "solved" the time problem by working from her house:

I have a downtown office but I also have a full office, with complete sets of files, in the basement: a large desk—one of those mammoth things—and three phones—one on every floor—I have the typewriter, the adding machine, I have everything I need there. . . . Of course this is something that has only developed over the years; at one time I used to spend more time down there. And I would say I don't go in more than once or twice or three times a month; more in the winter. Summers I don't go in at all if I can help it.

The most innovative solution for women who wanted to spend time at home with children was that established for two lawyers by David G. Trager U.S. attorney for the eastern district of New York. There, Bernard Fried Chief of the Criminal Division,* worked out a schedule with two assistant U.S. attorneys who each worked three-fifths time, as a way of fulfilling the office's commitment to increase the percentage of women and minorities.

Fried claimed the plan worked out well, except that one woman had become involved in a trial and was "here six weeks virtually day and night." Trial work is hard to cut up into part-time arrangements. Fried tried to find a solution. "We will try to make sure her trial load doesn't get too heavy and we will reassign cases, if she wants us to," he reported.

Seeking the Cooperation of Role Partners

Women lawyers learn that they must often depend on others in their role networks to make adjustments to help them budget their total role obligations. The "others" include employers, husbands, mothers, baby sitters, and colleagues. The adjustments include practical aid as well as encouragement and approval. Often more than one role partner is involved. A lawyer with a five-month-old baby commented:

There have been times when I've needed to work nights and I've called my mother. But the man I work for has said "go on home if you can't get her." The people here aren't oriented so that they'd ask me to foresake my child. If the worst came to worst, I guess I could always take my work home with me.

The assistant U.S. attorney who was given a three-fifths appointment was married to an assistant U.S. attorney in another district with trial duties to perform as well. How did they manage?

My husband and I were sort of balancing schedules on weekends. I would come in one day and he'd go into his office the other day, and when we started the trial I asked my regular baby sitter to come about 8:00 in the morning, then I'd have a teenager from the neighborhood come at 5:00 so our regular baby sitter could leave between 5:00 and 5:30, and then she'd stay till my husband could get home between 8:00 and 9:00. It was really a juggling act so my mother was coming for Thanksgiving vacation with my father, and we just asked her to stay on, and she stayed with us until a week before Christmas, when she took my

*Now Judge of the Criminal Court of Appeals in New York.

son with her, and we brought him back after Christmas, and did the same thing with the regular baby sitter and the local baby sitter until the verdict.

Whether one defined problems as solvable or not, managing the demands of work and of child care were issues that had to be confronted by the women and their husbands. The two necessities most women stressed were supportive husbands and good household help. They might also have added, but did not, an implication that ran through the interviews—the willingness to delegate responsibilities, not to insist on doing "everything," and perhaps to accept lower standards.

One mechanism for "coping" characteristic of the 1960s was joint law practice by husband and wife.[11] Although small partnerships of this type were diminishing in the 1970s and 1980s, a few women lawyers formed this kind of practice for at least a period of their careers. Husband-wife law practices, no matter what their other drawbacks, gave women flexibility of time, a supportive work environment in which obligations of child care could be shared with a husband, or minimally, a partner who understood intimately the pressures from home impinging on the work setting. Women could easily cut down the volume of work, if necessary, and husband and wife could cover for each other in these practices.

Although women who had left or limited their practices to care for children felt that lack of help was a major impediment to their career activity, it seemed obvious that they were also affected by current social values that stressed the importance of women personally caring for their children. One solution for this problem was to turn to the age-old universal source of assistance—the grandmother, and in some cases, grandfather.[12] In the group interviewed in the 1960s, two women attorneys, each of whom had four children under twelve years of age, shared households with their parents. Other women reported systematically turning to their parents, especially as backups to employed help. Most of those who did felt their parents were satisfactory substitutes for them. No one is faulted for delegating child care to grandparents.

Women lawyers interviewed in the 1970s were more ambivalent about household help than those from the earlier study. Some of the younger women complained that help was more expensive now, but more were loathe to use full-time housekeepers for practical or normative considerations. For some, living quarters were too small to maintain privacy, or they were uncomfortable with a "stranger" around, or they had been exposed to the view that maids were somehow incongruent with a philosophy of egalitarianism. However, women who used full-time help (and only a small number had help who lived with the family) argued that good surrogates could be found if one paid a decent wage, provided benefits that other workers enjoy, and treated the person respectfully.

Most of the women partners on Wall Street with children had full-time, live-in help. Those who did not had full-time child care in the home and lists of baby sitters who could be called on for work on Saturdays and Sundays, as well as evenings.

One lawyer interviewed envied a peer, a woman who

arrived pregnant, although she didn't realize it, and had the child after she'd been here nine months. She and her husband are both attorneys, but most important they have parents who live near. She has a nurse, but often their parents rush over and baby-sit when they have to work overtime and weekends when the nurse is off.

But calling in surrogates was not an acceptable answer for some women. Like the differences between the parents who feel it is right to send children to boarding school and the parents who feel it is criminal, the question of how much time to spend with children is an outgrowth of personality and social milieu. Many women faced the "normative dilemmas" characterized by Rhona and Robert Rapoport in their study of dual-career families.[13] Full-time careers for women, which precluded meeting child-care responsibilities according to the judgment of older traditions, still posed latent dilemmas for working women, covert uneasiness, perhaps even anxiety and guilt, although intellectually the pattern was approved.

For some lawyers, weekends represented most of their commitment to child care, and they viewed that as sufficient:

We don't work on weekends. We have absolutely made that commitment, except this weekend, my husband's got a trial starting Monday and he's just come out of another trial, so its really backed up.

Our weekends are almost always with Elise. Rarely do we do anything without her, so that is our time with her.

Others made sure their children's schedules were altered so they could spend time together:

People often question how late Judy stays up. She's always done that. That's just been the way it was. She used to wonder—maybe because she sees her friends, they're fed and ready to go to bed by 7:30. That's the time we begin to do whatever we do together. I think it works. I think kids get enough sleep.

Children

Other lawyers broke into long work days to spend dinner time and early evening with their children:

I'd run home to have dinner with the kids; read to them, put them to bed, run back to the office and work till midnight.

I'd go home and bring work to do after the children were asleep. My husband did the same thing.

One lawyer with a high demand private practice explained her no-guilt philosophy:

Every child has a need for his parent, which he expresses directly and sometimes through manipulation. Its a very valid need. But I have answered my son's needs in the following way. The fact of my work is something that's just an assumption in his life that he has to accept as part of his reality. I make no excuses for it. I intend to have no guilt behind it, because it's pointless. That's where it's at. The time I spend with him, I try as much as possible to make meaningful in the sense that I devote my attention to him. More importantly, I think he really understands that I love him and care for him. And more than that, I cannot give him. And if it works out well, it works out well. And if it doesn't, I've done the best as I was able. And I'm not going to second-guess myself.

At issue for some women was another quality identified by the Rapoports as "identity dilemma."[14] Women's views about being a "good person" often fixed on the quality and quantity of child care they were delivering. This tied into their conception of the motherhood role. The woman above had brought her role behavior and role conception into alignment. Women who could not do it, who thought there were aspects of child care that couldn't be delegated to hired help or husbands, had more anxiety and worked overtime at mothering.

I would stay up all night tie-dying a shirt so my boy would have a costume when he was supposed to be knight in a play. It never occurred to me to try and get someone else to do that.

When one of my children had a birthday I would work until all hours to make sure the party favors were special and there would be a home-baked birthday cake. I never wanted them to feel because I worked they would be deprived.

In a world where the standards of child care are not normatively defined in any specific way, women (and men) develop different criteria for their behavior. Far fewer younger women seem to feel they needed to make costumes or bake cakes; as with housekeeping and entertaining, they are willing to buy services to tend to children's needs, from party favors to boarding school.

How do some couples come to define their child-care commitments as adequate and others not? A lot has to do with the extent they are exposed to evaluations from friends and relatives or companions at work. Not only were many feminist lawyers more "traditional" with regard to norms specifying they should spend time with children, but they also were exposed to many discussions with their peers about child care as well as other personal relations. These discussions often contributed to their ambivalence.

In contrast, women in large firms worked in a milieu in which child care came up less as a topic of conversation. Women in some highly demanding practices had little time for casual friendships, telephone discussions of personal problems, or for popular magazines or television programs that would question their behavior. These women were insulated* by their work from sanctions that created guilt in other attorneys and other women. Ironically, the more they worked, the less guilt they were exposed to, since the rest of the world was less intrusive in their personal lives.

Insulation also had an impact on the role strain induced by children and the extent to which they experienced their parents' absences as deprivation or not. Of course, children can protest on a verbal or a non-verbal level. We asked the lawyers whether their children complained about the lack of time spent together. Not many said their children complained, or if they did, not many thought it was important to do anything about it. This lawyer's response to our question about whether her daughter felt deprived was typical:

I think she sees that more and more of the mothers are working now and are not bringing their children to school either. Most of the women who are going back to work, they're not getting into careers; they work, but it's not quite the same as her mother has. Sometimes she gets frustrated and says, "Take a day off" or "Why don't you quit." If she has a frustration, she's free to express it. I'm really not terribly concerned. That is not something I can consider making any decision based on. I'm just trying to make life as pleasant as I can for her within what I am doing, and I think there are ways to have fairly decent housekeepers if you can, decent schools, things to play with. Our weekends are almost always with her, rarely do we do anything without her, so she has her time. Except she doesn't want it sometimes.

*Insulation is a mechanism for alleviation of role strain. Robert K. Merton, *Social Theory and Social Structure* (New York: The Free Press, 1968) p. 429. One could also conceptualize the procedure as being "barriers to instrusion"—work being the barrier of women lawyers.

Children

Most women lawyers with children reported no special or severe problems with them. There were no reported patterns of poor school grades, chronic stomach trouble, or the host of ailments commonly conceived of as psychosomatic. Only a few reported that their children suffered from the serious problems one might expect ordinarily in any population of children: one was asthmatic; one was recovering from a car accident; another had dyslexia. Perhaps those mothers who faced problems that defied solution had long ago given up law. But most of the lawyer-mothers interviewed did not speak of their children as burdens, but spoke positively and sensibly, and sometimes glowingly about their accomplishments.

Norms and practices also had changed in the profession. We interviewed several lawyers in large firms who were nursing mothers and had gone back to work within a few months of their children's births. One had the nurse bring her baby to her at work, closed the door, nursed the baby, and sent it home. Her partners knew about it, she said, and no one said a word.

Another lawyer brought her baby to work on weekends and, while he was still an infant, made a bed for him in a large (legal-sized) file drawer. Male lawyers came in and out of the office, she reported, and never noticed the baby was there. Still another woman brought her baby to the office on Saturdays and put him in the bottom desk drawer. For most of the men, the presence of a baby in the office was so out of their frame of reference they never noticed those who were quiet. Those who did notice didn't object, partly because many were divorced and often had to come to work with their children on days when they had visits scheduled. Divorce has made some men more attentive fathers than they ever had been in their marriages, when they left child care entirely to their wives. An associate in a large firm commented, "I didn't dare bring my kids into the office on a Saturday until some of the new male partners started doing it."

A Wall Street associate described the daily pattern of one of her colleagues in the tax department:

"She rarely stays past 7:00 or 7:30, and then she'll take a cab home. Her baby was three months old when she started and she had a four-year-old. And she's pretty much told the head of her department that she wants to work 9:30 to 6:00, I mean, she's still nursing the baby so she has to be home at night—the baby's on a daytime bottle. But that's a tough schedule to maintain because there are times in most jobs that are demanding, when you do need to stick around—in law a lot more so than in others. But she has been holding so far.

She has stayed nights. One night she came back and brought the baby with her."

"How did they feel about that?"

"It was funny, several people came to see her in the office, and she had the baby sitting there and they didn't notice it. The baby was sitting in one of those little baby chairs, and she sort of had it on the desk, [pointing] on that side, and she was talking here, and she realized they did not notice the baby. Because the baby was just sitting there being quiet, and I guess was off angle. But it was really funny. One person saw her in the library with the baby and thought she had a doll. Another said to her the next day, 'What were you doing with a doll in the library, Emily?' She said, 'That wasn't a doll, that was my baby.' 'Oh,' he said."

But few women lawyers really approved of bringing children to the office; small children tend to run around and be disruptive. Some men asked their secretaries to take care of their children, and the women objected to this delegation of responsibility, saying they would never dream of doing it. One woman felt that Wall Street ought to have daycare centers, but this "radical" solution was never suggested by anyone else. She described one father's dilemma:

"I'll never forget one young man who had two children. He was getting divorced and brought his two kids in every Saturday. He worked all day Saturday and all of the support staff in the office had to watch his children because there was no place for them. The children would get on the telephone, tear things off other people's desks, run around shrieking through the entire floor of that staid downtown building. You can imagine what it was like. It was a cyclone!"

"What did the other partners say about this?"

"Everyone thought as soon as he got his divorce it would be over. No one took the next step and suggested, 'Why don't we set up some facilities.' They just viewed it as an idiosyncratic case."

One lawyer brought her eighteen-month-old child into the office during the week and kept her in a playpen. She was asked how clients felt about that.

It's a matter of selection. Those who mind don't come to me; those who don't mind aren't surprised. They just don't say anything. But as my daughter is getting older I think I won't be able to do that anymore.

Several lawyers were interviewed at home, with babies crying in the background or demanding attention. Feeling disoriented, I wondered how much of a solution this could be for anyone. To their credit, the women who conducted business with their children around seemed to direct attention to the business at hand and were less distracted than other mothers I have seen. Nevertheless, all viewed this as a stop-gap measure.

Husbands' Participation

How about sharing child care with husbands? Among women of all ages, and in all kinds of practice, most did not share child care equally with husbands. Many women lawyers are married to or living with men who share an egalitarian view of women's right to a career. The men have more ambivalence, however, about their own responsibility for sharing household responsibilities. Most work out some sort of schedule for housework when there are no servants. Child care is another matter. In the final analysis, men and women agree that the women will shoulder most of the physical and mental responsibility for child care.

This is obviously an outcome of socialization and structure that reinforce the sex division of labor with regard to child care. Most lawyer couples follow the usual American pattern in that the husband is older than the wife, he is usually (therefore) more established in his career, and also (therefore) making more money. Thus, when the question comes up about how to divide up child care, it is often reasoned that the wife will do it, not because she is the woman—for that would be an unthinkable sexist view in these modern times—but because he is, after all, the major breadwinner for the family, the demands on his time are greater and more important, and he, more than his wife, is ambitious. Often there is no discussion, and the wife just assumes it is her responsibility. Men are rarely put in a position where they have no choice but to take on major responsibility for child care because a wife refuses to do it.* Alice Rossi was probably right when she wrote in *Daedalus* that women will take on responsibility for child care and that most men will not.[15]

Reviewing the comments of women lawyers about their husbands, Judith Thomas, a colleague on this project, points out that any man who gets involved actively in caring for his children is regarded by his wife

*One of the men interviewed was left with two children when his wife decided to forge an independent life. His friends interpreted the wife's behavior as the pathological response of a mentally ill woman.

as special and as going beyond ordinary expectations.[16] This is true equally for women identified as feminist attorneys and for lawyers in conventional practices. She suggests that they face an ideological schizophrenia, in which they espouse a principle that says a father should be equally responsible for his children yet at the same time believe fathers who live up to the maxim are unusual and should get extra credit. Equal responsibility is something closer to an ideal than a real norm. A young attorney summed it up:

Ninety-nine cases out of a hundred, when there's a choice of a husband or wife curtailing their career because of children it will be the woman who does it. I don't know of any men who have given up careers because of their children. I don't resent it although it troubles me, because I have been so conditioned, I guess.

On the other hand, the husbands of these women certainly participated in child care. Their help ranged from the dedicated attention of one husband who spent hours every night helping his child with a reading problem, to another who took full responsibility for the children on the weekends when his wife was engrossed in a litigation problem, to those who had more minimal participation with the children.

What husbands did not seem to do, said one lawyer, was worry about their children, that is, about:

whether Johnny was seeing enough children his own age and making sure to call some parents and see if we couldn't get some of the kids together; or whether the clothes Jill had were the same kind as the other children were wearing at school; or even whether Amy wasn't a little withdrawn these days and what could we do—get her involved in a sports program or something.

And in a similar vein:

He shares a lot, but . . . he would be the first to admit that he does not go to the office and wonder if Jeremy's cold doesn't get worse in school. Or if Jeremy hasn't visited a friend in a while—I should call somebody up, because he needs that experience of having a long-time relationship with a child his age in our house. Or, Alison's birthday is coming up, and what should we do? Where shall I order the cake?

But yes, he helps me. He goes home, when I'm on trial he goes home every night and he takes the children on weekends. He did it all through law school. I could never have gotten through law school without him. He's 100 percent supportive. And he did wash, and he washed the floor when he wasn't making that much money to hire help. But still . . . I did more.

Children

Several women had agreements about sharing child care with their husbands before they had children. Those husbands did try to cooperate, although sometimes there were problems. One feminist lawyer said her husband "was no stranger to a diaper or the vacuum cleaner," but she wanted to be sure that child care did not interfere with his career. "I'm watchful," she said, "of what he needs to do to make sure others feel he is pulling his weight at work, and if necessary I ask less of him." Here, the role of the man as main breadwinner was established and the contribution of each partner in the marriage was considered in the equation.

In fact, the motherhood mystique, as noted before, seemed to be living most heartily among some of the more feminist women lawyers, although there were a number who were insistent that their husbands take charge of children some of the time. The young feminist lawyers who have babies seemed dedicated to personally taking care of them. Most of those interviewed in New York were in part-time practice, taking off several days a week to be home with children and eschewing full-time help. This decision rested in part on their desire to be with their children, in part on their distaste for employing domestic help. It was made possible by their ability to work part-time, a choice not available to most women who worked in larger firms. In consequence, these feminist lawyers may be sharing less of the responsibility with their husbands than other lawyers do because they put more time into child care.

Some of these women also expressed their commitment to motherhood symbolically and publicly. Several women in the feminist network carried their small babies about (sling-style in the new mode) to parties and public functions such as meetings and in some cases to their offices. One woman who knows many of them claims that the sense of commitment to causes which motivated them in law is now applied to motherhood, where their ideology is also strong.

The fathers of these children are also public in their attention to the children and often seem to be making a political statement that it is appropriate for them to be involved in the care of their children. The husbands and consorts (for some of the couples are not legally married, though they live in a nuclear, monogamous family unit) of these women are also committed to a feminist stance and understand that they are "on review" in public situations. Most of these women are married to or living with men who are also lawyers. One often will see a father holding his baby in a meeting, or rising to help the woman change the baby or feed it. Yet, in all cases I have seen or heard reported, it is clear that it is the mother who is primarily responsible for the baby, by choice, agreement, or simply *de facto,* with the father acting as an interested and proud partner.

In this group of radical or feminist lawyers, it is also clear that in most cases the women alone have curtailed their practice, in a repetition of the traditional pattern of male breadwinner and female caretaker. In one case, a couple decided to go into law practice together, which probably represents some limitation on the career possibilities of each, but both have decided that this is best for the family at the moment.

The young women who bring their children into their professional lives appear to be following a classic American pattern, but it is not the same. If the women are bringing babies to work and to professional association meetings and parties, it is with the cooperation of the men, who may also have to leave early if the child becomes obstreperous. This is not the same as women taking their babies to the public square or barn raising. In those cases, the women sat with other women and received help from their sisters and mothers and friends. In the modern version, they are taking the babies into an integrated world, and their men will be asked to provide assistance and support.

The constraints of the profession rather than lack of commitment seemed the most serious barrier to fuller participation of men in fathering. A feminist lawyer living with a lawyer older than she who had a family from a previous marriage described her companion as an "easy and flexible man." They had a small child. "It would be impossible if he weren't." She described his taking over care of the baby on a Thanksgiving weekend when she had to prepare a case for court the following Monday. "But what would happen," she was asked, "if your court deadline came when he was at work:" "He can't adjust his work life," she offered.

He's a partner in a firm and can't say "I'm sorry, I have to take care of the baby." I knew that before I had a child. And so that part of it is my responsibility. I wish I could live according to my beliefs that each parent should share equally in the care of the child. But, unfortunately, [P] can't do that.

She only worked in the office three days a week, and if she had to come in on a day when the baby sitter did not come in, the baby came to work too.

A feminist lawyer in practice with her very committed feminist husband explained her greater attentiveness to her child:

Children

Does [E] feel the same? Yes. But he wants to get more and more into corporate work, and whatever is necessary to do he will do even if it means seeing a little bit less of Alison. He says that, but it hasn't meant that yet. He has a lot of business trips that take him out of town. I feel a passionate and very positive need to be with her. He's a little more loose about it. I'm not only compulsive but I really want to be.

Of course, women often judge themselves by their responsiveness to children, and are reared to be sensitive to them, and men sometimes have to be reminded to be responsive, partly because reminders about children are not built into their conditioning or the structure of their lives.

I was particularly touched by that fact at a party given by a feminist lawyer at which the guests included an older, well-known, civil rights lawyer and a lawyer couple—the woman a feminist attorney and the husband a lawyer who had just won political office in New York. The couple had brought their six-month-old baby to the party and handed her to their friend by way of greeting. The older civil rights lawyer swept up the baby with a whoop, exclaimed over her, and then set her down and went on with his conversation. He so quickly and completely forgot the baby that he nearly stepped on her as he rose, gesturing to make a point. The women in the room, in contrast, had their eyes on the baby and were poised ready to rescue her, as one did in a way that did not embarrass the older man. He never was aware of the small drama going on at his feet. The child's father was called into service as the evening went on, to diaper the baby and later to leave early so the child could be put to bed. An important part of his presentation of self in this gathering was to show a participatory role in care of the baby. This was an ideologically committed group, and although it was clear they were ready to exclude the older man from evaluation, they were evaluating the younger man and he was aware of it.

Thus, there are changes to report. Young fathers (and some older ones) are becoming more sensitive to the demands of child-rearing and are no longer anxious to or allowed to place all responsibility for children on their wives. In social gatherings it is common to see lawyer couples—and other professional couples—share the comforting of fussing babies, taking turns at the dinner table if necessary. Most young men now do this automatically, although sometimes their wives must remind them. (Men don't remind women because that would be considered traditional and sexist.) One older radical lawyer, a "star" who had recently had a baby with his second wife, said in an interview that he had been a traditional father with his older children, but that now he was committed to sharing child care. He had a crib put in his office (at home) so that he could watch the child when there was no baby sitter and his wife was in court.

Most professional families today seem committed to an ideology of sharing, and this is true for couples of all social strata and lawyers in all types of practice. (Though women report that most men who work in highly demanding legal practices have traditional marriages in which the wives stay at home.) The hesitancy of young men about having children invariably comes from their sense that it would affect their careers. Ideology has moved men to commit themselves to sharing the work of children, although the precise share is still negotiable in most marriages.

Men's sharing is not the entire answer to the problem of child care because both partners have time pressures they have to meet. Without institutional supports we can look forward to men as well as women suffering from ambivalence and guilt. But men's participation is still minimal and surrogate institutions, the schools included, have not yet recognized the joint responsibility of husband and wife for children or even their own responsibility in an age when women are part of the work force.

Responsiveness of Other Institutions

Those outside the family who are involved in the lives of children—their teachers, school nurses, or doctors—generally assume the mother is the parent who will be available if there are problems to discuss about the child or if the child is ill. This is true even in cases where school personnel are acquainted with the fact that many mothers work.

One well-known attorney told me that when her son felt ill at school and went to the school nurse the nurse would call her to come and take him home. The child quickly realized he could get out of school when he complained of stomach aches. After a number of such calls, the lawyer told the nurse that she didn't see why the nurse always called her at work. "After all," she said, "the child has a father as well as a mother, and you can call my husband on some of these occasions." The school didn't call her after that, nor her husband, but rather managed to care for the child until normal dismissal time. The nurse obviously felt uncomfortable calling the father about a problem that she didn't feel was important enough to warrant disturbing him though she had felt comfortable disturbing the mother.

Schools in the United States are not geared to the needs of working mothers. There are no provisions made for care of children during school holidays, which are not holidays for working people. For sick children, schools are rarely permitted to administer the simplest medicines. Yet many parents would be relieved if a trained nurse administered aspirin or cough medicine (with a general waiver of liability) and a child could rest in an infirmary at the school. No rooms are usually set aside for a child with slight discomfort to lie down, or take a nap, which is all that might be needed.

The Workplace

Most workplaces are structured in ways that are maladaptive to the integration of private lives and work lives. Even those that are not "greedy institutions"[17] are nevertheless oriented toward convenience and principles of efficiency geared to the marketplace. Individuals are supposed to mold their lives to the needs of the occupational sector. This priority system makes it difficult for women to work, but it also makes it difficult for all members of the family. School and work vacations are not synchronized for married couples, or parents and children; school days and work days begin and end at different hours, and deviations from the pattern are regarded as abnormal. This is not only a "woman's problem," but a social problem. However, unlike the individual women who can manage strain by defining the many demands on them as "no problem," the problems of the many women who cannot do this remain unsolved by a society which ignores its problems and leaves to individuals the delicate coordination necessary to be productive at work and good parents and partners as well.

Conclusion:

Prognosis of Progress

THIS STUDY has shown how women have succeeded in making their way into the legal profession, despite the fact that they were not wanted, and how their presence has brought change to them and to the profession as a whole. But it will not have fully accomplished its purpose if it does not alert sociologists, policy makers, and informed readers to certain sociological findings.

First, sex-related characteristics have little or no bearing on a person's ability to engage in legal activity of any kind. Women have proved their talent for understanding and practicing law, whether by qualifying on the same law school entry tests or bar exams as men, or by winning cases or soundly advising clients on business matters. No special courses or programs or exceptions to standard practices have been needed to accomplish this, nor have any been employed.

Second, attitudes and orientations that social scientists and the public believed to be inherent and to account for women's self-selection into occupations other than law proved to be the easily changed choices of women who did not wish to confront discrimination in male-dominated domains. The opening up of opportunities proved immediately effective in creating interest. And, contrary to popular myths, it was not necessary for women to go through long years of "resocialization," retraining, or reorientation to prepare for their new roles.

Third, despite commonly held beliefs that women did not have the skills necessary to become effective lawyers, it was found that these skills were acquired primarily on the job, and women could not have possessed them because that training was barred to them in the past. This is no accident: by design, professions limit access to training to a

380

selected membership. It is only recently that the discriminatory practices of the professions have been acknowledged and myths about women's innate incapacities dispelled.

In considering the role of women in the legal profession in the recent past and near future, a paradox emerges: that of radical changes and the prevalence of old patterns. There have been radical changes in the number, character, and composition of women at the bar. At the beginning of this century, a mere five hundred women were lawyers in America, representing 1 percent of the profession. By mid-century, the proportion had crept to only 3 percent, and by 1980, women represented just 2 percent of all lawyers. But some predict that fully one third of legal professionals in the United States will be women by the end of this century.[1] The type of legal work women now can expect to do has also changed—it is both broader in the range of types of practice open to them (from public interest to corporate work) and broader in span (from associates' jobs to top partnerships) offering greater mobility within the profession.

Women lawyers have turned out not to be "the breed apart" men once feared would disrupt the customs and habits of centuries. Not only have women proved their ability to negotiate the combat zone of the courtroom and the choppy waters of corporate mergers and antitrust actions with interest and relish, but, having confronted the same challenges as men, they exhibit similar ambitions for recognition and money. Many women share their male colleagues' taste for courtroom drama and corporate machination, as well as their distaste for domestic relations practices and legal aid work for indigents.

Yet, a disproportionate number of women lawyers still do the legal work believed appropriate for women—namely, family and government law, public interest and defender work. Although these practices offer other satisfactions, their material rewards and professional prestige are lower. Many of these lawyers continue to share the larger cultural view that, even when they share the same occupational status, women not only have different interests than men, but tend to exhibit different behavior—behavior that is less materialistic and more dedicated to service. These views, which stereotype men as well as women, derive from a general assumption that all women share emotional characteristics because of their gender and social position as wives and mothers.

Of course women still fill the roles of wife and mother, and this presents them with common problems and affects their career profiles as lawyers. But the ways in which they deal with these issues vary from one lawyer to the next, and often they do not act according to the stereotypes. Many manage high-powered and demanding practices while bringing up small children; others opt for marriage without motherhood; and, of course, as in the past, a sizable proportion of them are single or divorced. And, like all working women today, they face a balancing act between demands of work and of home because of the scarcity of institutionalized assistance of the kind available in other countries, such as child-care centers and job-secure maternity leaves.

Women lawyers, no matter what their preference of practice or family situation, feel they are capable of matching the men in their profession in excellence, and few of the men would deny that they are as intellectually competent. In the last decade, women have demonstrated that, as with men, education, social background, and personal history are very important predictors of the kind of law practice they will select and, in the absence of discrimination on the basis of sex, even more important than gender.

The future of women in the legal profession cannot be viewed as a simple progression from exclusion to inclusion, with accession to all the rights, privileges, and responsibilities due to any true member of the profession. As in the past, women's future position in the legal world will not depend solely on women's own ambitions, interests, or legal abilities, but on how receptive others may be to these characteristics. It will depend on the disappearance of the stereotypes that define women as unlikely professional partners for men. It will also depend on the decisions of the men who guard the gates to the inner core of professional life at all strata of the bar, from courthouses in the towns and suburbs to the boardrooms of large corporate firms.

What is certain is that women are increasingly being welcomed into law as fledglings, as recruits, and as practitioners, but the welcome is still a reserved one in those spheres in which women compete with men for the choicest rewards. In the past decade, men have *relaxed* their hold on the legal profession, permitting women to enter, but they have not *released* their hold. The barriers once impenetrable can now be broken through. But for those women who strive to be at the top, the climb is not easy, and these women often find that they will be accepted only by following the pattern set by men. Many women are willing and able to pay the costs the most ambitious men pay as long as they can reap the same benefits. But there are signs of general disenchantment with this model of total commitment or of commitment which excludes other aspects of life.

Changes within the legal profession reflect views of the larger society, which itself is still riven with ambivalence, about women who play by the rules unequivocally. In the past, opposition was not only directed at women who combined professional life with family responsibilities, or those who for other reasons were not willing to commit themselves to the "greedy institution" that was the law. It was directed against all women without exception, lumping them together as if all had the same obligations and preferences. Opposition to women in the law has been only one manifestation of a segregation of the sexes typical of all societies and supported by notions that such segregation is fair, expedient, and natural to the social order. Stereotyped views have helped to rationalize women's exclusion from this mighty profession as they have helped to justify men's hold on it. Thus, even women who play by the rules, as we have seen, may have their actions "misinterpreted" by men who resist change.

But resistence to women may also be seen as one aspect of the process by which the privileged in society maintain their hold on important resources and by which groups maintain their boundaries, limiting membership to people of "like kind" and excluding those who are considered outsiders. Like other prestigious institutions, law has inherited legitimacy from the culture and the social institutions that made it objectionable for women to become lawyers. Thus in the past the profession could rely heavily on self-selection to screen out all women but those who were the heartiest, most dedicated, and in some cases, most oblivious to the prejudice directed against them. A growing awareness of the injustice of excluding people on the basis of such characteristics as sex, race, or religion, gathering momentum after the second World War and exploding in the 1960s, first in the case of blacks and then women, chipped away at the cultural and social foundations of the monopoly of white men. Self-selection no longer limited the demand for places in the protected society of professionals. As the justification corroded, changes began occurring in the system.

I have resisted using political rhetoric in this book, partly because such words as "domination," "oppression," and "patriarchy" are slogans often used by ideological movements of a certain cast, obscuring the complexity typical of the processes of social exclusion. However, it should be obvious to the reader that the book focuses on practices of domination over women in law and in society. Women began in the law as victims of harsh discrimination and, even with the shedding of obvious discrimination during the past decade, they are still having difficulty being considered true participants in the profession. The legal profession, of course, is not unique. Many groups in the larger society still are sympathetic to the denial of women's rights—witness the organized opposition to the Equal Rights Amendment—while subtler messages of the popular culture still convey to women that it is inappropriate for them to compete with men in the occupational sphere.

Stepping back for a moment, one sees that the social processes limiting membership in the legal profession to men from a confined sector of society have done more than exclude women. Limiting the lawyers has certainly limited the perspective and narrowed the clientele of the legal world as well. In the early 1970s, a lawyer recently graduated from Harvard Law School argued that the legal profession "had become the servant of the haves rather than the guardian of the have nots."[2] He went on to cite Chief Justice Stone's comment that "the best skill and capacity of the profession has been drawn into the highly specialized service of business and finance . . ." and has "tainted it with the morals and manners of the market place in its most anti-social manifestations."[3]

Certainly arguments in the 1960s and 1970s for a more broadly based and responsive legal profession, and the government programs that provided financial support for them, created a new potential. But no such optimism and idealism characterizes the opening of the 1980s. The norms of the recent past that directed people toward greater social consciousness now lie largely unattended. The ranks of the socially conscious—never large—have diminished. Yet among the strongest advocates for a more democratic judicial system are the women who moved into law in the 1970s. Even those young women whose commitment and interest in the corporate sector has grown to match that of the young men in the field have often dared to ask hard questions about the character of the work world they were entering.

I would argue that the presence of women in the legal world has challenged enduring assumptions about the right of a select few to dominate such an important gate to society's decision-making positions. Of course, women have not been merely "present"; they have encouraged and supported the entrance of other women into the law schools and judiciary and, once represented in these institutions, have called for reinterpretation of the law and its practice. And those women who have engaged in legal practice outside the economic power centers—in government, public interest, and feminist practices—have accomplished powerful changes by confronting the establishment head-on. These women represented a group, comprising over half the population, whose interests had not been adequately represented before. They raised issues whose importance had gone unrecognized in the past—

384

issues of equality of employment, equality in marriage and divorce, protection of victims of physical assault or rape, and guarantees of the right to enter normal business relations through access to credit and the negotiations of business affairs. Although certain male lawyers and judges were instrumental in the drive toward justice on these issues, the major battles were fought and won by women lawyers in the law schools, law practices, and centers devoted to human rights. They picked up the cudgels that the first women lawyers in the late 1800s had raised in advocacy for women's rights.[4]

Lodged as they are in a web of many interwoven roles, women, more than men, often take a view that extends beyond the traditional legal mentality—developed by law schools—that Edmund Burke, the British political theorist, claimed sharpened men's minds by making them narrow.[5] Many women lawyers today have not been permitted the luxury of narrowness because, as outsiders, they have been sensitized to the insular qualities of law practice. Certainly, many men possess breadth of vision and women are as capable as men of narrowness. But the outsider role, no matter what its pains and handicaps, does give a perspective different from that of insiders with vested interests in the existing structure. Furthermore, one positive consequence—among many negative ones—of stereotyping women as the guardians of social consciousness has been to provide a corps of relatively selfless advocates who are less subject than men to the pressures of proving their success through financial gain.

That women attend to social change of a particular kind neglected by men and that their work often tends toward a participatory mode of decision making increasingly valued today should make their contribution as lawyers not merely tolerated but sought out by the legal profession and the society that benefits from it. Just as women's contributions to the nurturing of children and the support of the family are welcomed and sought, so should be their contributions to a larger system of justice.

Yet it would be unwise and unfair to delegate to women the responsibility for changing the legal profession. The law will become more responsive to all members of society only when those in power concede that is the proper thing to do, and it will be accomplished only if the powerful are drawn from a broader pool than before—one that includes not only women but also men with ideals and talent. Furthermore, in the matter of simple justice, no one group ought to be burdened with the expectation of unilateral altruism. This position is usually both unappreciated and ineffective. That women lawyers have done well is not surprising to any but the prejudiced. I would predict that they not only will continue to do well, but that they will do so with a certain idealism and humanity, simply because those qualities are expected from them. But society must be prepared to redefine certain aspects of humane concern not as "women's work" but as the work of all.

Appendix: Methodology

THIS ANALYSIS of the changing role of women in the legal profession is based on fifteen years of research using a number of different "methods." It is hard to isolate method from general experience when one has spent so much attention on a research subject for a long period of time. Although I was engaged in other activities during this period, the subject of this study was on my mind. I changed methods in the course of those years, partly because it seemed appropriate to my goals and to the changing situation of lawyers. I was also exposed to different styles of work in sociology, work that did not replace earlier perspectives that had formed my sociological vision but that, I felt, added a richness and depth. I was also inspired by work that altered my final writing of this book, although it came too late to change earlier constructions of inquiry in ways that might have been useful in creating the final product. Nevertheless, as I have been advised by writers and researchers who appreciate the affliction known as "trouble in letting go," this work will have to stand for what it is.

Therefore, I shall try to indicate what has been the basis of the work and what directions the research took over the years of inquiry. The study began in 1963 when I started to think about doing a Ph.D. dissertation on women in male-dominated professions. By 1965 I had narrowed that focus to looking at women in the profession of law and began a qualitative study of women's experiences in the legal profession. The study was completed in 1968 and is reported in "Women and Professional Careers: The Case of the Woman Lawyer."[1] I continued to study the legal profession by keeping abreast of changes reported in the popular media and legal journals. I also spoke to hundreds of lawyers and law students, informally and in focused[2] interviews of lawyers clustered in certain specialities (focused interviews were conducted with 179 lawyers). After the initial study I did not sample according to any strict method. I did use a "snowball" technique within certain kinds of law practices.

The First Study

As I described in the introduction to this book, my first study was a portrait of women lawyers at a time when their situation was somewhat static and reflective of discrimination against women who sought high-commitment and high-prestige occupations. Women also self-selected themselves out of the professions because of cultural views that it was inappropriate for women to work in them. Because women lawyers had made unusual choices for their time, I engaged in a deviant case analysis,[3] following the approach used by Lipset, Trow, and Coleman in their study of democracy in the International Typographical Union[4]—a union that was an exception to the trend toward oligarchy. Because most women did not work in careers of high commitment or high prestige in the 1960s, and those who did so were regarded as peculiar or at least to be explained in some way, I studied women lawyers in the hope that they would provide clues to what made women's entry and participation in such careers possible.

For this purpose I planned and executed an interview study administered to a sample of women lawyers drawn from the New York section of the *Martindale-Hubbell Law Directory*. Focused interviews, conducted in 1965 and 1966, contained many open-ended and unstructured questions that were used to identify many areas of interest, among them problems of choice, aspiration and conflict, their experiences on their jobs, and the character of their personal lives.

This study centered primarily on interviews administered to a sample of women lawyers drawn from the five boroughs of New York City and Westchester and Suffolk counties, an area that could be covered within the study's limits of time and funding. The representativeness of the characteristics of the sample was checked by other methods of inquiry (for example, analysis of other studies, and statistical accounts) and information from other sources, among them news reports and interviews with administrative personnel and faculty at law schools in the New York area.

Nature of the Sample. The sample was drawn on as random a basis as possible in order to obtain a group of respondents with a broad spread of ages and types of law practice. I used every twentieth fe-male name from the New York section of the *Martindale-Hubbell Law Directory,* a semi-official source that lists all lawyers in the United States who have been admitted to the bar. (It should be noted that inclusion in Martindale is voluntary and therefore omits those people who prefer to remain unlisted.) The sample included lawyers working in a range of practices that included large city firms, corpora-tions, small suburban practices, and small practices in towns one to two hours' travel time from Manhattan.[5]

However, the study was limited by its geographical location in several respects. First, the substantial numbers of women lawyers in govern-ment service were under-represented since most practice in Washing-ton, D.C. I did, however, interview some women in city government. Second, the sample probably had a larger ethnic component than is true for the country at large, although studies done in other large cities also reflect the fact that a good proportion of the U.S. bar is ethnic in character. However, since most women lawyers come from urban areas and practice in them (New York City ranks first in concentration of women lawyers in the U.S.),[6] our geographic focus did not seem to interfere with the purposes of the study.

A total of fifty-four respondents were personally interviewed (and in most cases the interviews were tape-recorded). Ten women lawyers who were interviewed in a pilot stage of the study were not included in the statistical analysis of the study, but their insights and comments contributed to the overall picture of the legal profession. Informal inter-views were also conducted with twenty male lawyers—deans of several law schools, husbands of respondents, and lawyers I knew personally.

Second Study

The experiences of the three black lawyers who had been in my original sample led me to do a study of black professional women in 1972. I then interviewed more lawyers as well as women in other male dominated professions. The problems black women professionals faced because of their double negative status, as well as the special treatment accorded to them as a result of their situation, were reported in an article, "Posi-tive Effects of the Multiple Negative: Explaining the Success of Black Professional Women," in 1973.[7] I continued to study black women lawyers in the early 1970s, partially to identify their special problems but also because many of them, like the other young lawyers of their day, were working in the public interest sector or in other cause-associated work, for example the NAACP Legal Defense Fund. I was assisted in that effort by Gwyned Simpson, a graduate student at Teacher's College of Columbia University, who worked with me in following up the careers of the black women from the original study and then went on to do her own study of black women lawyers for her Ph.D. dissertation.[8] Her interviews, conducted in 1977, comprise some of the data on which the analysis in this book is based.

Early in 1971, in preparation for revising my dissertation for publication, I decided that it would be important to interview some younger women who had come into the law since the time I had done my study. I had no plans to do another study, merely to update my impressions. Lawyers whom I had met through women's movement activity supplied me with names of several activist associates and I spoke to them. They, in turn, identified others, and gradually a network emerged. These were lawyers who identified with the feminist movement and who were variously situated in small feminist practices, in public interest firms, or in radical and left "movement" legal practices serving political dissidents and the poor. In sociological terminology, this may be called a "snowball" technique (asking each person to recommend others of their "sort" I might interview).

As I mention elsewhere in this book, women lawyers are joined in networks. This is true not only in New York City, where I experienced a sense of closure after lawyers I was interviewing began to name people I had already seen or heard about, but also true nationwide, particularly for lawyers who identified in any way with women's rights issues and those who had been students together when womens organizations were prominent at law schools. Of course, the networks I located then and in a later wave of interviewing were centered in the big cities and revolved around both issue-oriented law and large corporate practice. To the extent that my sample is a product of these networks, it is biased toward them and away from practitioners in small towns or in small firms. However, as I pointed out earlier, the general trend in the United States is away from just those kinds of practices in which my study is deficient, and in which women have also been generally under-represented in the past.[9] Because one must impose limits on a study I decided not to include one major category of women lawyers —attorneys in corporations—house counselors—because lawyers in industry face somewhat different problems of practice and advancement than the traditional "free professional." That will remain for another time.

But the network approach has provided a sample of women in certain kinds of practice who could not otherwise be identified. There is no directory of lawyers in certain specialties or types of firms. Geographical location or alphabetical order would not provide a meaningful ordering to identify what was happening in a rapidly changing professional environment for the women within. My aim—to pick up the tone of a changing culture and the changing career courses of women within it—could be measured only in part by hard indicators such as statistics and interviews of a particular sample of lawyers at one point in time. The networks approach provided me with information about individual lawyers not only from their own point of view but from the views of others about them. Sometimes this resulted in a "Rashomon" effect when lawyers differed in their accounts of phenomena.* As I described earlier, sometimes men thought women colleagues were stiff in their relations when the women thought they were merely businesslike; and there was the woman who described her husband as looking like Robert Redford while the senior (male) attorney in the office said he looked like a mouse!

Some of the women I met in the early 1970s became my informants over the next eight years, and some became my friends, reporting on their own experiences and those of their associates as I continued to "revise" my earlier study. I was not a participant in the legal world, but I became more and more an observer, using, as Kai Erikson calls it, the professional lens of the sociologist.

I would like to think I was an ethnographer of the law world over that period of time, but that task would have been too broad for even a highly organized research group. But if I could not explore that whole world, I did visit one or another of the subcommunities of the legal world over the past decade and a half, like the anthropologist who visits his or her culture for field work for periods of time spread out over a number of years. In some I traveled extensively and that is reflected in my richer descriptions of them; of other legal environments I was more dependent on reports of lawyers in networks I had not caught in my net of associations or whose experiences I could not interpret as fully.

Like many other sociologists of my generation I have admired the use of sophisticated statistical methods and valued the kind of verification

---

Rashomon is the title of a Japanese art film, directed by Akiro Kurosawa (1951), which portrays four different versions of an encounter between a man and a women, variously seen as a seduction, a rape, and an affair by the separate participants and observers.

promised by use of this method. But such techniques are not always appropriate and, indeed, are dependent on simplification of variables, often missing the complexity of experience. My first study, although small and designed to pick up qualitative information, was subjected to statistical analysis, most of which is not reported in this book. Patterns certainly emerged in that study which I felt secure in reporting as a sociological generalization, although statistical tests of significance could not honestly be applied because of the small numbers involved. Since my techniques of exploration changed over the next years, I also had no points of comparison other than my growing knowledge of the system in which I was becoming a continuing explorer.

But as I, and my graduate students, have discovered through the years, the educated sociological eye does see patterns, experiencing the childlike wonder of "yes, there is a sociology," and this was the revelatory experience through many stages of interviewing as two or three of us would find that we were hearing about "that experience again" or seeing "that same type once more," not as an expression of boredom as those comments might imply, but in an exhilarating way of becoming reinforced in the knowledge that our "method" worked.

Numbers, however, play a part in this study. I did want to know actually how many women were coming into the law schools or the law firms or whatever, and those figures were hard to come by. I attempted to collect many of them myself, and at various times polled the law schools, the placement offices, and law firms to assess the magnitude of the increase. One answer to the questions of friends who still wonder why this book took so long to write is that my surveys seemed outdated even as the information was being collected, and I felt that the "revision" of the earlier study needed even further information to document each new event or breakthrough. Had I known what I would be finally able to establish as a cut-off date, I would have been spared a certain amount of the labor and expense of doing those surveys. Although the documentation is by no means perfect, or even adequate, by the late 1970s other scholars and various legal associations had collected some of the data which I had such trouble locating earlier.

Yet there are figures here that others have not collected and that are important to the study. We wish we had more; indeed, we look forward to the day when our desk-side computers will provide us with running accounts of the numbers of people in any environment we wish to study, broken down by sex, race, age, or any statistic of other variables of interest to the social analyst. Further, a digital type machine would indicate the changing numbers at the moment as (for example) a lawyer might change speciality or drop out of the profession, or come back in, or throw it all over, Gauguin-style, to paint in Tahiti.

I not only searched out data in the course of this work; at some points it came to me. By the 1970s there had been a change in my role as investigator from that in the mid-1960s, when I was a graduate student doing research for my dissertation. Not only had I climbed up the academic ranks during the 1970s, I had also established some reputation as a scholar of women's professional roles with the publication of a book, *Woman's Place: Options and Limits in Professional Careers,*[10] which included data and observations that had grown out of the early research on women attorneys. Thus, I became more visible and legitimated as a researcher. Not that I encountered many problems of access in the first study—women attorneys, for the most part, were open and cooperative, even eager to share their experiences—but I was asked to meet with various lawyers groups and widened my contacts with women in the profession who were familiar with my work and who were interested in discussing it and eager to contribute their own experiences to my formal interviews. My first such opportunity was an invitation to speak at one of the first conferences on women and the law, which was held at Rutgers University in 1970, organized by the women law students and Professor Ruth Bader Ginsburg, who is now Judge of the U.S. Court of Appeals for the District of Columbia. Judge Ginsburg, as my acknowledgment in the preface of this book expresses, was an important informant as well as reviewer of my analysis at various times in the past few years.

But I did turn to a somewhat more systematic data collection as I was moving the study of women lawyers to completion in the late 1970s. I made a more decided attempt to conduct ethnographic investigation of clusters of lawyers in several spheres I had not yet explored. A Guggenheim fellowship in 1976/77 made it possible to conduct more interviews and helped pay for research assistance. Susan Wolf, a graduate student in sociology who came to the Bureau of Applied Social Research some years before to work on her dissertation, and with whom I had spent many hours of discussion about women professionals, the women's movement, and the interpersonal relations of women, men, and children in general, came to work on the lawyer project.

The earlier decision to interview a cluster of women in movement and social interest law seemed again appropriate in order to see how women in the spheres of law I had looked at before were being affected by the women's movement; by "normalization," as more women were moving into the profession; by changes in the legal profession itself; and

393

by the course of the life cycle. It was one thing to interview young, idealistic lawyers in the 1970s when a spirit of crisis and challenge was high, but we wondered if their experience was the product of a brief, passing generation, to be short-lived in event and consequence.

Susan Wolf and I went back to a number of women who had served as informants before to see what had happened to them and to other women in their networks. Wolf also had personal contacts with movement lawyers, and with lawyers in a sphere I had not tapped before, the federal prosecutor's office.

District Attorneys and Prosecutors

In 1965 it was rare to run across a woman district attorney or a prosecutor in the city or state. In the 1970s there was a notable change in that situation. Women in the original study had often commented that women did not like, nor were they permitted, to play visible combatant roles, particularly as litigators. By the mid-1970s, becoming a prosecutor at the city, state, or federal level was not uncommon. In 1977 we interviewed women in the New York City prosecutor's office, and women in a federal attorney's office in New York, as well as male attorneys in that section.

Wall Street Firms

It also seemed to be the time to look at women's changing roles in the law practices of high prestige and power. Only a tiny number of women in my original study had any experience in the big firms, although many had tales of being refused employment in them. Now I knew that more women were being hired.

My initial leads to lawyers in the big firms came from professors in the law schools and through the feminist network. Many of these women had been classmates; some had worked together in the women's organizations; and some had moved from Wall Street to the feminist firms, or gone the reverse route. From 1976 to 1978, Susan Wolf and I interviewed men and women in the big firms at all levels.

Appendix: Methodology

The analysis of changes on Wall Street was a story in and of itself, and I wrote it up in a scholarly version and a popular version for a new magazine for executive women, *Savvy*.[11] This was especially fortunate, because the research staff at *Savvy* Magazine conducted telephone interviews with all the women partners in firms with more than ninety-five members, and thus attained brief biographies of the entire "universe" (in sociological terms) of women Wall Street partners (and also eliminated any sampling problems). After it was published, I sent copies of the magazine to all the partners and asked whether the article was a fair representation of their experiences. We heard from six of the thirty-four, all of whom agreed with the characterization. Of these, some volunteered to talk further about their experiences. By this time, Susan Wolf had left the project to join a law firm herself (as a para-legal) and I was assisted in the interviewing (and sorting out of ideas and impressions) by another research assistant, Judith Thomas. We were thus able to interview six more lawyers, ending this investigation in the summer of 1980.

Feminist Firms: Final Stage

Judith Thomas also conducted telephone interviews with lawyers who worked at public interest and feminist firms in New York, California, and Washington, D.C. The last bit of interviewing was done after a version of the chapter on feminist lawyers appeared in an American Bar Association magazine *The Student Lawyer*, under a title chosen by the editor, "The Short, Unhappy Life of the Feminist Law Firms" (May 1980). This title, of which I was unaware until receiving a copy of the published magazine, unfortunately highlighted the problems suffered by those firms (and in the manner identified as "labeling" oriented the reader to the material stressing defeat and away from the analysis stressing positive affects).

As a result, feminist firm lawyers from Washington and elsewhere communicated that they were alive and not unhappy. Although we felt that the book version of the analysis on feminist firms would more fully focus on the positive aspects of those practices as well as the negative, we wished to talk to more firms to clarify our assessment of the state of feminist law practice. Those interviews are included here and represent an example of what we might call "reflexive" methods, where responses of the people who are being studied feed back into the study, enriching and correcting the analysis of the researcher.

The Law Schools

In the mid-1960s there were few women law professors to interview. I spoke to two who claimed that there weren't any others to be found in New York City, and both of these held low-ranking posts. In April 1970 I attended the first National Conference of Law Women, held at New York University Law School, and a month later, the Rutgers Law School conference. Through the next decade I attended a number of conferences and made visits to law schools at various places in the country, contributing to an ethnographic sense of the changing climate in the law schools. I attended a conference at Vanderbilt University Law School in 1974. A month's sojourn at the MacDowell Colony in 1976 made it possible to interview women law faculty members at a new, small New Hampshire law school (Franklin Pierce) and also to hear the comments of women lawyers in the state of New Hampshire who engaged in a variety of practices. An invitation to speak at the University of Virginia Law School in 1977 made it possible to interview a woman faculty member there, and a group of sixty or seventy women students enrolled in a women and the law seminar. During that same year I also interviewed students and faculty at Harvard and Yale Law Schools. The next year, on a fellowship at the Center for Advanced Study in the Behavioral Sciences at Stanford, I interviewed professors and students at Stanford University and participated in some seminars that the students arranged. I also met with, and spoke to, professors from Boalt Hall (the law school of the University of California at Berkeley), from the law school of the University of California at Davis, and from some of the smaller schools in the San Francisco area.[12] Earlier, and through the past decade, I, with one or another research assistant,[13] interviewed administrators, law professors, and scores of students at Columbia, New York University, and New York law schools. In the law schools, I interviewed women professors both at the full professor level and those who were new appointees, including women teaching in the clinical programs, programs that barely existed in the mid-1960s, at the time of my first study.

Appendix: Methodology

Judges

The number of judges in the United States who are women are still few, at all levels of the bench, and I did not plan to include judges in my study. However, several judges had cropped up in my initial study, and several lawyers I had interviewed since then became judges in the decade which followed. I interviewed a few judges in the recent past who were either introduced to me by lawyer friends or who volunteered to be interviewed. Most of the material on judges in this book comes from other secondary sources—newspaper and magazine articles. Studying the routes to judgeship, the political aspects of selection in particular, is not considered in this study. I decided to limit attention to judgeship as an element in the reward system of law and an expression of the changing motivations of women to attain prestigious places in the legal profession as opportunity opened for them.

Follow-up of the Original Sample

A longitudinal study, one which interviews a group of people at one point in time and then later, is considered to be a good assessment of change. At the time of the second wave of interviews I did in the early 1970s, I did not plan to reinterview the women I had studied in the 1960s. I was more interested in the "new women" of law. I also felt that so many in the original sample were set in careers or close to the end of them (the median age was forty-seven) that I would not learn much from talking to them six years later. However, since the time of analysis was extending, I did decide to send a mail questionnaire to the original sample in 1976 to see how they had fared in the years intervening. I had mixed results from this questionnaire.

I received answers from only a third. Many of the questionnaires came back because the lawyers could not be located. Perhaps they had changed their names, or moved out of the state. An attempt to find them through the *Martindale-Hubbell Law Directory* or the New York Telephone book was unsuccessful. We assumed that some had died (I read the obituaries of a few); many probably retired. It is not unlikely that the lawyers whose careers were more precarious in 1965 went on to other work.

I did, however, reinterview some of the women to see how the women's movement had changed their work lives, and some of the women in lines of work where we thought there might be some change. I also followed the careers of some of the women in the original sample through newspaper and magazine accounts or by listening to their talks before law students' associations. By the 1970s, a number of those who were still visible had achieved greater prominence, and many asked to recall their experiences and act as role models to young women law students. I ran into some of these women at conferences and at social gatherings at which women professionals tend to meet, at parties by book publishers for books written by women, at fund-raising parties for the National Organization for Women or for the Equal Rights Amendment.

Although in the preceding discussion of method I have noted that men were part of the study, I think it might be useful to comment on what role men played. In some sense, they were viewed as part of the structure or landscape. As in the initial study, the interviewees were not systematically selected, but rather were part of the social network of the women (their role partners). Sometimes we interviewed women's peers in firms, or senior partners, male professors at the law schools, husbands who were also lawyers.

We wanted to interview women and men who were working in the same environments, not to match them according to performance criteria, but to learn whether men and women had different perceptions of what was going on in their working networks. We wanted also to get a sense of the problems men as well as women had in regard to promotion and to see how men worked and lived with women who were lawyers.

I also interviewed male lawyers I came into contact with socially, and sometimes asked them if they would be willing to be formally interviewed if they practiced at firms where the women I had questioned were lawyers. Sometimes I interviewed lawyer-husbands as they walked into the interview I was conducting with a wife at home, but we also arranged for separate interviews with the men when that seemed possible. Sometimes men who were hostile to the notion of women being lawyers demanded to be interviewed. Sometimes when male lawyers learned of my study in a social situation, they merely blurted out their opinions. I admitted all hearsay as evidence—which is inadmissible in the law, but not in ethnography.

At the point of writing this book I have a deeper appreciation for a style of investigation which attempts to dip into the complex web of social relations. It had always seemed to me more interesting to locate scores of "findings" which could only be suggestive rather than those which a more focused analysis could set in formal method. A sense that investigation loosely called "qualitative" was somehow less worthy than analysis called "quantitative" made me insecure in the early years of my work. I now not only have confidence that qualitative research can produce results that are as valid truths as those produced by the so-called "rigorous" methods, but a feeling akin to spiritual (but educated) belief that without investigations that attempt to capture the "feel" of an environment, and reality as experienced by the people in that environment, the more rigorous methods may well represent a statistically reliable but unintelligible "reality."

Of course, it is easier to follow a formula, to locate a sample of people, and poll them and analyze their responses. A physician can learn much about the patient by studying the electrocardiogram of the heartbeat, but any sophisticated doctor will note how important it is to know the history of the patient, the medical history of the parents, the pace of his or her life, and how they say they "feel." As medical people are once again discovering the "whole" patient, sociologists are once again discovering the whole society or whole community. Perhaps we shall never succeed in knowing it entirely, but we will try to do better by looking at its texture and dynamics in greater detail—by using more than one indicator to note its illness or health or merely how it is doing.

Notes

Introduction

1. "Employment of Women in Selected Occupations, Selected Years, 1950, 1960, 1970, and 1979," U.S. Department of Labor, Bureau of Labor Statistics, *Perspectives on Working Women: a Databook* October 1980, from Table 11, p. 10.

Chapter 1

1. Lawrence M. Friedman, *Law and Society* (Englewood Cliffs, N.J.: Prentice-Hall, 1977), p. 24.

2. Ibid., p. 27.

3. Robert D. Putnam, *The Comparative Study of Political Elites* (Englewood Cliffs, N.J.: Prentice-Hall, 1976).

4. Friedman, *Law and Society,* p. 26.

5. Magali Sarfatti Larson, *The Rise of Professionalism: A Sociological Analysis* (Berkeley: The University of California Press, 1977), p. 167.

6. See William J. Goode on the history of professional prestige upgrading in *Celebration of Heroes: Prestige as a Social Control System* (Berkeley: University of California Press, 1978).

7. See Philip Ross Cortney Elliott, *The Sociology of the Professions* (New York: Herder and Herder, 1972); Amitai Etzioni, *The Semi-Professions and Their Organization* (New York: Free Press, 1969); Lynda Lytle Holmstrom, *The Two-Career Family* (Cambridge, Mass.: Schenkman Publishing Co., 1972); R. L. Coser and G. Rokoff, "Women in the Occupational World: Social Disruption and Conflict," *Social Problems* 18, no. 4 (Spring 1971): 535–53; Rhona and Robert Rapoport, *Dual Career Families Reexamined: New Integrations of Work and Family* (New York: Harper and Row, 1977); and Talcott Parsons, "The Professions and Social Structure" in *Essays in Sociological Theory* (Glencoe, Ill.: Free Press, 1954, rev. ed.).

8. William J. Goode, "Community Within a Community: The Professions," *American Sociological Review* 22 (1957): 194–200; Eliot Friedson (ed.), *The Hospital in Modern Society* (New York: The Free Press, 1963); Rose Laub Coser, *Training in Ambiguity* (New York: The Free Press, 1979).

9. Everett Hughes, *The Sociological Eye* (Chicago: Aldine-Atherton, 1971).

10. Jerold S. Auerbach, *Unequal Justice: Lawyers and Social Change in Modern America* (New York: Oxford University Press, 1976).

11. Tom Goldstein, "A Dramatic Rise in Lawsuits and Costs Concerns Bar," the *New York Times,* May 18, 1977, p. 1, B9.

12. "Those # ★ X!!! Lawyers," *Time,* April 10, 1978, p. 56.

13. Friedman, *Law and Society,* p. 23.

14. Peter W. Bernstein, "The Wall Street Lawyers are Thriving on Change," *Fortune,* March 13, 1978, p. 104–12.

15. "Those # ★ X!!! Lawyers."

16. Tom Goldstein, "Law, Fastest Growing Profession, May Find Prosperity Precarious," *New York Times,* May 16, 1977, p. 35.

17. New York State leads all others with about 67,000 lawyers. See ibid., Table: "Growing Numbers of Lawyers."

18. Ibid., p. 35.

19. Bernstein, "The Wall Street Lawyers," p. 104.

20. Tom Goldstein, "Law Firms are Shifting from Wall Street," *New York Times,* March 7, 1974, pp. 41, 77.

21. Mark Green, "The High Cost of Lawyers."

22. Bernstein, "The Wall Street Lawyers." p. 105.

23. Goldstein, "Law, Fastest Growing Profession."

24. See Peter Blau, *Inequality and Heterogeneity* (New York: The Free Press, 1977) for the consequences of size of social relations in large organizations. See also Goode, "Community Within a Community."

25. Green, "The High Cost of Lawyers."

26. Bernstein, "The Wall Street Lawyers," p. 105.

27. These are findings reported by Altman and Weil, management consultants who surveyed 300 law firms across the country. The findings are summed up in an article "National Survey Studies Starting Salaries for 1975," *Harvard Law Record* 62, no. 1 (January 30, 1976).

28. Green, "The High Cost of Lawyers."

29. "Those # ⋆ X!!! Lawyers."

30. Robert K. Merton and Elinor Barber, "Sociological Ambivalence" in Robert K. Merton, ed., *Sociological Ambivalence and Other Essays* (New York: The Free Press, 1976), p. 26.

31. Ibid.

32. Ibid., pp. 28–29.

33. "The Troubled Professions—A Special Report," *Business Week,* August 16, 1976, p. 126.

34. "Those # ⋆ X!!! Lawyers."

35. Ibid.

36. Green, "The High Cost of Lawyers."

37. Ibid.

38. Lewis H. Goldfarb et ux petitioners v. Virginia State Bar et al. No. 74–70, October 6, 1978 (423 US 886, 46, L. Ed 2nd 118).

39. "Those # ⋆ X!!! Lawyers."

Chapter 2

1. Jerome Carlin, *Lawyers on Their Own: A Study of Individual Practioners in Chicago* (New Brunswick, N.J.: Rutgers University Press, 1962) and *Lawyers Ethics: A Survey of the New York City Bar* (New York: Russell Sage Foundation, 1966); Joel Handler, *The Lawyer and His Community: The Practicing Bar in a Middle Sized City* (Madison: University of Wisconsin Press, 1967); Erwin O. Smigel, *The Wall Street Lawyers: Professional Organization Man?* (New York: The Free Press, 1964); Jack Ladinsky, "Careers of Lawyers, Law Practice and Legal Institutions," *American Sociological Review* 27 (1963): 47 and "The Impact of Social Background of Lawyers on Law Practice and the Law," *Journal of Legal Education* 16 (1963): 127; Seymour Warkov and Joseph Zelan, eds., *Lawyers in the Making* (Chicago: Aldine Publishing Co., 1966).

2. Louis Auchincloss, *The Great World and Timothy Colt* (Boston: Houghton Mifflin Co., 1956); *Tales of Manhattan* (Boston: Houghton Mifflin Co., 1967); *The Partners* (New York: Warner Books, 1975).

3. Warkov and Zelan, *Lawyers in the Making.* Dan Clements Lortie, *The Striving Young Lawyer: A Study of Early Career Differentiation in the Chicago Bar* (Chicago: University of Chicago Library, 1958).

4. The survey of elite and semi-elite law schools showed that the great majority of entrants were products of urban areas, although an increasing number came from the suburbs. Robert Stevens, "Law Schools and Law Students," *Virginia Law Review* 59, no. 4 (April 1973): 551–707.

5. Joseph Zelan, "The Role of Occupational Inheritance," in *Lawyers in the Making,* p. 52.

Notes

6. Cynthia Fuchs Epstein, "Women and Professional Careers: The Case of the Woman Lawyer" (Ph.D. dissertation, Department of Sociology, Columbia University, 1968), Table IV-12, p. 111.

7. Ibid., Table IV-13, p. 112.

8. Rita Lynne Stafford, "An Analysis of Consciously Recalled Professional Involvement for American Women in New York State" (Ph.D. dissertation, School of Education, New York University, 1966).

9. Stevens, "Law Schools and Law Students," p. 602.

10. Warkov and Zelan, *Lawyers in the Making,* p. 90.

11. Joseph Zelan, "The Roles of Occupational Inheritance," in *Lawyers in the Making,* p. 52. But, it may be noted that sons of Jewish lawyers who score high on an Academic Performance Index tend to select medicine.

12. Lortie, *The Striving Young Lawyer.*

13. Carlin, *Lawyers Ethics.*

14. Epstein, "Women and Professional Careers," p. 93.

15. Stevens, "Law Schools and Law Students," p. 572, fn. 46.

16. Cynthia Fuchs Epstein, *Woman's Place: Options and Limits in Professional Careers* (Berkeley: University of California Press, 1970) p. 48.

17. James J. White, "Women in the Law," *Michigan Law Review* 65, no. 6 (April 1967): 1051–1122.

18. Epstein, "Women and Professional Careers," Table IV-6, p. 98.

19. Stevens, "Law Schools and Law Students," pp. 601–2.

20. Cynthia Fuchs Epstein, "Positive Effects of the Multiple Negative: Explaining the Success of Black Professional Women," *American Journal of Sociology* 78, no. 4 (January 1973).

21. Epstein, *Woman's Place,* p. 62.

22. Jean Lipman-Blumen, "The Vicarious Achievement Ethic and Non-Traditional Roles for Women." (Paper presented at the annual meeting of the Eastern Sociological Society, April 1973). See also Jean Lipman-Blumen and Harold J. Leavitt, "Vicarious and Direct Achievement Patterns in Adulthood," *Counseling Psychologist* 6, no. 1 (1976): 26–32.

23. Epstein, "Women and Professional Careers," p. 102, fn.

24. Stafford, "An Analysis of Consciously Recalled Professional Involvement," see footnote, p. 264.

25. Dorothy Thomas, ed., *Directory of Women Lawyers* (New York: Scarecrow Press, 1957).

26. Stafford, "An Analysis of Consciously Recalled Professional Involvement," p. 248.

27. I gave some thought to the role of negative role models after my first study of women lawyers was completed. I found this concept was also identified by Rue Bucher and Joan Stelling, *Becoming Professional* (Beverly Hills, Calif.: Sage Library of Social Research, Sage Publications, N. 46, 1971). See in particular Chapter 5, "Role Models and Self-Evaluation," for a wise discussion of the complex nature of role modeling for medical trainees.

28. Bucher and Stelling also identify the notion of the "partial role model." Ibid.

29. Epstein, "Women and Professional Careers," pp. 91–2.

30. D. Kelly Weisberg, "Barred from the Bar: Women and Legal Education in the United States, 1870–1890," *Journal of Legal Education* 28, no. 4 (1977): 495.

31. Ibid., p. 489.

32. Cynthia Fuchs Epstein, "Sex Role Stereotyping, Occupations and Social Exchange," *Women's Studies* 3 (1976): 183–94.

33. Susan Edmiston, "Portia Faces Life: The Trials of Law School," *Ms* 2, no. 10 (April 1974): 74.

34. Mirra Komarovsky, "Cultural Contradictions and Sex Roles," *American Journal of Sociology* 52 (November 1946): 184–9.

35. See Cynthia Fuchs Epstein on sex-typing of occupations in *Woman's Place: Options and Limits in Professional Careers* (Berkeley: University of California Press, 1970), pp. 152–66.

36. Janette Barnes, "Women and Entrance to the Legal Profession," *Journal of Legal Education* 23 (1970): 276–307.

37. Alan Blum and Peter McHugh, "The Social Ascription of Motives," *American Sociological Review* 36 (1971): 98–129.

38. Matina S. Horner, "Sex Difference in Achievement Motivation and Performance in Competitive and Non-Competitive Situations" (Ph.D. dissertation, University of Michigan, 1968).

39. Matina S. Horner, "Fail: Bright Women," *Psychology Today* 3, no. 36 (November 1969).

40. A more elaborate version of this discussion of the implications of the "fear of success" theory appears in Cynthia Fuchs Epstein, "Separate & Unequal: Notes on Women's Achievement," *Social Policy* 6, no. 5 (March/April 1976): 17–23. Matina Horner herself has found that many women scoring high on "fear of success" imagery in college turn out to have successful careers ten years later. (Personal communication, August 1980).

See also Eve Spangler and Ronald M. Pipkin, "Portia Faces Life: Sex Differences in the Professional Orientations and Career Aspirations of Law Students," Report from the Law Student Activity Patterns Project, Research Program in Legal Education of the American Bar Foundation, 1977.

41. Arlie Russell Hochschild, "Emotion Work, Feeling Rules and Social Structure," *American Journal of Sociology* 85, no. 2 (November 1979): 551–95.

42. James J. White, "Women in the Law," *Michigan Law Review* 65, no. 6 (April 1967): 1051–1122.

43. Robert Stevens, "Law Schools and Law Students," *Virginia Law Review* 59, no. 4 (April 1973): 551–707.

44. Ibid., p. 578.

45. Ibid., p. 579.

46. Ibid., pp. 617–8 n126.

47. Scott Turow, *One L.,* (New York: G.P. Putnam & Sons, 1977).

48. "Law a Second Career for Many Students in Class of 1982," NYU School of Law *Alumni Newsletter,* Spring 1980, p. 3.

Chapter 3

1. Martin Mayer, *The Lawyers* (New York: Harper and Row, 1967), pp. 78–81.

2. D. Kelly Weisberg, "Barred from the Bar: Women and Legal Education in the United States, 1870–1890," *Journal of Legal Education* 28, no. 4 (1977): 485.

3. Albie Sachs and Joan Hoff Wilson, *Sexism and the Law* (New York: Free Press, 1978) p. 96.

4. Ibid.

5. D. Kelly Weisberg, "Barred from the Bar," p. 485.

6. Ruth Bader Ginsburg, speech given at Harvard Law School, April 15, 1978. See "Women at the Bar—A Generation of Change," *University of Puget Sound Law Review* 2, no. 1 (Fall 1978): 1–14, p. 3.

7. Ibid.

8. Epstein, "Women and Professional Careers: The Case of the Woman Lawyer (Ph.D. dissertation, Department of Sociology, Columbia University, 1968)," Table V–2, p. 141 reports the period 1953–1967. The 1972 survey updated that material and expanded it to include schools throughout the country. It is unpublished.

9. Helene E. Schwartz, *Lawyering* (New York: Farrar, Straus & Giroux, 1975), p. 83.

10. Ruth Bader Ginsburg, "Women at the Bar—A Generation of Change," *University of Puget Sound Law Review* 2 (Fall 1978): 4.

11. *Harvard Law Record* (December 9, 1965): 7.

12. Cynthia Fuchs Epstein, "Positive Effects of the Multiple Negative: Explaining the Success of Black Professional Women," *American Journal of Sociology* 78, no. 4 (January 1973): 915–16.

13. Georgia Dullea, "More Women Now in Law and Medicine," the *New York Times,* January 15, 1975, p. 68.

Notes

14. Shirley R. Bysiewicz, "Women Penetrating the Law," *Trial* (November/December 1973): 27–28.
15. Ibid.
16. Robert Stevens, "Law Schools and Law Students," *Virginia Law Review* 59, no. 4 (April 1973): 572.
17. James P. White, "Is That Burgeoning Law School Enrollment Ending?" *American Bar Association Journal* 61 (February 1975): 202.
18. Frederick M. Hart and Franklin Evans, *Major Research Efforts of the Law School Admission Council* (Washington, D.C.: Law School Admission Council, 1976), p. 9.
19. Becky Wesley, "Female Enrollment Nears Thirty Percent," *Harvard Law Record* 67, no. 1 (1978): 3.
20. New York University *Alumni Bulletin,* Spring 1980, p. 3. (However, the Ashburn and Cohen chart only reports 39 percent. See Elizabeth A. Ashburn and Elena N. Cohen, *Women's Integration into Law Teaching* (American Bar Association, Section on Individual Rights and Responsibilities, November 1980), p. 60.
21. James J. White, "Women in the Law," *Michigan Law Review* 65, no. 6 (April 1967): 1051–1122.

Chapter 4

1. A film, *The Paper Chase,* chronicled the harsh style of teaching at Harvard University, which law students have since used as their standard to measure change (meaning an easier atmosphere).
2. Martin Mayer, *The Lawyers* (New York: Dell, 1966), p. 82.
3. I was informed of this "tradition" by lawyer Emily Jane Goodman.
4. Mayer, *The Lawyers,* p. 96.
5. Erving Goffman, *Stigma: Notes on the Management of Spoiled Identity* (Englewood Cliffs, N.J.: Prentice-Hall, 1963).
6. Harvard Law School Alumnae Directory, Classes of 1953–1977, Twenty-fifth Anniversary Celebration, Harvard University Law School.
7. Patricia A. Moore, "Teaching the Distaff Side," *Harvard Law Bulletin* 25, no. 3 (February 1974): 15.
8. Ibid.
9. Harvard Law School Alumnae Directory, Twenty-fifth Anniversary.
10. Ibid.
11. Ibid.
12. Ibid.
13. Helene E. Schwartz, *Lawyering* (New York: Farrar, Straus & Giroux, 1976), p. 85.
14. Harvard Law School Alumnae Directory.
15. Robert B. McKay, "Women and the Liberation of Legal Education," *Women Lawyers Journal* 57, no. 4 (Fall 1971): 141.
16. J. Jason, L. Moody, and J. Scheurger, "The Woman Law Student: The View from the Front of the Classroom," *Cleveland State Law Review* 24 (1975): 223–45.
17. Ibid., p. 239.
18. Janette Barnes, "Women and Entrance to Legal Profession," *Journal of Legal Education* 23 (1970): 291.
19. Most of the material on the women's activities at NYU is from "Women's Activism at N.Y.U. School of Law—A Brief History," *Law Women News* 1, no. 1 (March 1980).
20. Virginia B. Nordby, "The New Women and the Law Course at Michigan," *Law Quadrangle Notes* (The University of Michigan Law School) 19, no. 4 (Summer 1975): 10–15.
21. Fairlea Sheehy, "Students Plan Research in Women's Issues Law," *Harvard Law Record,* October 24, 1975, p. 2.
22. Lionel Tiger, *Men in Groups* (New York: Random House, 1969).
23. "Celebration 25: A Generation of Women at Harvard Law School," *Harvard Law School Bulletin* (Summer 1978): 4.

Chapter 5

1. Cynthia Fuchs Epstein, *Women and Professional Careers: The Case of the Woman Lawyer* (Ph.D. dissertation, Department of Sociology, Columbia University, 1968), p. 153.

2. James J. White, "Women in the Law," *Michigan Law Review* 65, no. 6 (April 1967): 1051.

3. In Jerold S. Auerbach, *Unequal Justice: Lawyers and Social Change in Modern America* (New York: Oxford University Press, 1976): p. 307.

4. William J. Goode, "Community Within a Community: The Professions," *American Sociological Review* 22 (April 1957): 194–200.

5. Kai T. Erikson, *Wayward Puritans: A Study in the Sociology of Deviance* (New York: John Wiley & Sons, 1966).

6. Erwin O. Smigel, "The Impact of Recruitment on the Organization of the Large Law Firms," *American Sociological Review* 25 (February 1960): 56–66; Jerome Carlin, *Lawyers Ethics: A Survey of the New York City Bar* (New York: Russell Sage Foundation, 1966); Jack Ladinsky, "Careers of Lawyers, Law Practice and Legal Institutions," *American Sociological Review* 28 (February 1963): 47–54; Robert K. Merton, George A. Reader, and Patricia Kendall (eds.), *The Student Physician: Introductory Studies in the Sociology of Medical Education* (Cambridge, Mass.: Harvard University Press, 1957).

7. Jerold S. Auerbach, *Unequal Justice: Lawyers and Social Change in Modern America* (New York: Oxford University Press, 1976); Mark Green, *The Other Government: The Unseen Power of Washington Lawyers* (New York: W.W. Norton, 1978).

8. Everett Hughes is an exception, having written a classic article "The Dilemma and Contradictions of Status" as far back as 1945, in which he located processes that relegate women and black professionals to "deviant" status. *American Journal of Sociology* 50 (March 1945): 353–59. I also deal with these issues in Cynthia Fuchs Epstein, *Woman's Place: Options and Limits in Professional Careers* (Berkeley: University of California Press, 1970).

9. Albie Sachs and Joan Hoff Wilson, *Sexism and the Law* (New York: The Free Press, 1978).

10. Ibid., p. 26.

11. Verna Elizabeth Griffin, *Employment Opportunities for Women in Legal Work* (Washington, D.C.: U.S. Department of Labor, U.S. Government Printing Office, 1958).

12. Bruce Abel, "The Firms—What Do They Want?" *Harvard Law Record* (December 12, 1963): 1.

13. Daniel J. Kucera, "Women Unwanted," *Harvard Law Record* 37, no. 9 (December 12, 1963): 1.

14. Abel, "The Firms—What Do They Want?"

15. This reflected and supported Lortie's finding that Catholic school graduates in Chicago are found in all types of law careers *except* the largest and most illustrious. Dan Clement Lortie, *The Striving Young Lawyer: A Study of Early Career Differentiation in the Chicago Bar* (Chicago: University of Chicago Press, 1958), p. 498.

16. Bradley Soule and Kay Standley, "Perceptions of Sex Discrimination in Law," *American Bar Association Journal* 59 (October 1973): 1144–47.

17. "Women Lawyers: Past, Present and Future" (recorded at the New York County Lawyers' Association, Special Committee on Women's Rights, November 1, 1978).

18. "Nanette Dembitz," *New York Times,* June 22, 1972, p. 46.

19. The "bathroom excuse" has been common in excluding women from various institutions. See Cynthia Fuchs Epstein, *Woman's Place: Options and Limits in Professional Careers* (Berkeley: University of California Press, 1970), p. 185.

20. Smigel, "The Impact of Recruitment," p. 57.

21. John Francis Dooling, Jr., "Working for Love of the Law," *New York Times,* August 1, 1977, p. 42.

22. Alumnae Directory, Classes of 1953–1977: "Celebration: A Generation of Women at Harvard Law School" (no date, no pagination).

23. Epstein, *Women and Professional Careers,* p. 164.

24. Cynthia Fuchs Epstein, "Positive Effects of the Multiple # Negative: Explaining the Success of Black Professional Women," *American Journal of Sociology* 78, no. 4 (January 1973): 912–35.

Notes

25. Epstein, *Women and Professional Careers,* p. 164.
26. Cynthia Fuchs Epstein, "Sex Role Stereotyping, Occupations and Social Exchange," *Women's Studies* 3 (1976): 185–94.
27. Richard Sennett and Jonathan Cobbs, *The Hidden Injuries of Class* (New York: Vintage Books, 1973).
28. Kai T. Erikson, *Everything in Its Path* (New York: Simon & Schuster, 1976), p. 255.
29. Joan E. Baker, "Employment Discrimination Against Women Lawyers," *American Bar Journal* 59 (September 1973): 1029–32.
30. Robert B. McKay, "Women and the Liberation of Legal Education," *Women Lawyers' Journal* 57, no. 4 (1971): 140–41.
31. A point underlined, according to McKay, in the case of Phillips v. Martin Marietta Corp. 91 Sup. Ct. 496 (1971) holding that female job applicants could not be rejected because they were mothers of preschool children since fathers of preschool children were not similarly disqualified. Ibid., p. 141.
32. 404 U.S. 71 (1971).
33. 421 U.S. 7 (1975).
34. Califano v. Goldfarb 430 U.S. 199 (1977), 222 (1977). Concurring Opinion.
35. Ibid; Craig v. Boren 429 U.S. 190 (1976).
36. Ruth Bader Ginsburg, "Women at the Bar—A Generation of Change," *University of Puget Sound Law Review* 2, no. 1 (1978): 1–14. The preceding paragraph on the Supreme Court is also based on this article (p. 10).

Chapter 6

1. U.S. Bureau of the Census, *1960 Subject Reports: Occupational Characteristics,* Final Report PC(2)–7A (Washington, D. C.: Government Printing Office, 1963), table 21, p. 277.
2. U.S. Bureau of the Census, *1970 Subject Reports: Occupational Characteristics,* Final Report PC(2)–7A (Washington, D. C.: Government Printing Office, 1973), table 43, pp. 693, 704.
3. U.S. Bureau of the Census, *1960 Subject Reports,* p. 277.
4. U.S. Bureau of the Census, *1970 Subject Reports,* pp. 693, 704.
5. John P. Heinz, Edward O. Laumann, Charles L. Cappell, Terence Halliday and Michael H. Schaalman, "Diversity, Representation, and Leadership in an Urban Bar: A First Report of a Survey of the Chicago Bar," *The American Bar Foundation Research Journal* 1976, p. 785.
6. Nick A. LaPlaca, "Class of 1975 Employment Report" National Association for Law Placement, 1976, cited in Christine White-Wiegner, "The Post-Law School Search" *Trial: the National Legal News Magazine* vol. 13., no. 3, (March 1977,): p. 24.
7. Nick A. LaPlaca, "Class of 1977 Employment Report," National Association for Law Placement, 1978. p. 11
8. Bureau of the Census, 1970 Subject Reports, pp. 693, 704.
9. LaPlaca, "Class of 1975 Employment Report," in White-Wiesner, p. 24.
10. LaPlaca, "Class of 1977 Employment Report," p. 10.
11. Ibid.
12. LaPlaca, "Class of 1975 Employment Report," in White-Wiesner, p. 24.
13. LaPlaca, "Class of 1977 Employment Report," p. 10.
14. Ibid.
15. LaPlaca, "Class of 1975 Employment Report," in White-Wiesner, p. 10.
16. LaPlaca, "Class of 1977 Employment Report," p. 9.
17. Letter from Robert C. Wheeler, Executive Coordinator of the California Young Lawyers Association, January 9, 1978, accompanying the "Report on Unemployment and Underemployment Among Attorneys Admitted Between January 1972 and June 1977," The California Young Lawyers Association, Employment Opportunities Committee, October 1977.
18. Yale information from Gwendolyn R. Hatchette, Associate Director for Placement, Yale Law School, February 13, 1979.
19. Columbia information from Columbia University Law Placement Office.

20. Harvard information from the Alumnae Directory, Classes of 1953–1977, (The Celebration, Alumnae Director, Harvard University), and Job Reports of . . . Eleanor Appel, Director of Placement, February 5, 1979.

21. LaPlaca, "Class of 1977 Employment Report." p. 13.

22. See E. Laumann and J. Heinz for a more elaborate definition and exploration of specialization in law, "Fields of Law Practice: Volume of Professional Activity, Intensity of Specialization, and Patterns of Co-Practice in a Major Urban Bar" (paper presented at ASA meeting, September 5–9, 1978, San Francisco, California).

23. Mark Green, *The Other Government: The Unseen Power of Washington Lawyers* (New York: W. W. Norton and Company, 1978), p. 21.

24. Ibid., p. 10.

25. These are the factors identified by lawyers interviewed by Hubert O'Gorman in his study of matrimonial lawyers, *Lawyers and Matrimonial Cases: A Study of Informal Pressures in Private Professional Practice* (New York: The Free Press of Glencoe, 1963).

26. Verna Elizabeth Griffin, *Employment Opportunities for Women in Legal Work* (Washington, D.C.: U.S. Department of Labor, U.S. Government Printing Office, 1958), p. 12.

27. James J. White, "Women in the Law," *Michigan Law Review* 65, no. 6 (April 1967): 1062–63.

28. White, "Women in the Law," p. 1062.

29. Nancy Young, "Alumnae," *Harvard Law School Bulletin* (December 1956): p. 13.

30. "Lawyers: The Perils of Portia—The Weaker Sex," *Time*, March 6, 1964, p. 48.

31. White, "Women in the Law," p. 1060.

32. Erving Goffman, *Stigma: Notes on the Management of Spoiled Identity* (Englewood Cliffs, N.J.: Prentice-Hall, 1963).

33. For a more complete analysis see Cynthia Fuchs Epstein, "Sex Roles Stereotyping, Occupations and Social Change," *Womens Studies* 3 (1976): 185–94.

34. John Kosa and Robert E. Coker, "The Female Physician in Public Health: Conflict and Reconciliation of the Professional and Sex Roles," *Sociology and Social Research* 49 (April 1965).

35. Epstein, *Woman's Place*.

36. William J. Goode, "Why Men Resist," *Dissent* 27, no. 1 (Winter 1980).

37. Talcott Parsons underscores the disruptive potential of women's engagement in the occupational world for the family in "Age and Sex in the Social Structure of the United States," in his *Essays in Sociological Theory* (Glencoe, Ill.: The Free Press, 1954), pp. 89–103.

Chapter 7

1. In a comment on my manuscript, "Women and Professional Careers: The Case of the Woman Lawyer" (Ph.d. dissertation, Columbia University, 1968).

2. Other protected settings have been husband-wife law partnerships, jobs in libraries, and jobs as assistants. See Epstein, "Women and Professional Careers," p. 161.

3. Jerold S. Auerbach, *Unequal Justice: Lawyers and Social Change in America* (New York: Oxford University Press, 1976).

4. U.S. Bureau of the Census, "Government Workers," *1970 Census of Population Subject Reports*, vol. 2, PC(2)–7 (Washington, D. C.: Government Printing Office, 1973), table 1, p. 7.

5. Epstein, "Women and Professional Careers," p. 181.

6. James J. White, "Women in the Law," *Michigan Law Review* 65, no. 6 (April 1967): 1051–122.

7. U.S. Bureau of the Census, *1970 Census of Population Subjects Reports* (Washington, D. C.: Government Printing Office, table 9, p. 102.

8. *Attorney Employment Fact Book*, as of December 31, 1978. U.S. Department of Justice, Office of Management and Finance, Central Management Offices Staff, Table 6.

9. Information provided by Glenn Stafford, U.S. Attorney's Office, in Washington, D.C., 1980.

Notes

Chapter 8

1. Sidney Zion, "Lawyers Begin Drive Against Poverty" *New York Times,* November 20, 1966 p. 83. Furthermore, this attitude was a "holdover from Elizabethan days," or so warned Edward V. Sparer, legal director of the Center of Social Welfare and Public Policy at Columbia University at a meeting sponsored by the NAACP Legal Defense and Education Fund, Inc., to develop new and imaginative legal remedies for poverty.
2. Tom Goldstein, "At 100 Legal Aid Strives to Live Within Budget," *New York Times,* March 6, 1976, p. 29.
3. Ibid.
4. Ibid., p. 3.
5. Michael Matza, "Another Defeat in the War Against Poverty," *The Student Lawyer* 8, no. 6 (February 1980): 19–42.
6. Goldstein, "At 100 Legal Aid Strives to Live Within Budget."
7. Stephen Wexler, "Practicing Law for Poor People," *Yale Law Journal* 1049 (1970): 212–13.
8. Jerold S. Auerbach, *Unequal Justice: Lawyers and Social Change in Modern America* (New York: Oxford University Press, 1976), p. 58.
9. Howard Erlanger, "Lawyers and Neighborhood Legal Services: Social Background and the Impetus for Reform," in *Lawyers and the Pursuit of Legal Rights,* ed. Joel F. Handler, Ellen Jane Hollingsworth, and Howard Erlanger (New York: Academic Press, 1978), p. 66.
10. Auerbach, *Unequal Justice,* p. 58.
11. Bernard Weintraub, "Legal Aid to Poor Arranged by City," *New York Times,* September 7, 1967.
12. Lacey Fosburgh, "Legal Aid Lawyers Vote to End Strike," *New York Times,* July 10, 1973.
13. Robert Hermann, Eric Single, and John Boston, *Counsel for the Poor: Criminal Defense in Urban America* (Lexington, Mass.: Lexington Books, 1977), p. 98.
14. Ibid.
15. Ibid., p. 98.
16. Lesley Oelsner, "400 Legal Aid Lawyers Go on Strike for Better Pay," *New York Times,* July 3, 1973, pp. 1, 24.
17. Ibid.
18. *Legal Aid Society Annual Report,* 1977–78, p. 22.
19. Gary Bellow, "Turning Solutions into Problems," *NLADA Briefcase* 34 (August 1977): 12.
20. Erlanger, "Lawyers and Neighborhood Legal Services," p. 26.

Chapter 9

1. Jan Hoffman, "Kris Glen: Defender of the Airways," *Ms,* February 19, 1978, p. 21.
2. Robert K. Merton and Elinor Barber, "Sociological Ambivalence," in *Sociological Ambivalence and Other Essays,* ed. Robert K. Merton (New York: Free Press, 1976), pp. 4–5.
3. Helene E. Schwartz, *Lawyering* (New York: Farrar, Straus & Giroux, 1976), p. 91, 126–28.
4. Laurie Johnson, "Two Law Firms Push Feminism-With All-Women Staffs," *New York Times,* February 17, 1973, p. 33.
5. Ibid.
6. Ibid.
7. On October 6, 1977, the *New York Times* reported the demise of the first women's law firm, Blank, Goodman, Rone, and Stanley.
8. Kai T. Erikson, *Everything in its Path* (New York: Simon & Schuster, 1976).
9. Joseph C. Goulden, *The Million Dollar Lawyers* (New York: Putnam, 1977).
10. Hubert O'Gorman, *Lawyers and Matrimonial Cases* (New York: The Free Press of Glencoe, 1963).

11. Goulden, *The Million Dollar Lawyers*.

12. William Goode, "A Theory of Role Strain," *American Sociological Review* 25 (August 1960): 483–96. Robert K. Merton, "Continuities in the Theory of Reference Groups and Social Structure," *Social Theory and Social Structure* (Glencoe, Ill.: The Free Press, [1949] 1957) pp. 281–386.

13. Cynthia Fuchs Epstein, *Woman's Place: Options and Limits in Professional Careers* (Berkeley: University of California Press, 1970).

14. Charles Page and Louis Killian suggested these comparisons to me at a seminar at the University of Massachusetts at Amherst, March 6, 1980, where I presented some of this material.

15. Rosabeth Moss Kanter, *Men and Women of the Corporation* (New York: Basic Books, 1977).

16. Pamela Fishman, "Interaction: The Work Women Do," *Social Problems* 25, no. 4 (April 1978): 397–406.

17. Cynthia Fuchs Epstein, "Sex Role Stereotyping, Occupations, and Social Exchange," *Women's Studies* 3 (1979): 185–94.

18. Howard Erlanger and Douglas Klegan, "Socialization Effects of Professional School: The Law School Experience and Student Orientations to Public Interest Concerns," *Law and Society Review* 13, no. 1 (Fall 1978): 253–74.

19. Eileen Moran, in a lecture to my seminar on "Women and the Power Structure," at the Graduate Center of CUNY, 1980.

Chapter 10

1. James J. White, "Women in the Law," *Michigan Law Review* 65, no. 6 (April 1967): 1062–3.

2. Jerome Carlin, *Lawyers Ethics: A Survey of the New York City Bar* (New York: Russell Sage Foundation, 1966).

3. Hubert O'Gorman, *Lawyers and Matrimonial Cases: A Study of Informal Pressures in Private Professional Practice* (New York: The Free Press of Glencoe, 1963).

4. Ibid., p. 108.

5. Ibid., p. 117.

6. An earlier and fuller description of husband-wife law partnerships may be found in Cynthia Fuchs Epstein, "Law Partners and Marital Partners, Strains and Solutions in the Dual-Career Family Enterprise," *Human Relations* 24, no. 6 (1971): 549–64.

7. D. Kelly Weisberg, "Barred from the Bar: Women and Legal Education in the United States, 1870–1890," *Journal of Legal Education* 28, no. 4 (1977): 485–507.

8. Letter from Kathryn S. Marshall of Marshall and Marshall, Ltd., Waukegan, Ill., September 25, 1978.

9. Ibid.

10. *National Law Journal* 2, no. 36 (May 19, 1980): 35.

Chapter 11

1. David Margolick and Jon Gelberg, "Few Dents in Male Dominance, Survey Finds," *National Law Journal* 2, no. 47 (August 4, 1980): 59.

2. Erwin O. Smigel, *The Wall Street Lawyer: Professional Organization Man*. (New York: The Free Press of Glencoe, 1964).

3. "The Legal Mating Dance," *Newsweek*, December 1, 1980, p. 111.

4. Erwin O. Smigel, "The Wall Street Lawyer Reconsidered," *New York*, August 18, 1969, pp. 36–41.

5. Robert K. Merton, has written about stereotypes in *Social Theory and Social Structure* rev. ed. (Glencoe, Ill.: The Free Press, 1957), 424–425. The reference to germs of truth is from "oral presentation" in lectures at Columbia University, 1963.

6. This perspective rests, of course, on the Thomas Theorem (so called by Robert K.

Notes

Merton) of W. I. Thomas: "If men define situations as real, they are real in their consequences." See p. 362 for further explanation.

7. Merton, "The Self-Fulfilling Prophecy," *Social Theory and Social Structure*, pp. 421–436.

8. Smigel, "The Wall Street Lawyer Reconsidered," *New York* August 18, 1969 pp. 36–41. and Cynthia Fuchs Epstein, "Women and Professional Careers: The Case of the Woman Lawyer" (Ph.D. dissertation, Department of Sociology, Columbia University, 1968).

9. My staff surveyed all thirty-two firms in August 1979 to obtain the latest figure on women partners.

10. Margolick and Gelberg, "Few Dents in Male Dominance," p. 59.

11. The figures are from my own study.

12. Tom Goldstein, "Law, Fastest-Growing Profession May Find Prosperity Precarious," *New York Times*, May 16, 1977, pp. 1, 35.

13. Mary Murphree, "Work Rationalization and Secretarial Stress: The Case of Wall Street Legal Secretaries." (Ph.D. dissertation, Department of Sociology, Columbia University, 1980).

14. Jerold S. Auerbach, *Unequal Justice: Lawyers and Social Change in Modern America* (New York: Oxford University Press, 1976).

15. Erwin O. Smigel, *The Wall Street Lawyer* (Bloomington: Indiana University Press, 1969), p. 370.

16. Jim Drinkhall, "Women Attorneys, Now Over 9% of Profession, Keep Making Gains in All Areas of Legal Work," *Wall Street Journal*, May 31, 1978. p. 46.

17. "Harriet Rabb, Scourge of Corporate Male Chauvinism," was the title of a *New York Magazine* article by Lindsey Van Gelder (June 26, 1978), chronicling the role of the Columbia project in the precedent-setting work in defeating discriminatory practice.

18. Arnold M. Laubasch, "Top Law Firm Bans Sex Discrimination," *New York Times*, May 8, 1977.

19. *Chicago Tribune*, June 12, 1972; *Chicago Daily News*, July 11, 1972.

20. Cynthia Ozick, "Puttermesser: Her Work History, Her Ancestry, Her Afterlife," *New Yorker*, May 9, 1977, pp. 38–44.

21. For example, Jonathan R. Cole, in *Fair Science: Women in the Scientific Community* (New York: The Free Press, 1979), p. 11, suggests that women may opt out of certain spheres of science.

22. Erving Goffman, *The Presentation of Self in Everyday Life* (New York: Doubleday Anchor, 1959).

23. Paul Hoffman, *Lions in the Street* (New York: Signet, 1973), p. 10.

24. Ibid., p. 10.

25. Rosabeth Moss Kanter, *Men and Women of the Corporation* (New York: Basic Books, 1977); Judith Long Laws, "The Psychology of Tokenism: An Analysis," *Sex Roles* 1 (1975): 51–67.

26. Cynthia Fuchs Epstein, *Woman's Place: Options and Limits in Professional Careers* (Berkeley: The University of California Press, 1970), chapters 5 and 6.

27. Following Merton's analysis in "Social Structure and Anomie," in *Social Theory and Social Structure*, pp. 131–60.

28. Rose Laub Coser, "Where Have All the Women Gone? Like the Sediment of a Good Wine, They Have Sunk to the Bottom," in *Access to Power: Cross National Studies of Women and Elites*, ed. Cynthia Fuchs Epstein and Rose Laub Coser (London: George Allen & Unwin, 1981), pp. 16–33.

29. Joseph C. Goulden, *The Superlawyers: The Small and Powerful World of the Great Washington Law Firms* (New York: Dell, 1971).

30. Mark Green, *The Other Government: The Unseen Power of Washington Lawyers* (New York: W.W. Norton and Company, 1978), p. 51.

31. Steven Brill, "Fulbright and Jaworski Settles Sex Discrimination Suit," *American Lawyer* (September 1979): 3,4.

32. "Updates," *American Lawyer*, (August 1980): 8.

33. Brill, "Fulbright and Jaworski," p. 4.

34. "Atlanta Law Firm Accused of Sex Discrimination," "Notes on People," *New York Times,* March 26, 1980, p. B7.

35. Tom Goldstein, "Law, Fastest-Growing Profession, May Find Prosperity Precarious."

36. Goldstein, "Law, Fastest Growing Profession May Find Prosperity Precarious," gives Cravath as an example of a New York firm with a three-to-one ratio, and Morgan, Lewis, and Bockius in Philadelphia, and Sidley and Austin in Chicago as examples of where the ratio is one-to-one.

37. Ibid.

38. Steven Brill, "Over the Hill at Thirty," *Esquire,* November 7, 1978, pp. 14, 15.

39. "In-house Lawyers Gain in Numbers and Stature at Many Companies," *Wall Street Journal,* April 22, 1980, p. 1. The article noted that "women chief legal officers remain a rare breed—only 47 among the 2,050 firms listed in a current directory."

40. Here I am referring to the phenomenon described by Erving Goffman in *Stigma: Notes on the Management of Spoiled Identity* (Englewood Cliffs, N.J.: Prentice-Hall, 1963.)

41. Smigel, *The Wall Street Lawyer,* pp. 114–16.

42. There is a growing literature on the impact of employment experiences on values, attitudes, and career goals. See Melvin Kohn and Carmi Schooler, "Occupational Experience and Psychological Functioning: An Assessment of Reciprocal Effects," *American Sociological Review* 38 (February 1973):97–118; Joanne Miller, Carmi Schooler, Melvin Kohn, and Karen Miller, "Women and Work: The Psychological Effects of Occupational Conditions," *American Journal of Sociology* 85 (July 1979): 66–94; and Jeylan T. Mortimer and Jon Lorence, "Work Experience and Occupational Value Socialization: A Longitudinal Study," *American Journal of Sociology* 84 (May 1979):1361–85.

43. Lewis A. Coser, *Greedy Institutions: Patterns of Undivided Commitment* (New York: The Free Press, 1974), p. 4.

44. Smigel, *The Wall Street Lawyer Reconsidered,* p. 75.

45. Louis Auchincloss, "Foster Evans on Lewis Boves," in *Tales of Manhattan* (Boston: Houghton Mifflin, 1967), p. 149.

46. Peter W. Bernstein, "The Wall Street Lawyers are Thriving on Change," *Fortune,* March 13, 1978, pp. 104–12.

47. Marilyn Machlowitz, "Newest Drug: Work," *Vogue,* March 1980, p. 327.

48. William J. Goode points out that over-producing, in other words, an over-fulfillment of the norms, creates prestige in *Celebration of Heroes: Prestige as a Social Control System* (Berkeley: University of California Press, 1979).

49. "Questions, The Bar Exams," *Time,* February 25, 1980, p. 44.

50. Cynthia Fuchs Epstein, "The Partnership Push," *Savvy,* March 1980, pp. 28–39.

51. David Margolick, "Wall Street's Sexist Wall," *National Law Journal,* 2, no. 47 (August 4, 1980): 58.

52. Margolick, "Partnership Picks Worry Women at Jenner & Block," 2, no. 36 *The National Law Journal* (May 19, 1980): 2.

53. Margolick, "Wall Street's Sexist Wall," p. 59.

Chapter 12

1. See the history of appointments in Benchmarks "Women in the Judiciary," chapter 24.

2. Donna Fossum, "Women Law Professors," *American Bar Foundation Research Journal* 4 (Fall 1980): 906.

3. Kenneth M. Davidson, Ruth Bader Ginsburg, and Herma Hill Kay, *Text, Cases and Materials on Sex-Based Discrimination* (St. Paul, Minn.: West Pub. Co., 1974), p. 885.

4. See Fossum, "Women Law Professors"; and D. Kelly Weisberg, "Women in Law School Teaching: Problems and Progress," *Journal of Legal Education* 30 (1979): 226–48.

5. Shirley R. Bysiewicz, "Women in Legal Education," *Journal of Legal Education* 25 (1973): 503.

6. Fossum, "Women Law Professors," p. 906.

7. Weisberg, "Women in Law School Teaching," p. 228.

8. Victor G. Rosenblum and Frances K. Kemans, *Making of a Public Profession* (Chicago, Illinois: American Bar Foundation, 1981). See also Donna Fossum, "Law Professors: A Profile of the Teaching Branch of the Legal Profession," *American Bar Foundation Research Journal* 3 (Summer 1980): 501–54.

9. Fossum, "Law Professors: A Profile of the Teaching Branch of the Legal Profession," *American Bar Association Journal* 3 (Summer 1980): 513.

10. Fossum, "Women Law Professors," p. 533.

11. Shirley R. Bysiewicz, "1972 AALS Questionnaire on Women in Legal Education," *Journal of Legal Education* 15 (1973): 508.

12. "University of Miami Names Soia Mentschikoff Law Dean," *Columbia Alumni Observer* 3, no. 2 (December 5, 1973).

13. Judith Younger, "My Life and Hard Times in Vanderbilt Hall," *The Commentator,* February 26, 1976; reprinted as "A Commentator Retrospective: From Miss to Ms.," October 13, 1976. Accompanying this article (original version) was another entitled "No Tenure for Younger: Bias is Charged."

14. Herma Hill Kay, "Legal and Social Impediments to Dual Career Marriages," *UCD Law Review* 12, no. 2 (Summer 1979): 207–25.

15. Martin Ginsburg joined the Columbia law faculty seven years after his wife, Ruth Bader Ginsburg.

16. Roberta Brandes Gratz, "Closeup: Lady of the Law," *New York Post,* May 31, 1966, p. 69.

17. Weisberg, "Women in Law School Teaching," p. 236.

18. Fossum, "Women Law Professors," p. 904.

19. Weisberg, "Women in Law School Teaching," p. 236.

20. Susan Edmiston, "Portia Faces Life: The Trials of Law School," *Ms,* April 1974, p. 74.

21. *New York Law School Bulletin,* 1976–1977.

22. Fossum, "Law Professors," p. 537.

23. Weisberg, "Women in Law School Teaching," p. 243.

24. Bysciewicz, cited in Ashburn and Cohen, "The Integration of Women Into Law Faculties," p. 58.

25. Mary G. Crawford, "Climbing the Ivy Covered Walls", *Ms,* November 1978, pp. 61–63.

26. Fossum, "Women Law Professors," p. 913.

27. Kenneth M. Davidson, Ruth Bader Ginsburg, and Herma Hill Kay, *Texts, Cases and Materials on Sex-Based Discrimination* (St. Paul, Minn.: West Publishing, 1974).

28. This follows the pattern that sex-typing reflects sex-ranking. See Cynthia Fuchs Epstein, *Woman's Place: Options and Limits in Professional Careers* (Berkeley: University of California Press, 1970).

29. In 1974, of the 158 ABA approved schools, approximately 63 (40 percent) had faculty listed under "Women and the Law" courses. In 1979, 70 schools (41 percent of 170) had such courses. Ashburn and Cohen, "The Integration of Women Into Law faculties."

Chapter 13

1. Cynthia Fuchs Epstein, "Women and Power: The Roles of Women in Politics in the U.S.," in *Access to Power: Cross National Studies of Women and Elites,* ed. Cynthia Fuchs Epstein and Rose Laub Coser (London: George Allen & Unwin, 1980), p. 139.

2. *Sourcebook: Tenth Annual National Conference on Women and the Law,* March 29–April 1, 1979, San Antonio, Texas.

3. Beverly Blair Cook, "Women Judges: The End of Tokenism, in *Women in the Courts,* ed. Winifred Hepperle and Laura Crites (Williamsburg, Va.: National Center for State Courts, 1978), pp. 84–105.

4. "Florence Allen, 82, First Woman on United States Appellate Bench, Dead," *New York Times,* September 14, 1966, p. 47.

5. Cook, "Women Judges: The End of Tokenism," p. 86.

6. Ibid.

7. Elizabeth Shelton, "Are the Scales Weighted Against Women Judges?" *The Washington Post,* September 19, 1965, p. F1.

8. Doris L. Sassower, "Women and the Judiciary: Undoing the Law of the Creator," *Judicature* 57, vol. 7 (February 1974): 282–288. Doris L. Sassower, "Women and the Legal Profession" (speech before the Women's Rights Group of New York University Law School, February 25, 1970).

9. Emily Jane Goodman, "The Boys on the Bench," *Juris Doctor* 8 (July/August 1976): 8–9.

10. Sharon Johnson, "Female Judges Push for More Women on the Federal Bench," *New York Times,* October 30, 1979, p. B11. See also Susan Tolchin, "The Exclusion of Women from the Judicial Process," *Signs: A Journal of Women in Culture and Society* 2, no. 4 (1977): 877–87.

11. Johnson, "Female Judges Push for More Women on Federal Bench."

12. *The Columbia Law Alumni Observer,* December 31, 1980.

13. Ibid.

14. *Sourcebook,* p. 155.

15. Stuart Taylor, Jr., "Carter Judge Selections Praised, but Critics Discern Partisanship," *New York Times,* October 3, 1980.

16. Ibid.

17. Cook, "Women Judges: The End of Tokenism," pp. 84–105.

18. "Florence Allen. . . . " *New York Times*

19. Goodman, "The Boys on the Bench," p. 8.

20. Ruth Hochberger, "Women in Law—How Far Have They Come?" *New York Law Journal* 177, no. 55 (March 22, 1977): 1.

21. "Embattled Chief Justice Rose Elizabeth Bird," *New York Times,* June 12, 1979, p. 12; "Coast Hearing Today on a Chief Justice" *New York Times,* March 7, 1977, p. 16; "Woman Chosen to Head Court: Rose Elizabeth Bird," *New York Times,* February 14, 1977, p. 48.

22. *New York Times,* February 14, 1977, p. 48.

23. *New York Times,* June 12, 1979, p. 12.

24. Joan Dempsey Klein, "Women Judges Join Together," *Judges Journal* 2 (Spring 1980).

25. Angel Castillo, "Women at Bar Parley Seek Rise in Status as Lawyers and Judges," *New York Times,* February 10, 1981, p. B9.

26. Ruth Marcus, "Few Women to Clerk at High Court," *National Law Journal* 2, no. 47 (August 4, 1980): 3, 12.

Chapter 14

1. Tom Goldstein, "ABA Marks Century as Voice for Lawyers," *New York Times,* June 11, 1978, p. 24.

2. Letter from Suzanne G. Manker, Information Service, American Bar Association, 1155 E. 60th Street, Chicago, Illinois, February 1, 1967.

3. Barbara Armstrong, "2997 Women Practice Law in the U.S., Still Find Going Tough, Survey Shows," *Harvard Law School Record* 13 (December 6, 1951): 1.

4. "A Woman Lawyer Alleging Sex Bias, Sues City Bar Unit," *New York Times,* February 27, 1971. p. 29

5. Telephone inquiry to Charles S. Vaccaro of the Metropolitan Trial Lawyers Association, March 5, 1981.

6. In a telephone inquiry at the Brooklyn Bar Association, a spokesperson reported this to me in 1968.

7. Armstrong, "2,997 Women Practice Law in U.S."

8. Computed from the 1966 *Year Book* of the New York County Lawyers' Association (New York: Correct Printing Company, Inc.), and the 1966 *Year Book* of the Association of the Bar of the City of New York (New York: (no publisher listed).

9. Computed from the 1972 *Year Book* of the Association of the Bar of the City of New York (New York: no publisher listed).

10. Angel Castillo, "Women at Bar Parley Seek Rise in Status as Lawyers and Judges," *New York Times,* February 10, 1981, p. B9.

11. Cynthia Fuchs Epstein, "Women and Professional Careers: The Case of the Woman Lawyer," (Ph.D. dissertation, Department of Sociology, Columbia University, 1968) p. 290.

12. Armstrong, "2,997 Women Practice Law in the U.S."

13. *The Daily Telegraph,* London, December 30, 1922, cited in Nellie Alden Franz, *English Women Enter the Professions*, Cincinnati, Ohio (privately printed by Columbia University Press, 1965), p. 277.

14. Henrietta Dunlop Stonestreet, "Women Lawyers vs. The Baltimore Bar Association," *Woman Lawyers' Journal* 19 (1931): 18.

15. Hubert M. Blalock, Jr., *Toward a Theory of Minority Group Relations* (New York: John Wiley and Sons, 1967), p. 51.

16. Ibid.

17. The comparison between women and blacks is effectively demonstrated by Helen Hacker in "Women as a Minority Group," *Social Forces* 30 (October 1951): 62.

18. Florence Joyce, "How We Started," in *75 Year History of National Association of Women Lawyers, 1899–1974: The First Seventy-Five Years,* ed. Mary H. Zimmerman (Chicago, Ill.: National Association of Women Lawyers, 1975) p. 13.

19. Armstrong, "2,997 Women Practice Law in the U.S.," p. 1.

20. Castillo, "Women at Bar Parley Seek Rise in Status," p. B9.

Chapter 15

1. I discuss the development of competence in professions in "Encountering the Male Establishment: Sex Status Limits on Women's Careers in the Professions," *American Journal of Sociology,* 75 (May 1970): 965–82; and in "Separate and Unequal: Notes on Women's Achievement," *Social Policy* 6, no. 5, (March/April 1976): 17–23. Lester Thurow claims most jobs are learned in the workplace in *Zero Sum Society: Distribution and the Possibilities for Change* (New York: Basic Books, 1980).

2. Cynthia Fuchs Epstein, "Ambiguity as Social Control: The Salience of Sex Status in Professional Settings" (Paper given at the Western Psychological Association, San Francisco, California, April 19, 1978).

3. Robert K. Merton and Elinor Barber, "Sociological Ambivalence," in *Sociological Ambivalence and Other Essays,* ed. Robert K. Merton (New York: The Free Press, 1976), p. 18.

4. Gregory Bateson, Don D. Jackson, Jay Haley, and John H. Weakland, "A Note on the Double Bind—1962" in *Communication, Family, and Marriage,* vol. 1, ed. Don D. Jackson (Palo Alto, Cal.: Science and Behavior Books, 1968) p. 58.

5. Robert S. Lynd, *Knowledge for What?* (Princeton, N.J.: Princeton University Press, 1939).

6. Merton and Barber, "Sociological Ambivalence," 1963.

7. Robert K. Merton, "The Self-Fulfilling Prophecy" [1948], *Social Theory and Social Structure* (Glencoe, Ill.: The Free Press, 1957) pp. 421–436.

8. Albie Sachs and Joan Hoff Wilson, *Sexism and the Law* (New York: The Free Press, 1978), pp. 96–97.

9. Although it has brought them secondary gains. See Erving Goffman, *Stigma: Notes on the Management of Spoiled Identity* (Englewood Cliffs, New Jersey: Prentice-Hall, 1963), p. 10.

10. Betty Friedan, *It Changed My Life: Writings on the Women's Movement,* (New York: Random House, 1976).

11. Ralph Nader has commented on the legal system in many public lectures and statements; Jerold S. Auerbach, *Unequal Justice: Lawyers and Social Change in Modern*

America (New York: Oxford University Press, 1976); Gary Bellow, "Turning Solutions into Problems," *National Legal Aid and Defense Association Brief Case,* 34 (August 1977): 106–10.

12. "Those # ★ X!!! Lawyers," *Time Magazine,* April 10, 1978, pp. 56–65.

13. Robert K. Merton, "The Self-Fulfilling Prophecy."

14. Merton and Barber, "Sociological Ambivalence," p. 21.

15. Elizabeth Janeway discusses the dynamics of risk-taking among men and the ways women are dissuaded from it in *Powers of the Weak,* (New York: Alfred A. Knopf, 1980).

16. Enid Nemy, *New York Times,* March 6, 1981, p. A15.

17. Cynthia Fuchs Epstein, "Encountering the Male Establishment: Sex Status Limits on Women's Careers in the Professions," *American Journal of Sociology* 75, no. 6 (May 1975): 965–82.

18. Robert K. Merton, "The Role Set: Problems in Sociological Theory," *British Journal of Sociology* 8 (June 1957): 106–20.

19. Robert K. Merton has, over the past several decades, developed the analysis of the dynamics of status sets to identify the circumstances in which the functionally irrelevant status would be activated.

20. Everett Hughes, "The Dilemmas and Contradictions of Status," *American Journal of Sociology* 50 (March 1945): 353–359.

21. Epstein, "Encountering the Male Establishment," pp. 965–82.

22. Frances Coles, "Women in Litigation Practice: Success and the Woman Lawyer" (Ph.D. dissertation, Department of Criminology, University of California, Berkeley, 1974).

23. David M. Margolick, "Wall Street's Sexist Wall," *National Law Journal* 2, no. 47 (August 4, 1980): 1.

24. Carey Winfrey, "In New York, Private Clubs Change with Times," *New York Times,* January 5, 1978.

25. Talcott Parsons, *Social System* (Glencoe, Ill.: The Free Press, 1952), Chapter 10.

26. David Margolick, "Wall Street's Sexist Wall," *National Law Journal* 2, no. 47, p. 58.

27. See the discussion of the protégée-sponsor relationship in Cynthia Fuchs Epstein, *Women's Place* (Berkeley: University of California Press, 1970), pp. 168–70.

28. This comprises the "role-set" of the status of lawyer according to sociological role theory. See Merton, *Social Theory and Social Structure.*

29. Rose Laub Coser, "Laughter Among Colleagues," *Psychiatry* 23 (February 1960): 81–90.

30. Pamela Fishman, "Interaction: The Work Women Do," *Social Problems* 25, no. 4 (April 1978): 397–406.

31. Nancy Henley, "Power, Sex, and Nonverbal Communication," *Berkeley Journal of Sociology* 18 (1973–74): 1–26.

32. Ibid.

33. Arlie Russell Hochschild, "Emotion Work, Feeling Rules and Social Structure," *American Journal of Sociology* 85, no. 3 (1979): 551–75.

34. Erving Goffman, "The Nature of Deference and Demeanor," *American Anthropologist* 58 (June 1956): 473–502. Reprinted in Goffman, *Interaction Ritual* (New York: Doubleday Anchor, 1967).

35. Howard Newby, "The Deferential Dialect," *Comparative Studies in Society and History* 17, no. 2 (April 1975): 159. The tradition includes Montaigne and Lord Chesterfield, as Robert K. Merton reminded me.

36. Newby, "The Deferential Dialect."

37. R. E. Park, *Race and Culture* [1911], (Glencoe, Ill.: The Free Press, 1950), p. 258. (Cited in Newby.)

38. Erving Goffman, "Deference and Demeanor," p. 63.

39. Howard Newby, "The Deferential Dialectic," p. 159.

40. Edward O. Laumann and John P. Heinz, "Specialization and Prestige in the Legal Profession: The Structure of Deference," *American Bar Association Research Journal* 100, no. 1 (Winter 1977): 179, 185.

41. My early study showed that in New York City, women attorneys, like men, tended to come from immigrant families. Cynthia Fuchs Epstein, "Women and Professional

Careers: The Case of the Woman Lawyer" (Ph.D. dissertation, Department of Sociology, Columbia University, 1968), pp. 91–92.

42. Joel F. Handler, Ellen Jane Hollingsworth, and Howard S. Erlanger, *Lawyers and the Pursuit of Legal Rights* (New York: Academic Press, 1970), p. 26.

43. I have heard complaints about this from women at women and the law conferences, but they generally have been second-hand. Laura Meyers, in a recent article, reports more specific examples. "Behind Closed Doors," *The Student Lawyer* 7, no. 3 (November 1978): 40–48.

44. Epstein, "Women and Professional Careers," p. 245.

45. Howard Erlanger and Douglas Klegan, "Socialization Effects of Professional Schools: The Law School Experience and Student Orientations to Public Interest Concerns," *Law and Society Review* 13, no. 1 (Fall 1978): 253–74. See Chapter 8, p. 121n for figures in this book.

46. Rose Laub Coser, "Laughter Among Colleagues," *Psychiatry* 23 (February 1960): 85.

47. Rosabeth Moss Kanter has underscored this phenomenon in *Men and Women of the Corporation* (New York: Basic Books, 1977).

48. Barrie Thorne and Nancy Henley report research in this area in *Language and Sex: Difference and Dominance* (Rowley, Massachusetts: Newbury House, 1975). Sidney M. Jourard and Jane E. Rubin, "Self-Disclosure and Touching: A Study of Two Modes of Interpersonal Encounter and Their Inter-Relation," *Journal of Humanistic Psychology*, 8, no. 1 (1968): 39–48.

49. Judith Blake and Kingsley Davis, "Norms, Values and Sanctions," in *Handbook of Modern Sociology*, ed. Robert E. L. Faris (Chicago: Rand McNally, 1964), p. 464.

Chapter 16

1. Lillian Kagan, "The Relationship of Achievement Success, Heterosexual Relationship Satisfaction, Sex-Role Orientation, and Self-Esteem in Male and Female Law School Students" (Ph.D. dissertation, School of Education New York University, 1976).

2. David Reisman, *The Lonely Crowd* (New Haven: Yale University Press, 1967).

3. Cynthia Fuchs Epstein, "Separate and Unequal: Notes on Women's Achievement," *Social Policy* 6, no. 5 (March/April 1976): 17–23.

4. C. Kronus, "Occupational U.S. Organizational Influences of Reference Group Identification," *Sociology of Work and Occupations*, 3, no. 3 (August 1976): 303–30.

5. Grace Baruch, Rosalind Barnett, and Caryl Rivers, "A New Start for Women at Midlife," *New York Times Magazine*, December 7, 1980, pp. 196–201.

6. Studying self-esteem, situational anxiety, and comfort during role playing, a social psychologist found that praised subjects were less stressed than those who were criticized. See Brenda Lee Sharp, "Stress Effects: The Assumption of Congruent and Incongruent Social Roles" (Ph.D. dissertation, Department of Social Psychology, Colorado State University, 1977).

7. Ellen Berscheid and Elaine Walster, "Physical Attractiveness," *Advances in Experimental Social Psychology* 7 (1974): 157–215.

8. Ibid., p. 165.

9. Robert Henry Stretch, "A Social Reinforcement Model of Successful Interpersonal Attraction" (Ph.D. dissertation, Department of Social Psychology, Purdue University, 1977).

10. Harriet Zuckerman has shared her insight on the confidence-inspiring attributes of large women.

11. Priscilla Tucker, "Legal Chic," *New York*, Sept. 19, 1977, p. 56.

12. Marilyn M. Machlowitz, "Dressing Up to the Executive Suite," *New York Times*, Sunday, Dec. 5, 1976, III, p. 3; and Jani Woolridge, "Dressing for the Top," *New York Times Magazine*, July 26, 1981, pp. 49–50.

13. "Practice Pointers; Attires for Women Attorneys," *Professional Responsibility Update*, release no. 8 (Los Angeles: Harcourt, Brace and Jovanovich, April 15, 1978), p. 32.

Notes

Chapter 17

1. Rose Laub Coser and Lewis A. Coser, "The Housewife and Her 'Greedy Family,'" in *Greedy Institutions* (New York: The Free Press, 1974), pp. 89–100.

2. Cynthia Fuchs Epstein, "Women and Professional Careers: The Case of the Woman Lawyer" (Ph.D. dissertation, Department of Sociology, Columbia University, 1965), pp. 313–14.

3. *National Survey of Working Women: Perceptions, Problems and Prospects,* Washington, D.C., National Commission on Working Women, June 1979.

4. I discuss this in "Role Transitions in the Changing Social Context" (Paper given at a conference on "The Dynamics of Stress: Women and Other people," College of Physicians and Surgeons of Columbia University, October 2, 1979).

5. Stephen R. Marks, "Multiple Roles and Role Strain, Some Notes on Human Energy, Time and Commitment," *American Sociological Review* 42, no. 6 (December 1977): 921–36.

6. Sam Sieber, "Toward a Theory of Role Accumulation," *American Sociological Review* 39 (August 1974): 567–78.

7. The discussion comes from Stephen R. Marks, "Culture, Humans Energy and Self-Actualization: A Sociological Offering to Humanistic Psychology," *Journal of Humanistic Psychology* 19, no. 3 (Summer 1979): 27–42; and "Multiple Roles and Role Strain."

8. Seiber, "Toward a Theory of Role Accumulation," p. 576.

9. Rose Laub Coser, "The Complexity of Roles As a Seedbed of Individual Autonomy," in Lewis A. Coser (ed.) *The Idea of Social Structure: Papers in Honor of Robert K. Merton* (N.Y.: Harcourt Brace Jovanovich, 1975), pp. 237–263.

Chapter 18

1. Betty Friedan, *The Feminine Mystique* (New York: W.W. Norton, 1963).

2. James J. White, "Women in the Law," *Michigan Law Review* 65, no. 6 (April 1967): 1065.

3. By 1970 only the proportion of women physicians who were married went up, while women lawyers and engineers had declined slightly, and a higher proportion of women lawyers were divorced.

4. Robert Reinhold, "Trend to Living Alone Brings Economic and Social Change," *New York Times,* March 20, 1977, p. 1.

5. Gwyned Simpson, "The Daughters of Charlotte Ray: An Investigation into the Process of Vocational Development in Black Female Attorneys" (Ph.D. dissertation, Teacher's College, Columbia University, 1979), p. 97.

6. Rita Lynne Stafford, "An Analysis of Consciously Recalled Professional Involvement of American Women in New York State" (Ph.D. dissertation, School of Education, New York University, 1966).

7. Mary Loretta Rosenlund and Frank Oski, "Women in Medicine," *Annals of Internal Medicine* 66 (1967): 1008–12; and M. C. Diamond, "Women in Modern Science," *Journal of the American Medical Women's Association* 18 (November 1963):891–96. See also John Kosa and Robert E. Coker, "The Female Physician in Public Health: Conflict and Reconciliation of the Professional and Sex Roles," *Sociology and Social Research* 49 (April 1965): 295 Women engineers tend to marry "in the late twenties to early thirties." *Profile of the Woman Engineer* (New York: Society of Women Engineers, May 1963), p. 3. R. L. Stafford found 60 percent of the doctors and 60 percent of the lawyers in her sample married after graduate school. Women in "female" professions tended to marry a little earlier. Percentages of those in educational administration and nursing administration who married after graduate school were 58.7 and 46.2 percent, respectively. Stafford, "An Analysis of Consciously Recalled Professional Involvement," p. 341.

8. Joseph Berger, Susan J. Rosenholtz, and Morris Zelditch, Jr., *Status Organizing Processes,* Technical Report no. 77, Stanford University, January 1980, mimeographed.

9. George Herbert Mead, *Mind, Self, and Society from the Standpoint of a Social*

Notes

Behaviorist (Chicago: University of Chicago Press, 1934); Erving Goffman, *Stigma* (Englewood Cliffs, N.J.: Prentice-Hall, 1963).

10. Dee Widemeyer, "Single Politicians: Public Life Without the Family Photograph," *New York Times,* December 29, 1979, p. 15.

11. For a discussion of "pools of eligibles" in marriage, see William J. Goode, *The Family* (Englewood Cliffs, N.J.: Prentice-Hall, 1964).

12. For statistics on the proportions and numbers of available men in the various age pools, see Cynthia Fuchs Epstein, "Positive Effects of the Multiple Negative: Explaining the Success of Black Professional Women," *American Journal of Sociology* 78, no. 4 (1973):912–35; and G. Simpson, "The Daughters of Charlotte Ray."

13. Fifty-one percent of Stafford's sample of eminent women lawyers in *Who's Who* were married to other lawyers. (R. L. Stafford, "An Analysis of Consciously Recalled Professional Involvement," p. 322).

14. Herma Hill Kay, "Legal and Social Impediments to Dual Career Marriages," *UCD Law Review* 12, no. 2 (Summer 1979): 207–25.

15. Stafford, "An Analysis," p. 217.

16. Debra Kaufman, "Associational Ties in Academe: Some Male and Female Differences," *Sex Roles* 4, no. 1 (1978): 9–12.

17. See Richard J. Lowry, ed., *Dominance, Self-Esteem and Self-Actualization: Germinal Papers of A. M. Maslow* (Monterey, Calif.: Brooks/Cole Publishers, 1973).

18. See Arlie Russell Hochschild, "Emotion Work, Feeling and Social Structure," *American Journal of Sociology* 85, no. 3 (November 1979): 551–75.

19. The references here are many: to cite only two—Edward Osborne Wilson, *Sociobiology: The New Synthesis* (Cambridge: Belknap Press of Harvard University, 1975); and Erik Erikson, *Childhood and Society* (New York: W. W. Norton, 1964).

20. Gary Becker, "A Theory of Marriage," *Journal of Political Economy Part I* 81, no. 4 (July/August 1973): 813–46.

21. Talcott Parsons, "Age and Sex in the Social Structure of the United States," *Essays in Sociological Theory* (Glencoe, Ill.: Free Press, 1958).

22. I am indebted to Alexander Epstein for this insight.

23. Only 7.3 percent of wives have earnings perceptibly higher than their husbands. Andrew Hacker, "Goodbye to Marriage," *New York Review of Books,* March 3, 1979. p. 27.

24. Daniel Patrick Moynihan, "The Negro Family: The Case for National Action," (Washington D.C.: U.S. Department of Labor, Office of Policy Planning and Research, 1965).

25. This insight was offered by Susan Wolf.

26. I am grateful to Susan Wolf for this observation.

27. Alan F. Blum and Peter McHugh, "The Social Ascription of Motives," *American Sociological Review* 36, no. 1 (February 1971): 98–109.

Chapter 19

1. Judy Klemesrud, "Survey Finds Major Shifts in Attitudes of Women," *New York Times,* March 13, 1980.

2. Gwen Kinkead, "On a Fast Track to the Good Life," *Fortune,* April 7, 1980, pp. 74–84.

3. Ibid.

4. Several scholars have shown that women who have become professionals are having fewer children than in the past. See Nan E. Johnson and C. Shannon Stokes, "Family Size in Successive Generations: The Effects of Birth Order, Intergenerational Change in Life Style and Family Satisfaction," paper presented at the Annual Meeting of the Rural Sociological Society, San Francisco, Calif., August 1975 (ERIC ed. 110237). C. Shannon Stokes and Alice P. Wrigley, "Sex Role Ideology, Selected Life Plans and Family Size Preferences" (paper presented at the Annual Meeting of the Southern Sociological Society, Washington, D.C., April 5–12, 1975); and Linda J. Waite, "Working Wives: 1940–1960," *American Sociological Review* 41 (February 1976): 65–80.

5. Cynthia Fuchs Epstein, "Women and Professional Careers: The Case of the Woman Lawyer" (Ph.D. dissertation, Department of Sociology, Columbia University, 1968).

6. Gwyned Simpson, "The Daughters of Charlotte Ray: An Investigation into the Process of Vocational Development in Black Female Attorneys" (Ph.D. dissertation, Teacher's College, Columbia University, 1979), p. 94–95.

7. Epstein, "Women and Professional Careers," pp. 127–28.

8. James J. White, "Women in the Law," *Michigan Law Review* 65, no. 6 (April 1967): 1065.

9. *Profile of a Woman Engineer* (New York: Society of Women Engineers, May 1963), p. 3.

10. Mary Loretta Rosenlund and Frank A. Oski, "Women in Medicine," *Annals of Internal Medicine* 66 (1967): 1010.

11. Cynthia Fuchs Epstein, "Law Partners and Marital Partners: Strains and Solutions in the Dual-Career Family Enterprise," *Human Relations* 24, no. 6 (1971): 549–64.

12. For data on this see Cynthia Fuchs Epstein, *Woman's Place: Options and Limits in Professional Careers* (Berkeley: University of California Press, 1970), p. 109. The discussion also gives fuller explanation for the reasons why this works well normatively.

13. Rhona Rapoport and Robert Rapoport, *Dual-Career Families Re-examined: New Integrations of Work and Family* (New York: Harper Colophon Books, 1977).

14. Ibid., p. 310.

15. Alice S. Rossi, "A Biosocial Perspective on Parenting," *Daedalus* 106, no. 2 (Spring 1977): 1–31.

16. Private communication.

17. Lewis A. Coser, *Greedy Institutions,* (New York: The Free Press, 1974).

Conclusion

1. Angel Castillo, "Women at Bar Parley Seek Rise in Status as Lawyers and Judges," *New York Times,* Feb. 10, 1980, p. B9.

2. Eric E. Van Loon, "The Law School Response: How to Sharpen Students' Minds by Making Them Narrow," in *With Justice for Some,* ed. Bruce Wasserstein and Mark J. Green (Boston: Beacon Press, 1970), p. 335.

3. Ibid., p. 336.

4. D. Kelly Weisberg, "Barred From the Bar: Women and Legal Education in the United States 1870–1890," *Journal of Legal Education,* 28, no. 4 (1977): 485–507. Arabella Mansfield, the first woman to be admitted to the bar in the United States, for example, was active in women's suffrage. Aleta Wallach, "Arabella Babb Mansfield (1846–1911)," *Women's Rights Law Reporter* (April 1974):3–5.

5. Cited in *The Mind and Faith of Justice Holmes, His Speeches, Essays, Letters and Judicial Opinions,* ed. Max Lerner (New York: The Modern Library, 1943), p. 36.

Appendix

1. Cynthia Fuchs Epstein, "Women and Professional Careers: The Case of the Woman Lawyer" (Ph.D. dissertation, Department of Sociology, Columbia University, 1968).

2. According to the protocol for the "focussed interview" as set out by Robert K. Merton and Patricia L. Kendall in "The Focussed Interview," *American Journal of Sociology,* 51, no. 6 (May 1946): 541–57.

3. The rationale and precedent for deviant case analysis is described in Ibid., pp. 6–7.

4. Seymour Martin Lipset, Martin Trow, and James Coleman, *Union Democracy* (Glencoe, Ill.: The Free Press, 1956).

5. Further details on the problems of obtaining a sample may be found in Epstein, "Women and Professional Careers," pp. 7–16.

6. Glenn Greenwood, *The 1961 Lawyer Statistical Report* (Chicago: American Bar Association, 1961); *Statistical Abstract of the United States,* U.S. Bureau of the Census,

Notes

87th ed. (Washington, D.C.: Government Printing Office, 1966), p. 159. Table 227 shows more women were lawyers in New York (1,204 to 41,075 men) than any other state in the year 1963.

7. Cynthia Fuchs Epstein, "Positive Effects of the Multiple Negative: Explaining the Success of Black Professional Women," *American Journal of Sociology* 78, no. 4 (January 1973): 912–35.

8. Gwyned Simpson, "The Daughters of Charlotte Ray: An Investigation into the Process of Vocational Development in Black Female Attorneys" (Ph.D. dissertation, Teachers College, Columbia University, 1979).

9. Martha Grossblat and Bette Sikes, *Women Lawyers: Supplementary Data to the 1971 Lawyer Statistical Report* (Chicago: American Bar Foundation, 1973).

10. Cynthia Fuchs Epstein, *Woman's Place: Options and Limits in Professional Careers* (Berkeley, California: University of California Press, 1970).

11. Cynthia Fuchs Epstein, "The Partnership Push," *Savvy,* March 1980, pp. 28–35.

12. Professor Ruth Bader Ginsburg, who was also a fellow at the center that year, thoughtfully introduced me to members of the legal academic community in the San Francisco area.

13. At various times from 1976 to 1980, Allan Grafman, Mary Murphree, Susan Wolf, and Judith Thomas performed able surrogate interviewing roles.

Name Index

Name Index

Name Index

Lifton, Robert Jay, *The Women in America*, 316*n*
Lincoln, Abraham, 272
Lipman-Blumen, Jean, 403*n*22
Lipset, Seymour Martin, 387; *Union Democracy*, 421*n*4
Llewellyn, Karl, 61, 80, 225
Lockheed, Marlaine E., 270*n*
Lockwood, Lorna, 240
Lomen, Lucille, 246*n*
Lorence, Jon, 412*n*42
Lortie, Dan Clements, 25; *The Striving Young Lawyer*, 402*n*3, 403*n*12, 406*n*15
Lowry, Richard J., *Dominance, Self-Esteem and Self-Actualization*, 419*n*17
Lynd, Robert, *Knowledge for What?*, 268, 415*n*5

McClelland, A. L., *Talent and Society*, 29*n*
Machlowitz, Marilyn, 412*n*47
McHugh, Peter, 404*n*37, 419*n*27
McKay, Robert B., 67, 405*n*15, 407*n*30–*n*31
McKelvey, Judith Grant, 225*n*
Magruder, Jeb, 216*n*
Mansfield, Arabella, 49, 420*n*4
Marcus, Ruth, 414*n*26
Margolick, David, 194*n*, 214, 216, 287, 410*n*1, *n*10, 412*n*51–*n*53, 416*n*23, *n*26
Marks, Steven R., 325, 418*n*5, *n*7
Marlatt, Frances, 51, 89
Marshall, Kathryn, 171–72, 178
Marshall, Thurgood, 229, 246*n*
Matthews, Burnita Shelton, 240
Matza, Michael, 409*n*5
Mayer, Martin, *The Lawyers*, 404*n*1, 405*n*4
Mead, George Herbert, 336; *Mind, Self, and Society*, 419*n*9
Meir, Golda, 346*n*
Mentchikoff, Soia, 60, 225
Merton, Robert K., 19, 112, 134, 144, 178–79, 215*n*, 226*n*, 247*n*, 271*n*, 272, 411*n*6, 416*n*18–*n*19, 417*n*35, 421*n*2; ambivalence, 266–67, 268; *Social Theory and Social Structure*, 266*n*, 271*n*, 370*n*, 410*n*12, 411*n*5, *n*7, *n*27, 416*n*7, *n*13, *n*28; *Sociological Ambivalence and Other Essays*, 144*n*, 266*n*, 402*n*30–*n*32, 409*n*2, 415*n*3, 416*n*6; status sets, 276*n*, 416*n*19; *The Student Physician*, 406*n*6; unanticipated consequences of social actions, 271*n*
Meyers, Laura, 417*n*43
Miller, Karen, 412*n*42
Miller-Lerman, Lindsey, 217
Miller, Steven, 187*n*
Millman, Marcia, *Another Voice*, 271*n*, 356*n*
Mitchell, Mildred B., 259*n*
Montaigne, Michel de, 416*n*35
Moore, Hoyt, 208*n*
Moore, Patricia A., 405*n*7, *n*8
Moran, Eileen, 161
Morgenthau, Robert, 116
Mortimer, Jeylan T., ix*n*, 412*n*42

Moynihan, Daniel Patrick, 346, 419*n*24
Murphree, Mary, 411*n*13
Murray, Pauli, 137–38

Nader, Ralph, 21, 122, 271
Nelson, Dorothy, 225*n*
Newby, Howard, 291, 416*n*35, *n*36, *n*39
Nixon, Richard, 176
Nizer, Louis, 146
Nordby, Virginia B., 405*n*20
Norton, Eleanor Holmes, 71, 185

O'Connor, Sandra Day, 84*n*, 240, 246
Oelsner, Lesley, 409*n*16, *n*17
O'Gorman, Hubert, 146, 164–65; *Lawyers and Matrimonial Cases*, 408*n*25, 410*n*3, *n*4, *n*5, *n*10
Oran, Ellen, 229*n*, 235
Oski, Frank H., 360, 418*n*7, 420*n*10
Ostroff, Ron, 179*n*
Owens, Elizabeth A., 227, 234, 235
Ozick, Cynthia, 411*n*20

Palmieri, Edmund L., 228
Park, R. E., 291; *Race and Culture*, 417*n*37
Parsons, Talcott, 144*n*, 152, 286; *Essays in Sociological Theory*, 139*n*, 401*n*7, 408*n*37, 419*n*21; *The Social System*, 144*n*, 416*n*25; *The Structure of Social Action*, 286*n*
Perkins, Frances, 346*n*
Peters, Ellen, 229
Pipkin, Ronald M., 41*n*, 41–42, 404*n*40
Piven, Francis Fox, 271*n*
Polier, Justine Wise, 239
Pollard, E. C., 105*n*
Powell, Lewis F., 246*n*
Pusey, Nathan, 52
Putnam, Robert D., *The Comparative Study of Political Elites*, 401*n*2

Rabb, Harriet, 66, 69–70, 108, 184–85, 186, 228
Rapoport, Rhona, 368, 369; *Dual Career Families Reexamined*, 401*n*7, 420*n*13–*n*14
Rapoport, Robert, 368, 369; *Dual Career Families Reexamined*, 401*n*7, 420*n*13–*n*14
Reagan, Ronald, 127, 285
Rehnquist, William H., 246*n*
Reinhold, Robert, 418*n*4
Reisman, David, 306, 316*n*; *The Lonely Crowd*, 415*n*2
Renne, Louise, 245
Ricker, Marilla, 239
Riggs, Bobby, 66

Name Index

Wise, Stephen, 239
Woods, Rose Mary, 216*n*
Wrigley, Alice P., 420*n*4

Young, Nancy, 408*n*29
Younger, Judith, 225, 413*n*13

Zelan, Joseph, *Lawyers in the Making*, 402
  *n*1, *n*3, *n*5, 403*n*10
Zeldich, Morris, 335
Zimmerman, Mary H., *75 Year History of
  National Association of Women Law-
  yers, 1899–1974*, 415*n*18
Zion, Sidney, 409*n*1
Zuckerman, Harriet, 418*n*10

Subject Index

Subject Index